Physics of Neural Networks

Series Editors:

E. Domany   J. L. van Hemmen   K. Schulten

Advisory Board:

H. Axelrad
R. Eckmiller
J. A. Hertz
J. J. Hopfield
P. I. M. Johannesma
P. Peretto
D. Sherrington
M. A. Virasoro

D0813329

Springer

Berlin
Heidelberg
New York
Barcelona
Budapest
Hong Kong
London
Milan
Paris
Santa Clara
Singapore
Tokyo

B. Müller   J. Reinhardt
M. T. Strickland

Neural Networks

An Introduction

Second Updated and Corrected Edition
With a MS-DOS Programm Diskette
and 105 Figures

 Springer

1995

Professor Dr. Berndt Müller

Department of Physics, Duke University
Durham, NC 27706, USA

Michael T. Strickland

Department of Physics, Duke University
Durham, NC 27706, USA

Dr. Joachim Reinhardt

Institut für Theoretische Physik
J.-W.-Goethe-Universität,
Postfach 11 19 32
D-60054 Frankfurt , Germany

Series Editors:

Professor J. Leo van Hemmen

Institut für Theoretische Physik
Technische Universität München
D-85747 Garching bei München
Germany

Professor Eytan Domany

Department of Electronics
Weizmann Institute of Science
76100 Rehovot, Israel

Professor Klaus Schulten

Department of Physics
and Beckman Institute
University of Illinois
Urbana, IL 61801, USA

ISBN 3-540-60207-0 2nd Ed. Springer-Verlag Berlin Heidelberg New York

ISBN 3-540-52380-4 1st Ed. Springer-Verlag Berlin Heidelberg New York

```
Library of Congress Cataloging-in-Publication Data
Müller, Berndt.
   Neural networks : an introduction / B. Müller, J. Reinhardt, M.T.
Strickland. -- 2nd updated and corr. ed.
      p.  cm. -- (Physics of neural networks)
   Includes bibliographical references and index.
   ISBN 3-540-60207-0 (alk. paper)
   1. Neural networks (Computer science)   I. Reinhardt, J.
(Joachim), 1952-   . II. Strickland, M. T. (Michael Thomas), 1969-
.  III. Title.  IV. Series.
QA76.87.M85  1995
006.3--dc20                                      95-24948
```

Typesetting: Camera ready copy from the authors.
SPIN 10120349   54/3144 – 5 4 3 2 1 0 – Printed on acid-free paper

In memory of
Jörg Briechle

Preface to the Second Edition

Since the publication of the first edition, the number of studies of the properties and applications of neural networks has grown considerably. The field is
now so diverse that a complete survey of its recent developments is impossible
within the scope of an introductory text. Fortunately, a wealth of material
has become available in the public domain via the Internet and the World
Wide Web (WWW). We trust that the reader who is interested in some of
the latest developments will explore this source of information on their own.
For those interested we have compiled a short list of Internet resources that
provide access to up to the date information on neural networks.

Neural-Network related WWW Sites

```
http://www.neuronet.ph.kcl.ac.uk:80/
```
       NEuroNet — King's College London
```
http://glimpse.cs.arizona.edu:1994/bib/
```
       Bibliographies on Neural Networks and More
```
http://http2.sils.umich.edu/Public/nirg/nirg1.html
```
       Neurosciences Internet Resource Guide
```
http://www1.cern.ch/NeuralNets/nnwlnHep.html
```
       Neural Networks in High Energy Physics
```
http://www.cs.cmu.edu:8001/afs/cs.cmu.edu/project
    /nnspeech/WorldWideWeb/PUBLIC/nnspeech/
```
       The Neural Net Speech Group at Carnegie Mellon
```
ftp://archive.cis.ohio-state.edu/pub/neuroprose/
```
       The Neuroprose preprint archive

Listservers

```
Connectionists-Request@cs.cmu.edu
```
       Connectionists — Neural Computation
```
neuron-request@psych.upenn.edu
```
       Neuron-Digest — Natural and Artificial Neural Nets
```
LISTSERV@UICVM.UIC.EDU
```
       Neuro1-L — Neuroscience Resources
```
mb@tce.ing.uniroma1.it (151.100.8.30)
```
       Cellular Neural Networks

Usenet Newsgroups

```
comp.ai.neural-nets
bionet.neuroscience
```

We have, therefore, resisted attempting to completely update the text and references with the latest developments and rather concentrated our efforts on eliminating two shortcomings of the first edition. We have added a separate chapter on applying genetic algorithms to neural-network learning, complete with a new demonstration program **neurogen**. A new section on back-propagation for recurrent networks has been added along with a new demonstration program **btt** which illustrates back-propagation through time. In addition, we have expanded the chapter on applications of neural networks, covering recent developments in pattern recognition and time series prediction.

One of us, Michael Strickland, would like to thank his wife Melissa Trachtenberg and all of his friends who helped him with proofreading and pep talks.

It is our hope that with the introduction of the new chapters and programs the second edition will be interesting and stimulating to new readers starting their own exploration of this fascinating new field of human science.

Durham and Frankfurt   *Berndt Müller*
May 1995   *Joachim Reinhardt*
                                                             Michael Strickland

Email addresses and homepages of the authors

```
http://www.phy.duke.edu/~muller/
        Berndt Müller — muller@phy.duke.edu
http://www.th.physik.uni-frankfurt.de/~jr/jr.html
        Joachim Reinhardt — jr@th.physik.uni-frankfurt.de
http://www.phy.duke.edu/~strickla/
        Michael Strickland — strickla@phy.duke.edu
```

 References to the demonstration programs are indicated in the main text by this "PC logo". We encourage all readers to do the exercises and play with these programs.

Preface to the First Edition

The mysteries of the human mind have fascinated scientists and philosophers for centuries. Descartes identified our ability to think as the foundation of ontological philosophy. Others have taken the human mind as evidence of the existence of supernatural powers, or even of God. Serious scientific investigation, which began about half a century ago, has partially answered some of the simpler questions (such as how the brain processes visual information), but has barely touched upon the deeper issues concerned with the nature of consciousness and the existence of mental features transcending the biological substance of the brain, usually encapsulated in the concept of the "mind".

Besides the physiological and philosophical approaches to these questions, so impressively presented and contrasted in the book by Popper and Eccles [Po77], studies of formal networks composed of binary-valued information-processing units, highly abstracted versions of biological neurons, either by mathematical analysis or by computer simulation, have emerged as a third route towards a better understanding of the brain, and possibly of the human mind. Long remaining – with the exception of a brief period in the early 1960s – a rather obscure research interest of a small group of dedicated scientists scattered around the world, neural network research has recently sprung into the limelight as a fashionable research field. Much of this surge of attention results, not from interest in neural networks as models of the brain, but rather from their promise to provide solutions to technical problems of "artificial intelligence" that the traditional, logic-based approach did not yield easily.[1]

The quick rise to celebrity (and the accompanying struggle for funding and fame) has also led to the emergence of a considerable amount of exaggeration of the virtues of present neural network models. Even in the most successful areas of application of neural networks, i.e. content-addressable (associative) memory and pattern recognition, relatively little has been learned which would look new to experts in the various fields. The difficult problem of position- and distortion-invariant pattern recognition has not yet been solved by neural networks in a satisfactory way, although we all know from experience that our brains can often do this. In fact, it is difficult to pinpoint any

---

[1] Neural networks are already being routinely used in some fields. For example, in high energy physics experiments they are being used as triggering software for complex detector systems.

X

technical problem where neural networks have been shown to yield solutions that are clearly superior to those previously known. The standard argument, that neural networks can do anything a traditional computer can, but do not need to be programmed, does not weigh as strongly as one might think. A great deal of thought must go into the design of the network architecture and training strategy appropriate for a specific problem, but little experience and few rules are there to help.

Then why, the reader may ask, have we as nonexperts taken the trouble of writing this book and devising the computer demonstrations? One motivation, quite honestly, was our own curiosity. We wanted to see for ourselves what artificial neural networks can do, what their merits are and what their failures. Not having done active research in the field, we have no claim to fame. Whether our lack of prejudice outweighs our lack of experience, and maybe expertise, the reader must judge for herself (or himself).

The other, deeper reason is our firm belief that neural networks are, and will continue to be, an indispensable tool in the quest for understanding the human brain and mind. When the reader feels that this aspect has not received its due attention in our book, we would not hesitate to agree. However, we felt that we should focus more on the presentation of physical concepts and mathematical techniques that have been found to be useful in neural network studies. Knowledge of proven tools and methods is basic to progress in a field that still has more questions to discover than it has learned to ask, let alone answer.

To those who disagree (we know some), and to the experts who know everything much better, we apologize. The remaining readers, if there are any, are invited to play with our computer programs, hopefully capturing some of the joy we had while devising them. We hope that some of them may find this book interesting and stimulating, and we would feel satisfied if someone is inspired by our presentation to think more deeply about the problems concerning the human mind.

This book developed out of a graduate level course on neural network models with computer demonstrations that we taught to physics students at the J.W. Goethe University in the winter semester 1988/89. The interest in the lecture notes accompanying the course, and the kind encouragement of Dr. H.-U. Daniel of Springer-Verlag, have provided the motivation to put it into a form suitable for publication. In line with its origin, the present monograph addresses an audience mainly of physicists, but we have attempted to limit the "hard-core" physics sections to those contained in Part II, which readers without an education in theoretical physics may wish to skip. We have also attempted to make the explanations of the computer programs contained on the enclosed disk self-contained, so that readers mainly interested in "playing" with neural networks can proceed directly to Part III.

Durham and Frankfurt
July 1990

Berndt Müller
Joachim Reinhardt

Table of Contents

Part II. Statistical Physics of Neural Networks

Part III. Computer Codes

Part I

Models of Neural Networks

1. The Structure
of the Central Nervous System

1.1 The Neuron

Although the human central nervous system has been studied by medical doctors ever since the late Middle Ages, its detailed structure began to be unraveled only a century ago. In the second half of the nineteenth century two schools contended for scientific prevalence: the *reticularists* claimed that the nervous system formed a continuous, uninterrupted network of nerve fibres, whereas the *neuronists* asserted that this neural network is composed of a vast number of single, interconnected cellular units, the *neurons*. As often in the course of science, the struggle between these two doctrines was decided by the advent of a new technique, invented by Camillo Golgi around 1880, for the staining of nerve fibres by means of a bichromate silver reaction. This technique was ingeniously applied by the Spanish doctor Santiago Ramon y Cajal in 1888 to disprove the doctrine of reticularism by exhibiting the tiny gaps between individual neurons. The modern science of the human central nervous system thus has just celebrated its first centennial![1]

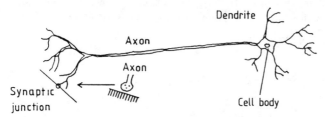

Fig. 1.1. Structure of a typical neuron (schematic).

The detailed investigation of the internal structure of neural cells, especially after the invention of the electron microscope some 50 years ago, has revealed that all neurons are constructed from the same basic parts, independent of their size and shape (see Fig. 1.1): The bulbous central part is called the cell body or *soma* ; from it project several root-like extensions, the

---

[1] Golgi and Ramon y Cajal shared the 1906 Nobel prize in medicine for their discoveries.

dendrites, as well as a single tubular fibre, the *axon*, which splits at its end into a number of small branches. The size of the soma of a typical neuron is about 10–80 μm, while dendrites and axons have a diameter of a few μm. While the dendrites serve as receptors for signals from adjacent neurons, the axon's purpose is the transmission of the generated neural activity to other nerve cells or to muscle fibres. In the first case the term *interneuron* is often used, whereas the neuron is called a *motor neuron* in the latter case. A third type of neuron, which receives information from muscles or sensory organs, such as the eye or ear, is called a *receptor neuron* .

The joint between the end of an axonic branch, which assumes a plate-like shape, and another neuron or muscle is called a *synapse*. At the synapse the two cells are separated by a tiny gap only about 200 nm wide (the *synaptic gap* or *cleft*), barely visible to Ramon y Cajal, but easily revealed by modern techniques. Structures are spoken of in relation to the synapse as *presynaptic* and *post-synaptic*, e.g. post-synaptic neuron. The synapses may be located either directly at the cell body, or at the dendrites, of the subsequent neuron, their strength of influence generally diminishing with increasing distance from the cell body. The total length of neurons shows great variations: from 0.01 mm for interneurons in the human brain up to 1 m for neurons in the limbs.

Nervous signals are transmitted either electrically or chemically. Electrical transmission prevails in the interior of a neuron, whereas chemical mechanisms operate between different neurons, i.e. at the synapses. Electrical transmission[2] is based on an electrical discharge which starts at the cell body and then travels down the axon to the various synaptic connections. In the state of inactivity, the interior of the neuron, the *protoplasm*, is negatively charged against the surrounding neural liquid. This resting potential of about -70 mV is supported by the action of the cell membrane, which is impenetrable for Na^+ ions, causing a deficiency of positive ions in the protoplasm (see Fig. 1.2).

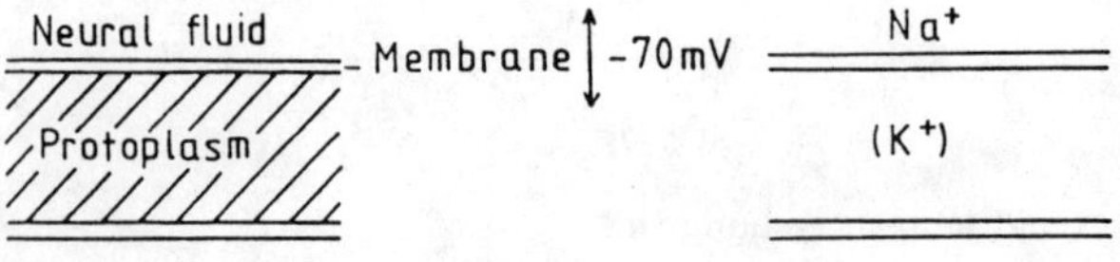

Neural fluid — Membrane $\downarrow$ -70mV $\quad Na^+$

Protoplasm $\quad (K^+)$

Fig. 1.2. Structure of an axon.

Signals arriving from the synaptic connections result in a transient weakening, or *depolarization*, of the resting potential. When this is reduced below -60 mV, the membrane suddenly loses its impermeability against Na^+ ions,

---

[2] Detailed studies of the mechanisms underlying electrical signal transmission in the nervous system were pioneered by Sir John Eccles, Alan Lloyd Hodgkin, and Andrew Huxley, who were jointly awarded the 1963 Nobel prize in medicine.

which enter into the protoplasm and neutralize the potential difference, as illustrated in the left part of Fig. 1.3. This discharge may be so violent that the interior of the neuron even acquires a slightly positive potential against its surroundings. The membrane then gradually recovers its original properties and regenerates the resting potential over a period of several milliseconds. During this recovery period the neuron remains incapable of further excitation. When the recovery is completed, the neuron is in its resting state and can "fire" again.

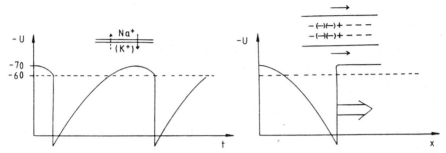

Fig. 1.3. Temporal sequence of activity spikes of a neuron (*left*), which travel along the axon as depolarization waves (*right*).

The discharge, which initially occurs in the cell body, then propagates along the axon to the synapses (see Fig. 1.3, right part). Because the depolarized parts of the neuron are in a state of recovery and cannot immediately become active again, the pulse of electrical activity always propagates in one direction: away from the cell body. Since the discharge of each new segment of the axon is always complete, the intensity of the transmitted signal does not decay as it propagates along the nerve fibre. One might be tempted to conclude that signal transmission in the nervous system is of a digital nature: a neuron is either fully active, or it is inactive. However, this conclusion would be wrong, because the intensity of a nervous signal is coded in the frequency of succession of the invariant pulses of activity, which can range from about 1 to 100 per second (see Fig. 1.4). The interval between two electrical spikes can take any value (longer than the regeneration period), and the combination of analog and digital signal processing is utilized to obtain optimal quality, security, and simplicity of data transmission.

The speed of propagation of the discharge signal along the nerve fibre also varies greatly. In the cells of the human brain the signal travels with a velocity of about 0.5–2 m/s. While this allows any two brain cells to communicate within 20–40 ms, which is something like a temporal quantum in the operation of the human central nervous system, it would cause unacceptably long reaction times for peripheral neurons connecting brain and limbs: a person would hit the ground before even knowing that he had stumbled. To increase the speed of propagation, the axons for such neurons are composed

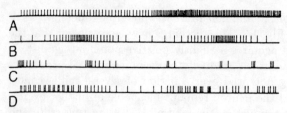

Fig. 1.4. Neuron as pulse-coded analog device: spike trains of some typical neural transmission patterns. The microstructure of the successive intervals is increasingly important from top to bottom (A–D), indicating messages of growing complexity (from [Bu77]).

of individual segments that are covered by an electrically insulating myelin sheath, which is interrupted from time to time at the so-called Ranvier nodes. The presence of an insulating cover causes the signal to propagate along the axon as in a wave guide from one Ranvier node to the next, triggering almost instantaneous discharge within the whole myelinated segment. This mode of propagation, called *saltatory conduction*, allows for transmission velocities of up to 100 m/s.

The discharge signal traveling along the axon comes to a halt at the synapses, because there exists no conducting bridge to the next neuron or muscle fibre. Transmission of the signal across the synaptic gap is mostly effected by chemical mechanisms. Direct electrical transmission is also known to occur in rare cases, but is of less interest here in view of the much lower degree of adjustability of this type of synapse. In chemical transmission, when the spike signal arrives at the presynaptic nerve terminal, special substances called *neurotransmitters* are liberated in tiny amounts from vesicles contained in the endplate (e.g. about 10^{-17} mol acetylcholin per impulse). The transmitter release appears to be triggered by the influx of Ca^{++} ions into the presynaptic axon during the depolarization caused by the flow of Na^+ ions. The neurotransmitter molecules travel across the synaptic cleft, as shown in Fig. 1.5, reaching the post-synaptic neuron (or muscle fibre) within about 0.5 ms. Upon their arrival at special receptors these substances modify the conductance of the post-synaptic membrane for certain ions (Na^+, K^+, Cl^-, etc.), which then flow in or out of the neuron, causing a polarization or depolarization of the local post-synaptic potential. After their action the transmitter molecules are quickly broken up by enzymes into pieces which are less potent in changing the ionic conductance of the membrane.

If the induced polarization potential δU is positive, i.e. if the total strength of the resting potential is reduced, the synapse is termed *excitatory*, because the influence of the synapse tends to activate the post-synaptic neuron. If δU is negative, the synapse is called *inhibitory*, since it counteracts excitation of the neuron. Inhibitory synapses often terminate at the presynaptic plates of other axons, inhibiting their ability to send neurotransmitters across the synaptic gap. In this case one speaks of presynaptic inhibition (see Fig.

1.6). There is evidence that *all* the synaptic endings of an axon are either of an excitatory or an inhibitory nature (*Dale's law*),[3] and that there are significant structural differences between those two types of synapses (e.g. the conductance for Na^+ and K^+ changes at excitatory synapses, that for Cl^- at inhibitory synapses).

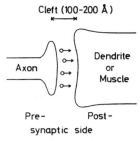

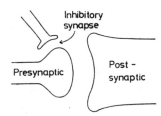

Fig. 1.5. Signal transmission at the synaptic cleft is effected by chemical neurotransmitters, such as acetylcholine.

Fig. 1.6. Presynaptic inhibition.

Under which condition is the post-synaptic neuron stimulated to become active? Although, in principle, a single synapse can inspire a neuron to "fire", this is rarely so, especially if the synapse is located at the outer end of a dendrite. Just as each axon sends synapses to the dendrites and bodies of a number of downstream neurons, so is each neuron connected to many upstream neurons which transmit their signals to it. The body of a neuron acts as a kind of "summing" device which adds the depolarizing effects of its various input signals. These effects decay with a characteristic time of 5–10 ms, but if several signals arrive at the same synapse over such a period, their excitatory effects accumulate. A high rate of repetition of firing of a neuron therefore expresses a large intensity of the signal. When the total magnitude of the depolarization potential in the cell body exceeds the critical threshold (about 10 mV), the neuron fires.

The influence of a given synapse therefore depends on several aspects: the inherent strength of its depolarizing effect, its location with respect to the cell body, and the repetition rate of the arriving signals. There is a great deal of evidence that the inherent strength of a synapse is not fixed once and for all. As originally postulated by Donald Hebb [He49], the strength of a synaptic connection can be adjusted, if its level of activity changes. An active synapse, which repeatedly triggers the activation of its post-synaptic neuron, will grow in strength, while others will gradually weaken. This mechanism of

---

[3] Sir Henry Dale shared the 1936 Nobel prize in medicine with Otto Loewi, who discovered the chemical transmission of nerve signals at the synapse.

synaptic *plasticity* in the structure of neural connectivity, known as *Hebb's rule*, appears to play a dominant role in the complex process of *learning*.

The release of neurotransmitters as well as their action at the receptor sites on the post-synaptic membrane can be chemically inhibited by substances such as atropin or curare, or their subsequent inactivation can be blocked, e.g. by cocaine. Similar, but less drastic changes of the synaptic efficacy are likely to occur naturally in the body, giving all synapses a certain degree of *plasticity*. Another point worth mentioning is that transmitter substance is randomly emitted at every nerve ending in quanta of a few 1000 molecules at a low rate. The rate of release is increased enormously upon arrival of an impulse, a single action potential causing the emission of 100–300 quanta within a very short time. However, even the random low-activity level results in small depolarization potentials, which may – from time to time – cause the spontaneous activation of the post-synaptic neuron.

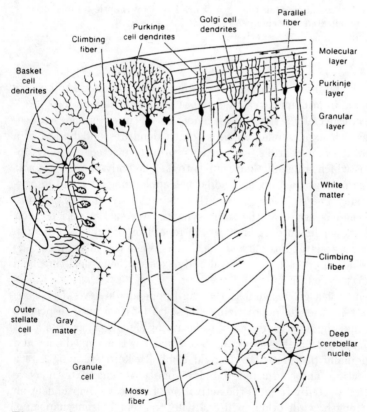

Fig. 1.7. Cross section through the cerebellar cortex, showing the multiply connected layered network of different types of neuron (from [Bu77]).

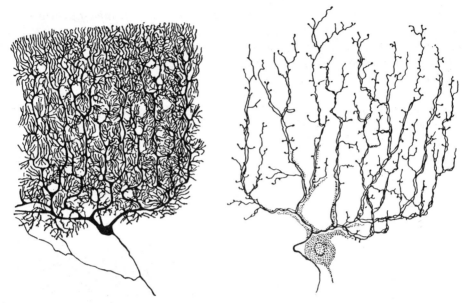

Fig. 1.8. Convergence and divergence in neural architecture: (*left*) Purkinje cell in the human cerebellum; (*right*) an axon ("climbing fibers") that intertwines with the dendritic tree of a Purkinje cell (from [Bu77]).

1.2 The Cerebral Cortex

The incredible complexity of the human central nervous system, especially of the human brain, rests not so much in the complexity and diversity of the single nerve cell, which is quite limited, as in the vast number of its constituent units, i.e. of the neurons and their mutual connections. However, it would be misleading to assume that all neurons in the brain are alike. There exist several very different types, distinguished by the size and degree of branching of their dendritic tree, the length of their axon, and other structural details. These types occur in different parts or layers of the cortex (a typical, schematic view of cortical structure is shown in Fig. 1.7), and their cooperation appears to be essential for the solution of complex cognitive tasks. But we repeat: all neurons operate on the same basic principles as explained above.

Connectivity in the central nervous system is characterized by the complementary properties of *convergence* and *divergence*. In the human cortex every neuron is estimated to receive converging input on the average from about 10 000 synapses. On the other hand, each cell feeds its output into many hundreds of other neurons, often through a large number of synapses touching a single nerve cell, as illustrated in Fig. 1.8. Perhaps the most impressive examples are found in the cerebellar cortex, where so-called *Purkinje cells* receive as many as 80 000 synaptical inputs, and a single *granule-cell* neuron connects to 50 or more Purkinje cells.

The total number of neurons in the human cortex is immense and can only be estimated. Complete counts in various small regions have yielded a virtually constant density of about 150 000 neurons per mm^2, which are distributed in layers over the full depth of the cortical tissue. Considering that the total area of the heavily folded cortex measures about 200 000 mm^2, one obtains a number of at least 3×10^{10} neurons. Some estimates even place the total number of neurons in the human central nervous system in the vicinity of 10^{11}. Combined with the average number of synapses per neuron, this yields a total of about 10^{15} synaptic connections in the human brain, the majority of which develop within a few months after birth (see Fig. 1.9).

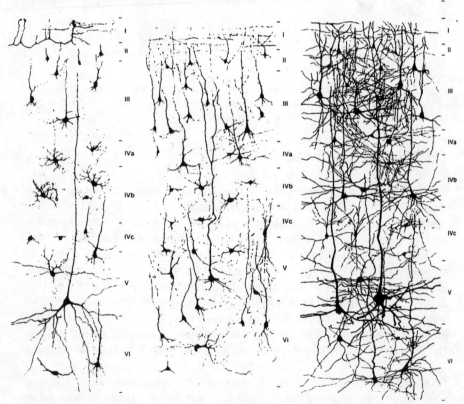

Fig. 1.9. Development of dendritic arborization in the human visual cortex, from (*left*) newborn, (*middle*) three-month-old, and (*right*) two-year-old infants (from [Bu77]).

All these synapses can vary in strength and precise location. Since the number of synaptic connections exceeds by far the amount of genetic information transmitted to the individual in its DNA, the detailed structure of the neural network in the human brain must be determined by a combination of rather general principles and information gathered or generated by the brain itself. In fact, it is well established that certain brain functions, such as the ability to perceive optical structures, are not innate. If a kitten is blindfolded right after birth for several weeks, its sense of optical pattern discrimination does not develop, although the sensitivity of the retina and the nerves transmitting optical perceptions to the brain are fully operative. However, the neural connectivity of the visual centre remains highly deficient, even if the cat is continuously exposed to optical experience at a later stage in its life.[4]

It is an interesting question in what respect the human central nervous system differs from that of other higher animals. Here it is important to relate the size of the brain to the total size of the animal. Studies in vertebrate animals have shown that the weight of the brain E grows with body weight P for species at a comparable level of evolution. Quantitatively, a quite strict relation of the form [Ch83]

$$\log E = a \log P + c \tag{1.1}$$

with $a \approx 0.6$ was found (see Fig. 1.10). Since the surface of a body grows as $P^{2/3}$, one may suspect that the brain size grows in proportion to the body surface, corresponding to the value $a = \frac{2}{3}$. This result is not too much of a surprise, because the amount of external sensory information that the brain has to process increases roughly in proportion to the surface of the body. Vertebrates belonging to different levels of evolution differ, not so much in the density of neurons, but in the magnitude of the constant of proportionality c in the relation (1.1). For example,

$$
\begin{aligned}
c_{\text{ape}}/c_{\text{rodent}} &= 11.3 \;, \\
c_{\text{man}}/c_{\text{rodent}} &= 28.7 \;,
\end{aligned}
\tag{1.2}
$$

i.e. the brain of an average human weighs roughly thirty times as much as that of a hypothetical rodent with the same body weight.

We may thus reach the tentative conclusion that the superiority of the human brain has its root in its increased complexity, especially in a certain overabundance of neurons, rendering some neurons free to perform tasks not associated with the immediate reaction to, or storage of, sensory information. This hypothesis is supported by the fact that there are large regions in the human cortex which do not serve any apparent, clear-cut purpose,

---

[4] One cannot help wondering about the ethical value of these and other experiments with animals, although they are often essential to the progress of brain research. Similarly, a major part of the knowledge about the human brain has been derived from experience with head injuries sustained by soldiers during the many wars of the last century. The prospect of learning about the brain through computer simulations appears particularly attractive and beneficial in this context.

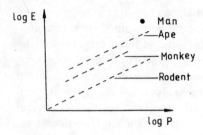

Fig. 1.10. Relation between brain size (E) and body weight (P) for various animals in comparison to man.

but which somehow contribute to the general organization and modulation of brain activities. The path to an understanding of the mode of operation of the brain in its various functions, culminating in the still-unexplained state of *consciousness*, must thus lead through the study of the properties of complex systems built of simple, identical units. This book attempts to give an introduction to the first steps taken by research scientists into the uncharted territory of complex neural networks. It will become clear that, in spite of the remarkable results obtained during the past decade especially, we are still very far away from the ultimate aim: to understand the working of the human brain.

Owing to lack of space (and the limited expertise of the authors) the description of the central nervous system has been somewhat cursory. We could not do justice to the large body of neurobiological and physiological data accumulated during the last few decades. Many important details are known about the diversity of neurons and synapses in the brain which we did not mention and which are usually also ignored by theoretical neural-network modelers. Nevertheless, even the simplified schematic and uniform "connectionist" units, upon which we will bestow somewhat indiscriminately the name "neurons" in the remainder of this book, offer a surprisingly rich structure when assembled in a closely interconnected network. The reader interested in the biology of the brain is encouraged to consult books dealing specially with this subject, e.g. [Bu77, Co72, Ec77, Ka85, Sh79]. For the less scientifically minded reader, who is interested more in the general principles than in the technical details, we suggest the following books: [Wo63, Po77, Ch83]. There are also some resources on the Internet that could prove useful:

WWW Sites

http://http2.sils.umich.edu/Public/nirg/nirg1.html
          Neurosciences Internet Resource Guide
http://ric.uthscsa.edu/
          BrainMap — Visual Brain Interface
http://www.loni.ucla.edu/
          International Consortium for Brain Mapping
http://www.bbb.caltech.edu/GENESIS
          Genesis — Neural Simulator

2. Neural Networks Introduced

2.1 A Definition

Neural network models are algorithms for cognitive tasks, such as learning and optimization, which are in a loose sense based on concepts derived from research into the nature of the brain. In mathematical terms *a neural network model is defined as a directed graph with the following properties*:

1. A state variable n_i is associated with each node i.
2. A real-valued weight w_{ik} is associated with each link (ik) between two nodes i and k.
3. A real-valued bias ϑ_i is associated with each node i.
4. A transfer function $f_i[n_k, w_{ik}, \vartheta_i, (k \neq i)]$ is defined, for each node i, which determines the state of the node as a function of its bias, of the weights of its incoming links, and of the states of the nodes connected to it by these links.

In the standard terminology, the nodes are called *neurons*, the links are called *synapses*, and the bias is known as the *activation threshold*. The transfer function usually takes the form $f(\sum_k w_{ik} n_k - \vartheta_i)$, where $f(x)$ is either a discontinuous step function or its smoothly increasing generalization known as a sigmoidal function. Nodes without links toward them are called *input* neurons; *output* neurons are those with no link leading away from them. A *feed-forward* network is one whose topology admits no closed paths. It is our aim to study such models and learn what they can (and cannot) do. Better than from any lengthy explanation of the concept of a neural network defined above, we will understand its meaning through a brief survey of its historic evolution.

2.2 A Brief History of Neural Network Models

In 1943 Warren McCulloch and Walter Pitts [Mc43] proposed a general theory of information processing based on networks of binary switching or decision elements, which are somewhat euphemistically called "neurons", although they are far simpler than their real biological counterparts. Each one of these elements $i = 1, \ldots, n$ can only take the output values $n_i = 0, 1$, where $n_i = 0$ represents the resting state and $n_i = 1$ the active state of the elementary unit. In order to simulate the finite regenerative period of real neurons, changes in the state of the network are supposed to occur in discrete time steps $t = 0, 1, 2, \ldots$. The new state of a certain neural unit is determined by the influence of all other neurons, as expressed by a *linear* combination of their output values:

$$h_i(t) = \sum_j w_{ij} n_j(t) \, . \tag{2.1}$$

Here the matrix w_{ij} represents the synaptic coupling strengths (or *synaptic efficacies*) between neurons j and i, while $h_i(t)$ models the total post-synaptic polarization potential at neuron i caused by the action of all other neurons. h_i can be considered the *input* into the neural computing unit, and n_i the *output*. The properties of the neural network are completely determined by the functional relation between $h_i(t)$ and $n_i(t + 1)$. In the simplest case, the neuron is assumed to become active if its input exceeds a certain threshold ϑ_i, which may well differ from one unit to the next. The evolution of the network is then governed by the law

$$n_i(t + 1) = \theta(h_i(t) - \vartheta_i) \, , \tag{2.2}$$

where $\theta(x)$ is the unit step function, i.e. $\theta(x < 1) = 0$ and $\theta(x > 1) = 1$.

McCulloch and Pitts showed that such networks can, in principle, carry out any imaginable computation, similar to a programmable, digital computer or its mathematical abstraction, the Turing machine. In a certain sense the network also contains a "program code", which governs the computational process, namely the coupling matrix w_{ij}. The network differs from a traditional computer in that the steps of the program are not executed sequentially, but in parallel within each elementary unit. One might say that the program code consists of a single statement, i.e. the combination of the equations (2.1) and (2.2). The extreme reduction of the program is compensated by the substitution of a vast number of processing elements (10^{11} in the human brain!) for the single processing unit of a conventional, sequential electronic computer.

The designer of a McCulloch–Pitts-type neural network now faces the problem of how to choose the couplings w_{ij} so that a specific cognitive task is performed by the machine. Here we use the word "cognitive task" in a generalized sense: it can mean any task requiring digital or, as we shall see later, analog information processing, such as the recognition of specific optical or acoustical patterns.[1] This question was addressed in 1961 by Eduardo

---

[1] The terms "neural computation" and "neural computer", which are frequently used in this context, are somewhat misleading: neural networks do not appear to be very adept at solving classical computational problems, such as adding, multiplying, or dividing numbers. The most promising area of application concerns cognitive tasks that are hard to cast into algebraic equations.

Caianiello [Ca61], who gave a *"learning"* algorithm that would allow the determination of the synaptic strengths of a neural network. This algorithm, called the *mnemonic* equation by Caianello, incorporates in a simple way the basic principle of Hebb's learning rule.

Around 1960 Frank Rosenblatt and his collaborators [Ro62] extensively studied a specific type of neural network, which they called a *perceptron*, because they considered it to be a simplified model of the biological mechanisms of processing of sensory information, i.e. perception. In its simplest form, a perceptron consists of two separate layers of neurons[2] representing the input and output layer, respectively, as illustrated in Fig. 2.1. The neurons of the output layer receive synaptic signals from those of the input layer, but not vice versa, and the neurons within one layer do not communicate with each other. The flow of information is thus strictly directional; hence one speaks of a *feed-forward* network.

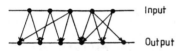 Input

Output

 Input

Output

Fig. 2.1. Simple perceptron, consisting of two layers of neurons. The input neurons feed into the neurons of the output layer, but not vice versa.

Fig. 2.2. The exclusive-OR function cannot be represented on a two-layer perceptron.

Rosenblatt's group introduced an iterative algorithm for constructing the synaptic couplings w_{ij} such that a specific input pattern is transformed into the desired output pattern, and even succeeded in proving its convergence [Bl62a]. However, M. Minsky and S. Papert [Mi69] pointed out a few years later that this proof applies only to those problems which can, in principle, be solved by a perceptron. What made matters worse was that they showed the existence of very simple problems which cannot be solved by any such two-layered perceptron! The most notorious of these is the "exclusive-OR" (XOR) gate, which requires two input neurons to be connected with a single output neuron (see Fig. 2.2) in such a way that the output unit is activated if, and only if, one of the input units is active. The logic XOR gate is a standard problem easily solved in computer design, and thus the result of Minsky and Papert represented a severe blow to the perceptron concept.[3]

---

[2] Technically, even the simple perceptron was a three-layer device, since a preprocessing layer of sensory units was located in front of the first layer of computing neurons, which is here called the "input" layer. The connections between the sensory units and the following (input) layer of neurons were not adjustable, i.e. no learning occurred at this level.

[3] In fact, the XOR problem can be easily solved by a feed-forward network with *three* layers, but no practical algorithm for the construction of the w_{ij} of such generalized perceptrons was known at the time.

Another very fruitful development began when W. Little [Li74] pointed out the similarity between a neural network of the type proposed by McCulloch and Pitts and systems of elementary magnetic moments or *spins*, see Fig. 2.3. In these systems, called *Ising models*, the spin s_i at each lattice site i can take only two different orientations, *up* or *down*, denoted by $s_i = +1$ (up) and $s_i = -1$ (down). The analogy to a neural network is realized by identifying each spin with a neuron and associating the upward orientation $s_i = +1$ with the active state $n_i = 1$ and the downward orientation $s_i = -1$ with the resting state $n_i = 0$.

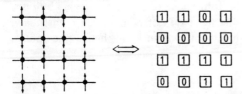

Fig. 2.3. A one-to-one correspondence exists between lattices of magnetic spins with two different orientations (Ising systems) and McCulloch–Pitts neural networks. A spin pointing up (down) is identified with an active (resting) neuron.

These ideas were further developed by Little and Gordon Shaw [Li78] and by John Hopfield [Ho82] who studied how such a neural network or spin system can store and retrieve information. The Little and Hopfield models differ in the manner in which the state of the system is updated. In Little's model all neurons (spins) are updated synchronously according to the law (2.1), whereas the neurons are updated sequentially one at a time (either in a certain fixed order or randomly) in the Hopfield model. Sequential updating has a considerable advantage when the network is simulated on a conventional digital computer, and also for theoretical analysis of the properties of the network. On the other hand, it holds the essential conceptual disadvantage that the basic feature of neural networks, namely the simultaneous operation of a large number of parallel units, is given up. Neurons in the human brain surely do *not* operate sequentially, this being precisely the reason for the brain's superiority in complex tasks to even the fastest existing electronic computer.[4]

The analogy with spin systems became especially fruitful owing to advances in the understanding of the thermodynamic properties of disordered systems of spins, the so-called *spin glasses*, achieved over the past decade. In order to apply these results to neural networks it is necessary to replace the deterministic evolution law (2.2) by a *stochastic* law, where the value

---

[4] On the other hand, it is highly questionable whether biological assemblies of neurons can operate *synchronously*, except in very special cases. Most likely, reality is somewhere intermediate between the two extremes, where precisely being a sharply debated subject [Cl85].

of $n_i(t+1)$ is assigned according to a probabilistic function depending on the intensity of the synaptic input h_i. This probability function contains a parameter T that plays the role of a "temperature". However, T is not to be understood as modeling the physical temperature of a biological neural network, but as a formal concept designed to introduce stochasticity into the network and thus to allow for the application of the powerful techniques of statistical thermodynamics. In the limit of vanishing "temperature" the deterministic McCulloch–Pitts model is recovered.[5] Studies, which will be discussed in the course of this book, have shown that stochastic evolution can render the network less susceptible to dynamic instabilities and may thus improve the overall quality of its operation.

Input layer

Hidden layer

Output layer

Fig. 2.4. A feed-forward network with one internal ("hidden") layer of neurons.

In recent years the interest in layered, feed-forward networks (perceptrons) has been revived. This development was initiated by the (re-)discovery of an efficient algorithm for the determination of the synaptic coupling strengths in multilayered networks with *hidden layers*, see Fig. 2.4. The power of this method, initially suggested by Werbos [We74] and now known as *error back-propagation*, was recognized around 1985 by several groups of scientists [Cu86, Pa85, Ru86a]. This learning algorithm is based on a simple but very effective principle: the synaptic strengths w_{ij} are modified iteratively such that the output signal differs as little as possible from the desired one. This is achieved by application of the gradient method, which yields the required modifications δw_{ij}. Since the operation of the network corresponds to a highly nonlinear mapping between the input and the output – the step function in (2.1) is nonlinear – the method must be applied many times until convergence is reached.

Error back-propagation is a particular example of a much larger class of learning algorithms which are classified as *supervised learning*, because at each step the network is adjusted by comparing the actual output with the desired output. Not only are such algorithms most probably not implemented in biological neural networks, but they also suffer because they are applicable only when the desired output is known in detail. For this reason also other concepts of "learning", i.e. adjustment of the synaptic strengths, are studied intensively, such as strategies based on reward and penalty, evolution and

---

[5] The stochastic evolution law introduced by Little may actually represent a real feature of biological neurons, viz. that nerve cells can fire spontaneously without external excitation, leading to a persistent noise level in the network.

selection, and so on. We shall discuss some of these developments in detail and study their strengths and weaknesses in the following chapters.

2.3 Why Neural Networks?

Historically, the interest in neural networks has two roots: (1) the desire to understand the principles on which the human brain works, and (2) the wish to build machines that are capable of performing complex tasks for which the sequentially operating, programmable computers conceived by Babbage and von Neumann are not well suited. Since the recent surge of interest in neural networks and "neural computing" is mainly based on the second aspect, we begin our discussion with it.[6]

2.3.1 Parallel Distributed Processing

Everyday observation shows that the modest brains of lower animals can perform tasks that are far beyond the range of even the largest and fastest modern electronic computers. Just imagine that any mosquito can fly around at great speed in unknown territory without bumping into objects blocking its path. And a frog's tongue can catch these insects in full flight within a split second. No present-day electronic computer has sufficient computational power to match these and similar accomplishments. They typically involve some need for the recognition of complex optical or acoustical patterns, which are not determined by simple logical rules.

Great efforts were made in the past two decades to solve such problems on traditional programmable computers. One product of these efforts was the emergence of the techniques of *artificial intelligence* (AI). While the products of AI, which are more appropriately described as expert systems, had a number of impressive successes, e.g. in medical diagnosis, they are much too slow to perform the analysis of optical or speech patterns at the required high rate. Moreover, the concept of AI is based on formal logical reasoning, and can thus only be applied when the logical structure of a certain problem has been analyzed.

Another source of dissatisfaction is that the speed of solid-state switching elements, the basic units of electronic computers, has reached a point where future progress seems to be limited. In order to accelerate computational tasks further one has therefore turned to the concept of *parallel processing*, where several operations are performed at the same time. Three problems are encountered here. The first is that the basic unit of a traditional computer,

---

[6] The two goals are sometimes hard to reconcile: the excitement generated by the recent progress in "neural computing" has an innate tendency to distract interest from research into the principles governing the human mind. A critical view has been voiced by F. Crick [Cr89].

the central processing unit (CPU), is already a very complex system containing hundreds of thousands of electronic elements that cannot be made arbitrarily cheap. Hence there is a cost limit to the number of CPUs that can be integrated into a single computer. Secondly, most problems cannot be easily subdivided into a very large number of logically independent subtasks that can be executed in parallel. Finally, the combination of a large number of CPUs into a single computer poses tremendous problems of architectural design and programming which are not easily solved if one wants to have a general-purpose computer and not a machine dedicated to one specific task.

The inadequacies of logic-based artificial intelligence and the speed limitations of sequential computing have led to a resurgent interest in neural networks since about 1980. One now considers neural networks as prototype realizations of the concept of *parallel distributed processing* [PDP86]. In contrast to the great complexity of the CPU of a traditional electronic computer the elements ("neurons") of a neural network are relatively simple electronic devices, which contain only a switching element and adjustable input channels. Even with presently available technology it is not unimaginable to arrange tens of thousands of binary decision elements on a single silicon chip. Many thousands of such identical chips might be integrated into a single *neural computer* at a reasonable cost. Thus neural computers with a billion or more neuron-like elements appear technically feasible, and a system approaching the complexity of the human brain (about 10^{11} neurons) does not belong entirely to the realm of science fiction.

The main difficulties along this road lie in the implementation of a sufficient number of synaptic connections and in the limited knowledge of appropriate strategies for synaptic-strength selection, commonly called *learning*. Especially the second of these problems, i.e. the choice of network topology and synaptic efficacies, will concern us here, although we will also briefly discuss some technical aspects of electronic "neurons". The main point here is that one would like to keep the network architecture and the learning protocol as general as possible, so that they can be applied to a large class of problems, e.g. in pattern recognition. In particular, since one would like to apply neural computing to problems whose logical structure is relatively poorly understood, details of this structure should be irrelevant for the learning algorithm. Accordingly, most of the learning strategies that have been extensively studied are based on rather general concepts, such as output optimization and strategies of reward or evolution.

Neural networks seem to be particularly well suited to approaching problems in subsequent steps of increasing detail, i.e. in hierarchical steps. The emphasis here is more on the geometrical or topological aspects of the problem than on logical relations between its constituent parts. For this reason the term *geometrical*, as opposed to the usual *analytical*, approach to computing has been coined. An important aspect of problem solving in hierarchies is that the network is potentially capable of *classification* and *generalization*.

Although traditional methods of AI can readily solve these tasks if the classification criteria or generalization rules are known, e.g. in the case of expert systems, they have difficulty establishing such relations on their own. This is different in the case of neural networks, which have been shown to be able to learn and generalize from examples without knowledge of rules. Particularly impressive examples are networks that have learned to pronounce speech or to analyze protein structure (see Chapt. 7).

At the present state of development, traditional computers are still superior to neural networks in performing tasks or solving problems whose structure has been thoroughly analyzed. However, it is only fair to state that often many man-years of analysis are required before the problem is sufficiently well understood to be accessible to computational treatment. On the other hand, a neural network may be able to learn from a large number of examples within a few hours of synaptic-strength adjustment. Moreover, since a very large number of computing elements work in parallel, the execution – once the correct synaptic connections have been established – may be much faster than on a normal computer. This requires, of course, that the neural network runs on dedicated electronic hardware; the remark does not apply to simulations on sequentially operating computers.

Finally, a potentially important advantage of neural networks is their high degree of *error resistivity*. A normal computer may completely fail in its operation if only a single bit of stored information or a single program statement is incorrect. This translates into a high risk of failure if the system cannot be continuously serviced. In contrast, the operation of a neural network often remains almost unaffected if a single neuron fails, or if a few synaptic connections break down. It usually requires a sizable fraction of failing elements before the deterioration is noticeable. In view of the rapidly increasing use of electronic data processing in vital areas this is an attractive feature of neural computing, which may become of practical importance.

A major disadvantage of neural networks is the broad lack of understanding of *how* they actually solve a given cognitive task. Our present ignorance stems from the fact that neural networks do not break a problem down into its logical elements but rather solve it by a holistic approach, which is hard to penetrate logically. When the logical structure of the solution of the problem is not known anyway, this represents an advantage (see above), but it becomes a disadvantage when one wants to determine whether the solution provided by the neural network is correct. The sole presently known method of testing the operation of a neural network is to check its performance for individual test cases, a not very enlightening technique.[7] No one knows how to judge the performance of a neural network knowing only its architecture, and

---

[7] Indeed, testing the correctness of conventional (von Neumann) computer programs is a difficult problem, as well. For example, there is no general algorithm that checks in a finite time whether a given program always terminates or not. But a variety of clever testing techniques allows us to check the accuracy of computer programs with a very high probability of being correct.

it is almost impossible to determine what task the network actually performs from the pure knowledge of the synaptic efficacies.

2.3.2 Understanding How the Brain Works

The second source of motivation, and for many scientists the more profound one, is the desire to understand better how the human brain processes information. Here neural network models can help in two ways: (1) they permit the investigation, by computer simulation, of processes that cannot be studied in the live brain; (2) they permit the modeling of abstract concepts and mechanisms involved in certain brain functions which cannot be directly deduced from observations.

Today, a great deal is known about the basic building block of the brain, i.e. the neuron, and the structure and operation of small ensembles of neurons has been studied in detail. Perhaps the most famous and ground-breaking study was that by Hubel and Wiesel on the striate visual cortex of the cat [Hu62, Hu65, Hu88]. Their finding that certain neurons are responsible for the detection of specifically oriented optical patterns, and that adjacent neurons detect patterns with a slightly different angle of orientation, posed a challenge to discover the mechanisms for topological-structure formation in the brain. Owing to the long duration of the structuring process it is not possible to investigate it in the biological environment. Neural network models permit the study of various hypotheses in simulation. After pioneering work in this field by C. von der Malsburg [Ma73], Stephen Grossberg [Gr76], Shun-ichi Amari [Am80, Am90] and Teuvo Kohonen [Ko82, Ko84], we now know several very effective mechanisms for topological structure formation in the brain.

Physiological observations inside the brain are generally able to yield information about the operation of neurons connected in a fixed configuration, but hardly about the detailed temporal evolution of the connectivity structure among neurons. In particular, synaptic plasticity cannot be studied *in situ*. Simulations of neural networks provide a useful way of probing theories of synaptic plasticity and studying its role in certain brain functions.

An area where neural networks have greatly contributed to our understanding of brain function is memory. Storage of information in the brain must be very different from that in traditional computers, which store information by location and in binary code. Only a tiny fraction of the incoming sensory information is permanently stored, and this selection depends on its relation to the already available memory, as everyone knows from daily experience. Similarly, the recall of stored information by the brain relies more on associations with other stored data than on the order in which the memory was acquired. Neural-network simulations have allowed researchers to study how and under what conditions such an *associative* or *content-addressable memory* works, and how much information it can store. Some surprising results of such studies are, e.g., that an associative memory can work better in the presence of a certain level of internal noise, or with a certain degree of "forget-fulness". Neural-network models involving nonlinear synaptic-plasticity laws have provided a better understanding of the working of short-term memory, which is limited in capacity, but where old memories are continually super-seded by new ones.

A question that has intrigued generations of psychologists is the role of dreaming in the operation of the brain. A modern view, originally advanced by Crick and Mitchison [Cr83], is that the purpose of dream sleep (REM sleep) is the elimination of undesirable states of memory.[8] It seems that the intensity of dreaming is not related to the amount of stored memories, but rather to the degree of synaptic plasticity of the brain which culminates right after birth. Neural-network simulations have lent credibility to this theory by showing that the detection and elimination of spurious memories is possible in a specific mode of operation, where the network receives no input from the outside, similar to the way the brain works during sleep.

One particularly fascinating aspect of the brain is that it is a reliable system built from intrinsically unreliable units, the neurons [La89]. More-over, it is able to process extremely noisy sensory data and extract reliable information from them to an astonishing degree.[9] Are these aspects interre-lated, i.e. is the ability to ignore internal noise necessary to cope with a high level of external noise? Studies of neural networks support this hypothesis. A related observation, that even extensive lesions of the brain often hardly affect the mental ability of a person, is also illuminated by neural-network studies. Neural nets are found to deteriorate "gracefully", i.e. their perfor-mance decreases only very gradually, even if a large fraction of the synaptic connections is severed.

Another interesting result from neural-network studies sheds light on the remarkable ability of the human brain to classify and categorize, which un-derlies much of our abstract thinking. It was found that the stability of stored patterns in associative-memory models is often arranged in a hierarchical or-der. Shallow, broad stability valleys branch into deeper but narrower ones, which fragment into even finer structures, and so on. This type of hierarchy (called *ultrametricity*) was originally observed in models of random magnetic materials, so-called spin-glass models, which were later found to be equivalent to a broad class of neural-network models.

---

[8] This view is supported by the observation that the daily duration of dream sleep gradually decreases during the life of a person, and that the longest periods of dreaming occur in newly born, or even unborn, babies. If this hypothesis is correct, attempts to remember the contents of dreams, as advocated by Freud and his followers, would be counterproductive rather than helpful!

[9] For example, focusing attention, one can follow the words of a single speaker in a room where many people talk simultaneously, often far exceeding the speaker's voice level. Among researchers of the auditory process this is known as the "cock-tail party effect".

Besides these schematic attempts to represent higher brain functions by artificial neural networks many studies have been carried out modeling the properties of small selected groups of neurons that perform well-defined functions, especially in simple organisms. Examples are *central pattern generators* of invertebrate animals (for an overview see [Se89]). The formal theory of such simple neuronal networks is usually a straightforward matter of electrical-circuit analysis, and rarely involves deeper concepts of theoretical physics. Here we shall not pursue this line of research further, since it is primarily of biological interest.

2.3.3 General Literature on Neural Network Models

For further reading we have listed several monographs which contain a more comprehensive presentation of the subject or emphasize different aspects of this rapidly developing research field.

Two classic monographs are the book of Kohonen [Ko84], and the two volumes by the PDP research group on parallel distributed processing [PDP86]. The textbooks of Hertz, Krogh, and Palmer [He91] and Haykin [Ha94b] provide excellent overviews of the subject material, and the more hands-on programming books of Masters [Ma94c] and Freeman [Fr94] give practical advice most other texts lack. The book by Mézard, G. Parisi, and M.A. Virasoro [Me87] contains many reprints of important original research articles on spin-glass theory, while the reprint volume of original neural-network articles edited by Shaw and Palm [Sh88] emphasizes the development before 1980 and thus provides an interesting historical perspective. For more historical perspective we also suggest the two volumes by Hecht-Nielsen [He90] and Anderson et al. [An90]. We would also like to mention Amit's book [Am89a] which gives a comprehensive treatment of the physics of neural network models.

3. Associative Memory

3.1 Associative Information Storage and Recall

Associative memory, i.e. storage and recall of information by association with other information, may be the simplest application of "collective" computation on a neural network. An information storage device is called an *associative memory*, if it permits the recall of information on the basis of a partial knowledge of its content, but without knowing its storage location. One also speaks of *content-addressable* memory.

With traditional computers the recall of information requires precise knowledge of the memory address, and a large part of data-base management is concerned with the handling of these addresses, e.g. in index files. However, it is obvious that our human memory cannot be organized in this way. If we try to recall the name of a certain person, it is completely useless to know that this was the 3274th name we encountered during our life. Rather, it helps a lot, if we know that the name starts and ends with the letter "N", even if this may be far from unique. If we remember, moreover, that the name reminded us of a famous English scientist, the successful recall "NEWTON" is virtually guaranteed. Other associations, such as the time or place where we last met the person in question, can serve a similar purpose.

This should not be taken to imply that address-oriented information storage has no place in the human memory. Often it is possible to recall a specific event by remembering all related memories in the order in which they occurred. Or one can recall an appointment by going through the day's schedule in mind. Nonetheless, it is clear that associations play an essential role in our memory processes, as exemplified by the "cribs" which are often useful for memorizing words or numbers that are hard to remember otherwise. Without associative memory, most crossword puzzles would remain unsolved.

Here we define an associative memory as follows.[1] Assume that p binary patterns containing N bits of information, ν_i^μ ($i = 1, \ldots, N$; $\mu = 1, \ldots, p$), are stored in the memory. If now a new pattern n_i is presented, that stored pattern ν_i^λ is to be recalled which most strongly resembles the presented

---

[1] The associative-memory model was studied by several researchers in the early 1970s [An72, Na72, Ko72, Am72]. It has enjoyed great popularity since Hopfield [Ho82] drew the analogy to physical spin systems.

pattern. This means that ν_i^λ and n_i should differ in as few places as possible, i.e. that the mean square deviation

$$H_\mu = \sum_{i=1}^{N} (n_i - \nu_i^\mu)^2 \tag{3.1}$$

is minimal for $\mu = \lambda$. H_μ is called the *Hamming distance* between the patterns n_i and ν_i^μ. In principle, this problem is easily solved on a standard digital computer by computing all values H_μ and then searching for the smallest one. However, if the memory contains many large patterns, this procedure can take quite a long time. We are therefore led to ask the question whether the patterns can be stored in a neural network of N elements (neurons) in such a way that the network evolves from the initial configuration n_i, corresponding to the presented pattern, into the desired configuration ν_i^λ under its own dynamics.

In order to formulate the problem, it is useful to invoke the analogy between neurons and Ising spins, mentioned in Sect. 2.2, replacing the quantities ν_i, n_i by new variables σ_i, s_i, which are defined as

$$s_i = 2n_i - 1 , \qquad \sigma_i = 2\nu_i - 1 . \tag{3.2}$$

The new variables take the values ± 1, instead of 0 and 1. The squared deviations are then expressed as

$$(n_i - \nu_i^\mu)^2 = \frac{1}{4}(s_i - \sigma_i^\mu)^2 = \frac{1}{4}\left[s_i^2 - 2s_i\sigma_i^\mu + (\sigma_i^\mu)^2\right] = \frac{1}{2}(1 - s_i\sigma_i^\mu) , \tag{3.3}$$

and the search for the minimum of H_μ is then equivalent to the search for the *maximal* value of the function

$$A_\mu(s_i) = \sum_{i=1}^{N} \sigma_i^\mu s_i \qquad (\mu = 1, ..., p) \tag{3.4}$$

for a given pattern s_i. A_μ is called the (Euclidean) *scalar product* or *overlap* of the N-component vectors σ_i^μ and s_i.

According to Hopfield [Ho82] the dynamical evolution of the state of the network is now defined as follows. The individual neurons i are assigned new values $s_i(t+1)$ in some randomly chosen sequence. The new values are computed according to the rule

$$s_i(t+1) = \mathrm{sgn}[h_i(t)] \equiv \mathrm{sgn}\left[\sum_{j=1}^{N} w_{ij}s_j(t)\right] , \tag{3.5}$$

where the instantaneous output values of all neurons are to be taken on the right-hand side. The temporal evolution proceeds, as already mentioned in the previous chapter, in finite steps, i.e. t takes only the discrete values $t = 0, 1, 2, \ldots$. Equation (3.5) corresponds to the rule (2.2), rewritten in the new variables s_i. Whether the self-coupling term with $j = i$ occurs on the right-hand side of (3.5) remains open for now; we can always drop it by requiring that the diagonal elements of the matrix w_{ij} vanish: $w_{ii} = 0$.

3.2 Learning by Hebb's Rule

The immediate problem consists in choosing the synaptic coupling strengths w_{ij} given the stored patterns σ_i^μ, such that the network evolves from the presented pattern s_i into the most similar stored pattern by virtue of its own inherent dynamics. For a start, we will give the solution to this problem and then prove that it really serves its purpose. In order to do so we have to show, first, that each stored pattern corresponds to a stable configuration of the network and, second, that small deviations from it will be automatically corrected by the network dynamics.

We begin with the simple, but not trivial, case of a single stored pattern σ_i. The state corresponding to this pattern remains invariant under the network dynamics, if $h_i = \sum_j w_{ij}\sigma_j$ has the same sign as σ_i. This condition is satisfied by the simple choice

$$w_{ij} = \frac{1}{N}\sigma_i\sigma_j \tag{3.6}$$

for the synaptic strengths. To see this, we calculate the synaptic potentials h_i, assuming $s_i(t) = \sigma_i$. By virtue of the relation $(\sigma_i)^2 = (\pm 1)^2 = 1$ we have[2]

$$h_i \equiv \sum_j \sigma_j = \frac{1}{N}\sum_j \sigma_i(\sigma_j)^2 = \frac{1}{N}\sigma_i\sum_j 1 = \sigma_i . \tag{3.7}$$

Inserting this result into the evolution law (3.5), we find, as required,

$$s_i(t+1) = \mathrm{sgn}(h_i) = \mathrm{sgn}(\sigma_i) = \sigma_i = s_i(t) . \tag{3.8}$$

Now assume that the network would start its evolution not exactly in the memorized state σ_i, but some of the elements had the wrong value (-1 instead of $+1$, or vice versa). Without loss of generality we can assume that these are the first n elements of the pattern, i.e.

$$s_i(t=0) = \begin{cases} -\sigma_i & \text{for } i = 1, \ldots, n \\ +\sigma_i & \text{for } i = n+1, \ldots, N \end{cases} . \tag{3.9}$$

Then the synaptic potentials are

$$\begin{aligned} h_i(t=0) &= \sum_j w_{ij}s_j(t=0) = \frac{1}{N}\sigma_i\sum_j \sigma_j s_j(t=0) \\ &= \frac{1}{N}\sigma_i[-n + (N-n)] = \left(1 - \frac{2n}{N}\right)\sigma_i , \end{aligned} \tag{3.10}$$

i.e. for $n < \frac{N}{2}$ the network makes the transition into the correct, stored pattern after a single update of all neurons:

$$s_i(t=1) = \mathrm{sgn}[h_i(t=0)] = \mathrm{sgn}\left(1 - \frac{2n}{N}\right)\mathrm{sgn}(\sigma_i) = \sigma_i . \tag{3.11}$$

---

[2] If we had excluded the self-coupling of neuron i, the sum would run over only $(N-1)$ terms, and the appropriate normalization factor in (3.6) would be $\frac{1}{N-1}$.

One also says that the stored pattern acts as an *attractor* for the network dynamics.

We are now ready to consider the general case, where p patterns $\sigma_i^1, \ldots, \sigma_i^p$ are to be stored. Generalizing (3.6), we choose the synaptic strengths according to the rule

$$w_{ij} = \frac{1}{N} \sum_{\mu=1}^{p} \sigma_i^\mu \sigma_j^\mu , \qquad (3.12)$$

which is usually referred to as *Hebb's rule*. If $\sigma_i^\mu = \sigma_j^\mu$, i.e. if neurons i and j are both active or dormant for a given pattern, a positive contribution to w_{ij} results. If this occurs for the majority of patterns, the synapse becomes excitatory. On the other hand, if $\sigma_i = -\sigma_j$ for most patterns, we obtain $w_{ij} < 0$, i.e. the synaptic connection becomes inhibitory. When we now select one specific pattern ν for the initial state of the neural network and study its stability, the synaptic potentials are given by

$$
\begin{aligned}
h_i^{(\nu)} &= \sum_j w_{ij} \sigma_j^\nu = \frac{1}{N} \sum_{\mu=1}^{p} \sigma_i^\mu \sum_j \sigma_j^\mu \sigma_j^\nu \\
&= \frac{1}{N} \left[\sigma_i^\nu \sum_j \sigma_j^\nu \sigma_j^\nu + \sum_{\mu \neq \nu} \sigma_i^\mu \sum_j \sigma_j^\mu \sigma_j^\nu \right] \\
&= \sigma_i^\nu + \frac{1}{N} \sum_{\mu \neq \nu} \sigma_i^\mu \sum_j \sigma_j^\mu \sigma_j^\nu .
\end{aligned}
\qquad (3.13)
$$

The first term is identical with the one in (3.7), obtained for a single stored pattern. In the case that the stored patterns are *uncorrelated*, one sees that the second term contains in all $N(p-1)$ randomly signed contributions (± 1). Hence, according to the laws of statistics, for large N and p its value will typically be of size $\frac{1}{N}\sqrt{N(p-1)} = \sqrt{(p-1)/N}$.[3] As long as $p \ll N$, i.e.

---

[3] This is a consequence of the *central-limit theorem*: the sum of a large number, N, of independent random variables will obey a Gaussian distribution centered at N times the mean value and having a variance which is $\sqrt{N}$ times the variance of the original probability distribution. For our special case this can be deduced as follows (see, e.g., [Wi86a]). Let $\xi_i, i = 1, 2, \ldots$, be discrete random variables taking the values $\xi_i = \pm 1$ with equal probability. The task is to find an expression for the probability distribution $p(x, N)$ of the sum $x = \sum_i^N \xi_i$. Clearly we have

$$p(x, 1) = \frac{1}{2}\left[\delta(x-1) + \delta(x+1)\right] = \frac{1}{2}\delta(|x| - 1) .$$

Furthermore, if we know $p(x, N-1)$, we obtain $p(x, N)$ by the convolution

$$
\begin{aligned}
p(x, N) &= \frac{1}{2} \int dz \, p(z, N-1) \delta(|x - z| - 1) \\
&= \int dz \, p(z, N-1) p(x - z, 1) .
\end{aligned}
$$

when the number of stored patterns is much smaller than the total number of neurons in the network, the additional term will most likely not affect the sign of the complete expression $h_i^{(\nu)}$, which alone determines the reaction of the ith neuron. The smaller the ratio p/N, the greater the likelihood that the pattern ν is a stable configuration of the network:

$$s_i(t=1) = \operatorname{sgn}\left[h_i^{(\nu)}(t=0)\right] = \operatorname{sgn}(\sigma_i^\nu) = \sigma_i^\nu \,. \tag{3.14}$$

Again, if n neurons start out in the "wrong" state, a combination of the previous considerations yields

$$
\begin{aligned}
h_i &= \left(1 - \frac{2n}{N}\right)\sigma_i^\nu + \frac{1}{N}\sum_{\mu\neq\nu}\sigma_i^\mu\sum_j\sigma_j^\mu\sigma_j^\nu \\
&= \left(1 - \frac{2n}{N}\right)\sigma_i^\nu + O\!\left(\sqrt{\frac{p-1}{N}}\right).
\end{aligned}
\tag{3.15}
$$

Under the condition $n, p \ll N$ we will still have $\operatorname{sgn}(h_i) = \sigma_i^\nu$, i.e. the network configuration will converge to the desired pattern within a single global update. However, if the number of stored patterns is comparable to the number N of neurons, the second, randomly distributed term in (3.15) becomes of order one, and the patterns can no longer be recalled reliably. As we shall prove in Part II with the help of powerful mathematical methods borrowed from statistical physics, this undesirable case occurs when the number p of stored random patterns exceeds 14% of the number N of neurons, i.e. $p/N < 0.14$.

An explicit expression is obtained by means of the Fourier transform

$$\tilde{p}(k,1) \equiv \int \mathrm{d}x\, p(x,1)\exp(\mathrm{i}kx) = \frac{1}{2}\left[\exp(\mathrm{i}k) + \exp(-\mathrm{i}k)\right] = \cos(k)\,.$$

Since the Fourier transformation converts a convolution of two functions into the product of the Fourier transforms, i.e. $\tilde{p}(k,N) = \tilde{p}(k,N-1)\tilde{p}(k,1)$, we find that

$$p(x,N) = \frac{1}{2\pi}\int \mathrm{d}k\, \tilde{p}(k,N)\exp(-\mathrm{i}kx) = \frac{1}{2\pi}\int \mathrm{d}k\, \cos^N(k)\exp(-\mathrm{i}kx)\,.$$

An approximate expression for large N can be found by noting that this limit corresponds to small values of k, where we can expand the cosine function:

$$\cos^N(k) \approx \left(1 - \frac{1}{2}k^2\right)^N \approx \exp\left(-\frac{1}{2}Nk^2\right),$$

Finally

$$p(x,N) \approx \frac{1}{2\pi}\int \mathrm{d}k\,\exp\left(-\mathrm{i}kx - \frac{1}{2}Nk^2\right) = (2\pi N)^{-1/2}\exp\left(\frac{-x^2}{2N}\right),$$

showing that the distribution has the width $\Delta x \propto \sqrt{N}$.

This limit becomes significantly higher if the various stored patterns happen to be orthogonal to each other, i.e. if their scalar products satisfy the conditions

$$\frac{1}{N} \sum_i \sigma_i^\mu \sigma_i^{\mu'} = \delta_{\mu\mu'} \;. \tag{3.16}$$

Obviously, the disturbing "noise" term in (3.15) vanishes in this case, and hence it is possible to store and retrieve N patterns.[4] As we shall see later (in Chapt. 10), there exist "learning" protocols, i.e. rules for the choice of the synaptic connections, which are superior to Hebb's rule and permit the storage of up to $2N$ retrievable patterns, and which even work in the presence of correlations. However, Hebb's rule is by far the simplest and most easily implementable of learning rules, and thus will remain of great practical importance. It also serves as the basis for more sophisticated concepts of learning, to be discussed in Chapt. 10.

Use the program ASSO (see Chapt. 22) to learn and recall the first five letters of the alphabet: A–E. Choose the following parameter values: (5/5/0/1) and (1/0/0/1;2), i.e. sequential updating, temperature and threshold zero, and experiment with the permissible amount of noise. Which letter is most stable?

3.3 Neurons and Spins

3.3.1 The "Magnetic" Connection

By denoting the two possible states of a neuron by the variables $s_i = +1$ (active neuron) and $s_i = -1$ (resting neuron), we have already exposed the analogy between a neural network and a system of atomic magnetic dipoles or *spins*, which can be oriented in two different directions (Ising system [Is25], see Fig. 2.3). We shall now show how this analogy can be advantageously exploited in view of the remarkable progress made in studies of the physical properties of magnetic systems.

Atoms interact with each other by inducing a magnetic field at the location of another atom, which interacts with its spin. The total local magnetic field at the location of an atom i is given by $h_i = \sum_j w_{ij} s_j$, where w_{ij} is the dipole force, and the diagonal term $j = i$ (self-energy) is not included in the sum. Newtons law "action = reaction" ensures that the coupling strengths w_{ij} are symmetric: $w_{ij} = w_{ji}$. In realistic cases the strength of interaction

---

[4] There are at most N different orthogonal patterns!

between two atoms falls off with distance r, because the dipole force falls as $1/r^3$. In simple models of magnetic-spin systems one therefore often retains only the interaction between atoms that are nearest neighbors. If all w_{ij} are positive, the material is *ferromagnetic*; if there is a regular change of sign between neighboring atoms, one has an *antiferromagnet*. If the signs and absolute values of the w_{ij} are distributed randomly, the material is called a *spin glass*. The ferromagnetic case corresponds to a neural network that has stored a single pattern. The network which has been loaded with a large number of randomly composed patterns resembles a spin glass.

In order to describe the properties of a spin system it is useful to introduce an *energy functional $E[s]$*:

$$E[s] = -\frac{1}{2} \sum_{ij}^{i \neq j} w_{ij} s_i s_j . \qquad (3.17)$$

This is always possible for symmetric couplings $w_{ij} = w_{ji}$; any antisymmetric contribution would cancel in the double sum in (3.17).[5] We can include the diagonal terms $i = j$ in (3.17) without prejudice since these contribute, because of $(s_i)^2 = (\pm 1)^2 = 1$, only a constant term to the energy, which does not depend on the orientation of the spins (or on the states of the neurons):

$$E[s] = -\frac{1}{2} \sum_{ij} w_{ij} s_i s_j + \frac{1}{2} \sum_i w_{ii} . \qquad (3.18)$$

This additional constant does not influence the dynamical evolution of the system and can be dropped.

If the couplings w_{ij} are determined according to Hebb's rule (3.12), and if the configuration of the system corresponds to a stored pattern σ_i^ν, the energy functional takes the value (considering $(\sigma_i^\mu)^2 = 1$)

$$\begin{aligned} E[\sigma^\nu] &= -\frac{1}{2N} \sum_{ij} \sum_\mu \sigma_i^\mu \sigma_j^\mu \sigma_i^\nu \sigma_j^\nu \\ &= -\frac{1}{2N} \left[N^2 + \sum_{\mu \neq \nu} \left(\sum_i \sigma_i^\mu \sigma_i^\nu \right)^2 \right] . \end{aligned} \qquad (3.19)$$

Assuming that the patterns are uncorrelated and the values $\sigma_i^\mu = \pm 1$ are equally likely, one finds that the term $\sum_i \sigma_i^\mu \sigma_i^\nu$ ($\mu \neq \nu$) is of order $\sqrt{N}$, and therefore the energy becomes

$$E[\sigma^\nu] \approx -\frac{1}{2N} \left[N^2 + O((p-1)N) \right] \approx -\frac{N}{2} + O\left(\frac{p-1}{2} \right) . \qquad (3.20)$$

The influence of all other patterns ($\mu \neq \nu$) causes a slight shift of the total energy, since it is typically caused by fluctuations around the ground state

---

[5] It is remarkable that Hebb's rule (3.12) always leads to symmetric synaptic couplings, which is no longer true for more general learning rules.

of a system. If the spins of a few atoms have the wrong orientation, the fluctuating term is replaced by another fluctuating term, which does not result in an essential modification of its magnitude. However, the term with $\mu = \nu$ suffers a substantial change: if the first n spins are incorrectly oriented ($s_i = -\sigma_i^\nu$ for $i = 1, ..., n$), its value changes to

$$\sum_{ij} \sigma_i^\nu \sigma_j^\nu s_i s_j = \left(\sum_i \sigma_i^\nu s_i\right)^2 = (N - 2n)^2 , \tag{3.21}$$

i.e. the energy of the configuration rises in proportion to the extent of its deviation from the stored pattern:

$$E[s] \approx -\frac{1}{2N}(N - 2n)^2 + O\left(\frac{p-1}{2}\right) = E[\sigma^\nu] + 2n - \frac{2n^2}{N} . \tag{3.22}$$

The memorized patterns are therefore (at least local) minima of the energy functional $E[s]$.

A more detailed investigation shows that the energy functional has, in addition, infinitely many other local minima. However, all these spurious minima are less pronounced than those given by the stored patterns σ_i^μ; hence these correspond to *global* minima of the energy surface $E[s]$, at least for moderate values of the parameter $\alpha = p/N$, which indicates the degree of utilization of storage capacity. An example of local, not global, minima is given by network configurations that are linear combinations of three stored patterns:

$$\pm \sigma_i^\mu \pm \sigma_i^\nu \pm \sigma_i^\lambda . \tag{3.23}$$

We shall study this example as well as more general cases in Part II, when we investigate the stability properties of Hopfield networks in the framework of statistical physics. There we shall find that the superposition states (3.23) become unstable when the deterministic activation law (3.5) of the neurons is replaced by a stochastic activation condition, which we discuss in the next chapter.

Finally, it is worth noting that Hebb's rule can be "derived" on the basis of an argument requiring the energy functional to attain minima at those configurations which have maximal resemblance to the stored patterns. For a given pattern σ_i, the overlap with any other configuration of the spin system (or neural network) cannot exceed unity:

$$\frac{1}{N}\sum_i \sigma_i s_i \leq 1 , \tag{3.24}$$

where equality holds only if $s_i = \sigma_i$ for all i. The maximality of the overlap function is converted into minimality of the energy functional given by

$$E[s] = -\frac{1}{2N}\left(\sum_i \sigma_i s_i\right)^2 = -\frac{1}{2N}\sum_{ij}(\sigma_i \sigma_j)s_i s_j . \tag{3.25}$$

A comparison with (3.17) immediately yields Hebb's rule (3.6). The argument is easily translated to the case of several stored patterns by postulating that the energy functional is given by a sum over expressions of the form (3.25):

$$E[s] = -\frac{1}{2N} \sum_{\mu=1}^{p} \left(\sum_i \sigma_i^{\mu} s_i \right)^2 = -\frac{1}{2N} \sum_{ij} \left(\sum_{\mu=1}^{p} \sigma_i^{\mu} \sigma_j^{\mu} \right) s_i s_j \ . \tag{3.26}$$

3.3.2 Parallel versus Sequential Dynamics

The energy function of a physical system is closely related to its dynamics. It is easy to see that the energy function $E[s]$ (3.17) continually decreases with time in the case of (random) sequential updating of the neuron states (the Hopfield model). That is to say, the contribution of a given neuron i to the energy (3.17) at time t is

$$E_i(t) = -s_i(t) \left[\sum_{j \neq i} w_{ij} s_j(t) \right] = -s_i(t) h_i(t) \ . \tag{3.27}$$

If the state of the neuron i is updated according to the law (3.5) its contribution to the energy becomes

$$\begin{aligned} E_i(t+1) &= -s_i(t+1) \left[\sum_{j \neq i} w_{ij} s_j(t) \right] = -\text{sgn}\left[h_i(t) \right] h_i(t) = -|h_i(t)| \\ &\leq -s_i(t) h_i(t) = E_i(t) \ . \end{aligned} \tag{3.28}$$

In other words, the energy contribution of a given neuron never increases with time. Since the energy functional $E[s]$ is bounded below, this implies that the network dynamics must reach a stationary point that corresponds to a (local) minimum of the energy functional. The network thus always ends up in a state of *equilibrium*, which is stable against changes in the state of any single neuron.

The random sequential updating of spin states was first considered as a model for the nonequilibrium statistical physics of magnetic systems by R. Glauber [Gl63]; one therefore also uses the term *Glauber dynamics* to describe the sequential mode of operation of a neural network. The approach to equilibrium becomes particularly important in the context of network models based on stochastic neurons, which are the subject of Chapt. 4. The Glauber dynamics then ensures – for an appropriate neural activation function – that the network enters a state of thermodynamic equilibrium, which can be studied with methods derived from statistical physics.

The argument given above does not apply to the *synchronous* mode of operation of the neural network (the Little model). Since all neurons assume new states in parallel and at the same moment, the contribution of an individual neuron to the energy function cannot be considered in isolation. In this case it is more appropriate to consider the stability function[6] [Br88c]

---

[6] This is the *Lyapunov function* of the system.

$$E_{\mathrm{L}}(t) = -\sum_{i,j} w_{ij} s_i(t) s_j(t-1) , \tag{3.29}$$

which depends on the state of the entire network at two subsequent moments of its dynamical evolution. $E_{\mathrm{L}}[s(t); s(t-1)]$ cannot be regarded as an energy functional, since it depends on the states of the neurons at two different times, i.e. it is nonlocal in the time variable. However, for symmetric synaptic weights $w_{ij} = w_{ji}$ it is again easy to show that $E_{\mathrm{L}}(t)$ decreases monotonously with time:

$$
\begin{aligned}
E_{\mathrm{L}}(t+1) &= -\sum_{i,j} w_{ij} s_i(t+1) s_j(t) = -\sum_{i} \mathrm{sgn}\big[h_i(t)\big] h_i(t) \\
&= -\sum_{i} |h_i(t)| \leq -\sum_{j} h_j(t) s_j(t-1) = E_{\mathrm{L}}(t) .
\end{aligned} \tag{3.30}
$$

Since, on the other hand,

$$E_{\mathrm{L}}(t+1) - E_{\mathrm{L}}(t) = -\sum_{i,j} w_{ij} s_i(t) \big[s_j(t+1) - s_j(t-1)\big] , \tag{3.31}$$

the network must eventually settle into a steady state with the same network configuration being repeated every other time step: $s_i(t+1) = s_i(t-1)$. This means that the network can either reach a steady state, as in the Hopfield model, or permanently cycle between two different configurations [Br88c, Ma89b].

The property of admitting cycling behavior as opposed to the unavoidable approach to a steady state is an attractive aspect of the parallel synchronous updating mode, especially if one wants to associate "trains of thought" with such *reverberations* of the neural network [Cl85]. Of course, symmetrical synaptic weights lead only to trivial cycling patterns (two-cycles), but networks with asymmetric synapses can exhibit very complex cycling behavior, with different cycle lengths possibly prevailing in separate regions of a large neural network [Cl85, Ik87]. The relevance of such phenomena, e.g. for the human short-term memory, is debated [Cl88], since synchronicity does not naturally arise in extended biological neural systems because of variations in the time required for signal transmission along axons of different length.

 Use the program ASSO (see Chapt. 22) to learn and recall the first five letters of the alphabet: A–E. Choose the following parameter values: (5/5/0/1) and (0;1;2/0/0/1), i.e. temperature and threshold zero, and experiment with the updating prescription: 0 = random; 1 = sequential; 2 = synchronous.

3.4 Neural "Motion Pictures"

Temporal sequences of recalled memories form a very important aspect of our mental capacity. They enable us to comprehend the course of events and actions, which otherwise would consist of meaningless single pictures. Periodic sequences of neural impulses are also of fundamental importance for the control of motor body functions, such as the heartbeat, which occurs with great regularity almost three billion times during an average person's life. Irregularities in the cardiac rhythm are fortunately rare, but cause grave concern if they do occur. The questions how neural networks can sustain highly periodic activity for a long period of time and what makes them fail under certain conditions are thus of vital interest.

It is clear from the very beginning that here we have to study the properties of networks with aysmmetric synaptic connections, because periodic activity cannot occur in the presence of thermodynamic equilibrium, toward which all symmetric networks develop. It turns out that the transition from a Hopfield–Little network with its stable-equilibrium configurations to a network exhibiting stationary periodic activity is surprisingly simple. One only has to modify Hebb's rule (3.12) in the following manner:

$$w_{ij} = \frac{1}{N} \sum_{\mu=1}^{p} \sigma_i^{\mu+1} \sigma_j^{\mu} \,, \tag{3.32}$$

where the patterns σ_i^{p+1} and σ_i^1 are identified in order to get temporal periodicity. In order to see how the synaptic rule (3.32) works, we assume that the network configuration is just given by one of the patterns, $s_i(t) = \sigma_i^{\nu}$. Assuming that the patterns are orthogonal, the synaptic polarization potential for the next update is

$$
\begin{aligned}
h_i(t) &= \sum_j w_{ij} \sigma_j^{\nu} = \frac{1}{N} \sum_{\mu} \sigma_i^{\mu+1} \sum_j \sigma_j^{\mu} \sigma_j^{\nu} = \frac{1}{N} \sum_{\mu} \sigma_i^{\mu+1} N \delta_{\mu\nu} \\
&= \sigma_i^{\nu+1} \,.
\end{aligned}
\tag{3.33}
$$

As a consequence the network immediately makes a transition into the next pattern, $s_i(t+1) = \sigma_i^{\nu+1}$, i.e. it performs "heteroassociation" instead of autoassociation. (Note that this mechanism works only in the synchronous, parallel updating mode.)

This rapid, undelayed transition between patterns is not always desired. In the biological context it would imply that one activation pattern would replace another on the scale of the elementary time constant of neural processes, i.e. patterns would change every few milliseconds. This is much too short for many processes, such as motor actions or chains of thoughts, where the characteristic time constants lie in the range of tenths of seconds. How is it possible to stretch the temporal separation between different patterns, to obtain a kind of "slow-motion" effect?

One way to achieve this delay is to add a stabilizing term to the expression (3.32) for the synaptic connections. Again Hebb's rule (3.12) does the trick, if we take the synapses as

$$w_{ij} = \frac{1}{N} \sum_{\mu} \sigma_i^\mu \sigma_j^\mu + \frac{\lambda}{N} \sum_{\mu} \sigma_i^{\mu+1} \sigma_j^\mu .$$ (3.34)

If we set $\lambda < 1$, the pattern initially remains stable, because the stabilizing Hebb term dominates; for $\lambda > 1$ the pattern is unstable and gives way to the following patterns in the next updating step. In order to expand the time between transitions, one only has to make the parameter λ time dependent in an appropriate way. One could start out with a small value of λ, let it grow until it triggers the change of patterns, then immediately reduce it to its initial small value to keep the next pattern stable, and so on.

A better, and less artificial, way to slow down the network is to introduce different types of synaptic connections[7] with stabilizing and transition-inducing tasks, characterized by different lengths of their relaxation time [Kl86, So86a]. We denote the rapidly relaxing, stabilizing ("fast") synapses by

$$w_{ij}^F = \frac{1}{N} \sum_{\mu} \sigma_i^\mu \sigma_j^\mu ,$$ (3.35)

and the slowly relaxing, change-inducing ("slow") synapses by

$$w_{ij}^S = \frac{\lambda}{N} \sum_{\mu} \sigma_i^{\mu+1} \sigma_j^\mu .$$ (3.36)

The total post-synaptic polarization potential at time t is defined as

$$h_i(t) = \sum_j \left[w_{ij}^F s_j(t) + w_{ij}^S \bar{s}_j(t) \right] ,$$ (3.37)

where the "slow" synapses w_{ij}^S average over a number of previous configurations of the network:

$$\bar{s}_j(t) = \int_{-\infty}^t dt' \, G(t - t') s(t') .$$ (3.38)

Here $G(t)$ plays the role of a delay function, which is normalized according to $\int_0^\infty dt \, G(t) = 1$. Standard choices for this function are $G(t) = \delta(t - \tau)$, representing a definite well-defined time delay (Fig. 3.1a), and $G(t) = \tau^{-1} \exp(-t/\tau)$, describing a gradual decay of the post-synaptic potential with a characteristic time constant τ (Fig. 3.1b).

---

[7] We note that it is possible to generate slow sequences of patterns without introducing an explicit temporal order through time-delayed synapses. Directed transitions between stored patterns can also be induced by the action of thermal noise [Bu87b]. The operation of networks at finite temperatures will be discussed in the next chapter.

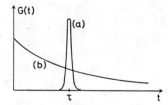

Fig. 3.1. Two possible choices (*curves a and b*) for the synaptic delay function $G(t)$.

For the exponential delay function the interesting range of λ is between 1 and 2. The temporal distance between transitions of patterns diverges at the lower boundary ($\lambda = 1$), and decreases to the value $t_0 = \tau \ln 2$ at ($\lambda = 2$). An interesting special case is that of a single pattern σ and the synapses

$$w_{ij}^F = \sigma_i \sigma_j, \qquad w_{ij}^S = -\lambda \sigma_i \sigma_j , \tag{3.39}$$

which cause the network to oscillate between the pattern and its inverse. In this manner one can construct neural networks that permanently cycle between states of high and low activity.

In addition to the possibility of going through periodic cycles of p patterns or of getting stuck at one of the patterns (if λ is too small), a further kind of temporal behavior has been observed for the case of the exponential synaptic delay function $G(t)$. If the parameter λ is chosen sufficiently large then the network switches to *chaotic time evolution*, tumbling through a set of distorted patterns in a highly irregular fashion [Ri88a].

Amit [Am88] has proposed a variant network model, where the transition between patterns does not occur automatically, but can be triggered from outside by a small perturbation. Here λ must take a value slightly below 1, so that each individual pattern remains stable. In addition to the internal synapses w_{ij}^F and w_{ij}^S, which connect different neurons, each neuron receives external signals through synapses w_i'. A random signal $w_i' = \rho \xi_i$ of strength $\rho > 1 - \lambda$ causes the network to jump into some other configuration, which is generally unrelated to any of the stored patterns σ_i^μ. The trick is that in the next step – when the external signal has disappeared – the fast stabilizing synapses have lost all memory of the previous pattern, whereas the slow synapses still transmit their influence and force the network to make a transition into the subsequent pattern.[8]

Neural networks of a similar structure, containing synaptic connections with different time constants, probably play a crucial role in the nerve centres that control motor activities. In biological studies reaction times for synaptic excitation have been measured which can differ by up to a factor of 20. Simple networks designed to model patterns of neural activity in motor centres (central pattern generators – CPG) of invertebrate animals have exhibited surprisingly close parallels to the physiological observations on which they were based [Se89].

---

[8] Note that the reaction of the network to an external stimulus is very fast in this model, occurring on the elementary neural time scale of a few milliseconds.

We note that it is possible to generate slow sequences of patterns without introducing an explicit temporal order through time-delayed synapses. Directed transitions between stored patterns can also be induced by the action of thermal noise [Bu87b]. The operation of networks at finite temperature will be discussed in the next chapter.

Yet another model that lets the network go through temporal sequences has been proposed by Horn and Usher [Ho89b]. Here the time dependence is not caused by the delay of the synaptic connections. Instead, the neuron activation function is endowed with an intrinsic time dependence by introducing *dynamical threshold parameters* $\vartheta_i(t)$ which depend on the history of the neuron activation. These thresholds are assumed to grow when a neuron has been active for a prolonged time interval, mimicking the effect of *fatigue*. Specifically [Ho89b] assume that the effective local field is determined by

$$h_i(t) = \sum_j w_{ij} s_j(t) - bR_i(t) \tag{3.40}$$

where the dynamical threshold function follows the time evolution law

$$R_i(t) = \frac{1}{c} R_i(t-1) + s_i(t) . \tag{3.41}$$

If a neuron is constantly active, $s_i = 1$, the corresponding synaptic strength variable grows to the saturation value $\vartheta_i \to bc/(c-1)$. If the parameters b and c are chosen suitably these thresholds may exceed the local fields induced by the synaptic couplings thus destabilizing the activation pattern. Depending on the choice of the parameters and on the synaptic matrix (which may contain heteroassociative couplings of the type (3.42)) such a network can undergo periodic oscillations as well as random or regular jumping between stored patterns. (Note that this works only at finite temperature). If this model has relevance to the operation of higher brain functions (which is, of course, questionable) one might say that trains of thought are induced by fatigue.

 Use the program ASSCOUNT (see Chapt. 23) to learn and recite the three numbers 0, 1, 2. Choose the following parameter values: (3/3/0/1) and the standard set on the second screen. Use the space bar to switch from one number to the next.

4. Stochastic Neurons

4.1 The Mean-Field Approximation

We now consider a simple generalization of the neural networks discussed in the previous chapter which permits a more powerful theoretical treatment. For this purpose we replace the deterministic evolution law (3.5)

$$s_i(t+1) = \text{sgn}\big[h_i(t)\big] \equiv \text{sgn}\left[\sum_{j=1}^{N} w_{ij} s_j(t)\right] \tag{4.1}$$

for the neural activity by a stochastic law, which does not assign a definite value to $s_i(t+1)$, but only gives the probabilities that $s_i(t+1)$ takes one of the values $+1$ or -1. We request that the value $s_i(t+1) = \pm1$ will occur with probability $f(\pm h_i)$:

$$\text{Pr}\big[s_i(t+1)\big] = f\big[h_i(t)\big] , \tag{4.2}$$

where the activation function $f(h)$ must have the proper limiting values $f(h \to -\infty) = 0, f(h \to +\infty) = 1$. Between these limits the activation function must rise monotonously, smoothly interpolating between 0 and 1. Such functions are often called *sigmoidal* functions. A standard choice, depicted in Fig. 4.1, is given by [Li74, Li78]

$$f(h) = \left(1 + e^{-2\beta h}\right)^{-1} , \tag{4.3}$$

which satisfies the condition $f(h) + f(-h) = 1$. Among physicists this function is commonly called the Fermi function, because it describes the thermal energy distribution in a system of identical fermions. In this case the parameter β has the meaning of an inverse *temperature*, $T = \beta^{-1}$. We shall also use this nomenclature in connection with neural networks, although this does not imply that the parameter β^{-1} should denote a physical temperature at which the network operates. The rule (4.3) should rather be considered as a model for a stochastically operating network which has been conveniently designed to permit application of the powerful formal methods of statistical physics.[1] In the limit $\beta \to \infty$, or $T \to 0$, the Fermi function (4.3) approaches the unit step function $\theta(h)$:

---

[1] A certain degree of stochastic behavior is in fact observed in biological neural networks. Neurons may spontaneously become active without external stimulus

$$\lim_{\beta \to \infty} f(h) = \theta(h) , \tag{4.4}$$

and the stochastic neural network goes over into our original deterministic network. Hence all results obtained for stochastic networks can be extrapolated to the deterministic case.

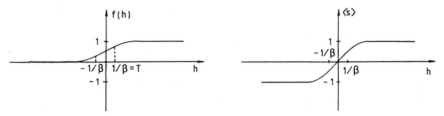

Fig. 4.1. The Fermi function.   **Fig. 4.2.** The function (4.5).

The state of a given neuron in a stochastic network is not of particular relevance, because it is determined randomly. However, the *mean activity* of a neuron, or the mean orientation of a spin in the magnetic analogy (i.e. the *mean magnetization*), is an interesting quantity. Let us first consider a single neuron; its mean activity is (brackets $\langle \cdots \rangle$ indicate an average)

$$\langle s \rangle = (+1) f(h) + (-1) f(-h) = \left(1 + e^{-2\beta h}\right)^{-1} - \left(1 + e^{+2\beta h}\right)^{-1}$$

$$= \frac{1 - e^{-2\beta h}}{1 + e^{-2\beta h}} = \tanh(\beta h) . \tag{4.5}$$

In the limit $\beta \to \infty$ one obtains $\langle s \rangle \to \mathrm{sgn}(h)$, i.e., depending on the sign of the local field h, the neuron is either permanently active or permanently dormant (see Fig. 4.2).

The evolution of a single neuron i in a network composed of many elements is difficult to describe, since its state depends on all other neurons, which continually fluctuate between $+1$ and -1. That even remains true if we are only interested in the mean activity of the ith neuron. This is determined by the value of the synaptic potential h_i, which depends, however, on the actual instantaneous states s_j of the other neurons and *not* on their mean activities. The difficulty is a result of the nonlinearity of the probability function $f(h)$, which does not permit one to take the average in the argument of the function. In such cases one often resorts to a technique called the *mean-field approximation*, which is defined as the exchange of the operations of averaging and evaluating the function $f(h)$:

---

or if the synaptic excitation does not exceed the activation threshold. This phenomenon does not, however, appear to be a simple thermal effect, rather it is a consequence of random emission of neurotransmitters at the synapses.

$$\langle f(h_i)\rangle \longrightarrow f(\langle h_i\rangle) = f\left(\sum_j^N w_{ij}\langle s_j\rangle\right). \tag{4.6}$$

The mean activity of a neuron is then computable as before:

$$\langle s_i\rangle = \langle f(h_i)\rangle - \langle f(-h)\rangle \longrightarrow \tanh\left(\beta\sum_j^N w_{ij}\langle s_j\rangle\right). \tag{4.7}$$

This is still a system of nonlinear equations with N unknowns $\langle s_i\rangle$, but these have become deterministic rather than stochastic variables. Since the mean-field approximation is more readily accessible to intuition for magnetic-spin systems than for neural networks, we shall base the following discussion mainly on this analogy.

4.2 Single Patterns

The solution of the system of equations (4.7) depends, of course, on the synaptic strengths w_{ij}. We begin by considering the simplest case $w_{ij} = \frac{1}{N}$ for all i, j, which corresponds to a ferromagnetic spin system.[2] In this case the mean magnetization is

$$\langle s_i\rangle = \tanh\left(\beta\frac{1}{N}\sum_j\langle s_j\rangle\right) \tag{4.8}$$

and, because the right-hand side does not depend on i,

$$\langle s\rangle \equiv \langle s_i\rangle = \tanh\left[\beta\langle s\rangle\right]. \tag{4.9}$$

We have thus managed to reduce the evolution law to a single equation. Since the slope of the function $\tanh(\beta x)$ at the origin $(x = 0)$ equals β and falls towards both sides, we have to distiguish the two cases illustrated in Fig. 4.3. For $\beta < 1$ ($T > 1$; Fig. 4.3, *left*), (4.9) admits only the trivial solution $\langle s\rangle = 0$, i.e. all spins point up or down with equal probability. (In the neural network this means that all neurons are as often active as they are resting.) The average magnetization vanishes, as it must in a ferromagnet at high temperature. The temperature $T = 1$ corresponds to the Curie point, above which the spin system becomes demagnetized.

For $\beta > 1$ ($T < 1$, right part of figure) there are altogether three solutions, namely $\langle s\rangle = 0$, and $\langle s\rangle = \pm S_0$, where $S_0 > 0$ satisfies the equation

$$S_0 = \tanh(\beta S_0). \tag{4.10}$$

---

[2] Our discussion here actually applies to the general case of a single stored pattern σ_i, which can be seen as follows: the transformation $s_i \to s_i' = \sigma_i s_i$ always maps this into the ferromagnetic case $\sigma_i' = \sigma_i\sigma_i = +1$. Such a transformation is called a *gauge transformation*, since it corresponds to a reinterpretation of the meaning of "up" and "down" at each lattice site. The transformed Hebbian synaptic connections are $w_{ij}' = \sigma_i w_{ij}\sigma_j = \sigma_i(N^{-1}\sigma_i\sigma_j)\sigma_j = 1/N$.

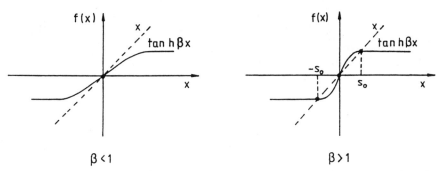

Fig. 4.3. Solutions of the equation $x = \tanh(\beta x)$ for $\beta < 1$ (*left*) and $\beta > 1$ (*right*).

In order to find out which solution is realized, we must check for stability against small fluctuations. The condition for a stable solution is that the reaction of the mean orientation of the spin against a small deviation $\delta\langle s \rangle$ from the solution be smaller than the deviation itself:

$$\delta\big[\tanh(\beta\langle s \rangle)\big] = \frac{\partial \tanh(\beta\langle s \rangle)}{\partial \langle s \rangle}\delta\langle s \rangle < \delta\langle s \rangle \,. \tag{4.11}$$

From the graphical representation (Fig. 4.3, *right*) it is obvious that in the case $\beta > 1$ this condition is satisfied for the two nonvanishing solutions, but not for the trivial solution, because the slope of the function $\tanh(\beta\langle s \rangle)$ is

$$\frac{\partial \tanh(\beta\langle s \rangle)}{\partial \langle s \rangle} = \frac{\beta}{\cosh^2(\beta\langle s \rangle)} \begin{cases} = \beta > 1 & \text{for } \langle s \rangle = 0 \\ < 1 & \text{for } \langle s \rangle = \pm S_0 \end{cases} . \tag{4.12}$$

We have thus found the following important result: at high temperatures $T > 1$ the fluctuations dominate and the network does not have a preferred configuration; at low temperatures $T < 1$ the network spontaneously settles either into a preferentially active ($S_0 > 0$) or into a preferentially resting ($S_0 < 0$) state. The equal degree of stability of one state and its complementary configuration is a characteristic property of networks with a symmetric activation function satisfying the relation $f(h) + f(-h) = 1$. According to (3.17) the configurations $\{s_i\}$ and $\{-s_i\}$ always have the same value of the "energy" E.

It is useful to study the behavior of the network in the vicinity of the critical temperature $T = 1$. Slightly below this threshold $|S_0| \ll 1$, and we can expand in powers of S_0:

$$S_0 = \tanh(\beta S_0) \approx \beta S_0 - \frac{1}{3}(\beta S_0)^3 + \dots \,, \tag{4.13}$$

with the nonvanishing solutions $S_0 \approx \pm\sqrt{3(\beta - 1)}$. For very small temperatures, i.e. $\beta \gg 1$, one finds that S_0 very rapidly approaches ± 1, like $S_0(\beta \gg 1) \approx \pm(1 - 2e^{-2\beta})$. Already for $\beta = 3$ one finds $S_0 = \pm 0.995$, i.e. on average only one neuron out of 400 deviates from the correct value. The behavior of $S_0(\beta)$ is illustrated in Fig. 4.4. At the point $\beta = 1$ $(T = 1)$ the function is not differentiable, indicating the phase transition between a magnetized and an unmagnetized phase. The mean magnetization shows a fractional power-law singularity $S_0 \propto (\beta - \beta_c)^{1/2}$ near the critical point $\beta_c = 1$, which is characteristic for phase transitions.

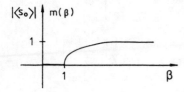

Fig. 4.4. The spontaneous "magnetization" S_0, or m, as a function of the inverse temperature $\beta = 1/T$.

4.3 Several Patterns

We now turn to the general situation, when the synaptic couplings are fixed according to Hebb's rule (3.12) for several patterns. The mean-field equation (4.7) then takes the form

$$\langle s_i \rangle = \tanh\left(\frac{\beta}{N} \sum_j^N \sum_\mu \sigma_i^\mu \sigma_j^\mu \langle s_j \rangle \right). \tag{4.14}$$

This relation does not have an obvious solution. However, we recall that every single stored pattern represents a stable configuration for a deterministic network. It is, therefore, not unreasonable to make the *ansatz* that $\langle s_i \rangle$ resembles one of the stored patterns, except for a normalization factor:

$$\langle s_i \rangle = m\sigma_i^\nu. \tag{4.15}$$

Inserting this into (4.14) and following the same considerations as with (3.13), we obtain:

$$m\sigma_i^\nu = \tanh\left[\beta m\sigma_i^\nu + \frac{\beta m}{N} \sum_{\mu \neq \nu} \sigma_i^\mu \sum_j \sigma_j^\mu \sigma_j^\nu \right]$$

$$= \tanh\left[\beta m\sigma_i^\nu + \beta m O\left(\sqrt{\frac{p-1}{N}} \right) \right]. \tag{4.16}$$

On account of $\sigma_i^\nu = \pm 1$, the normalization factor m is determined in the limit $N \to \infty$ and $p \ll N$ by the same equation (4.10) as S_0:

$$m = \tanh(\beta m). \tag{4.17}$$

Again we find the same fixed points as for a single stored pattern: for $T > 1$ one has $m = 0$, and the time-averaged network configuration does not resemble one of the stored patterns. One might say that the network is "amnesic".

For $T < 1$ one has $m = \pm S_0$ and the average configuration of the network points toward one of the stored patterns. Since these take only the values (± 1), the patterns can be uniquely (up to an overall sign) recovered from the averaged state $\langle s_i \rangle$ of the network. However, if one takes a snapshot of the network at any given moment, the probability that a neuron is found in its "correct" state is only $\frac{1}{2}(1 + m)$; i.e. the network (almost) never reproduces the pattern without error. As already mentioned in connection with the function $S_0(\beta)$, shown in Fig. 4.4, the analytical behavior of the function $m(T)$ at $T = 1$ is characteristic of a *phase transition,* as it often occurs in many-body systems. In the magnetic analogy it corresponds to the transition from a magnetized to an unmagnetized phase. We shall study this phase transition in Part II with help of the methods of statistical thermodynamics.

Numerical simulations as well as statistical analysis show that the number of stored patterns relative to the number of neurons, $\alpha = p/N$, plays a similar role as the parameter T. When the storage utilization α is increased starting from zero, i.e. if more and more patterns are stored, the recall quality of the network deteriorates slightly. However, when α approaches the *critical capacity* $\alpha_c \approx 0.138$, the network suddenly and catastrophically fails to recall any of the memorized patterns. Up to this point, the relative error $(1-m)/2$ of the time-averaged recall states remains small, actually less than 2 percent (see Fig. 4.5). In other words the networks suddenly jumps from almost perfect memory into a state of complete confusion.[3] The understanding of this surprising result by means of methods developed in the statistical theory of spin-glass systems, to be discussed in Part II, has kindled the interest of theoretical physicists in neural networks and in brain models in general.

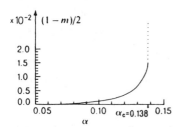

Fig. 4.5. The quality of memory recall deteriorates with increasing storage density $\alpha = p/N$ and breaks down at $\alpha \approx 0.138$.

---

[3] A second, much more restrictive approach can be taken to define the memory capacity. One may require *error-free recall,* which means that the stored patterns have to be attractors of the network, without there being any errors introduced by interference among them. The number of patterns that can be perfectly memorized is much smaller than $\alpha_c N$, namely

$$p \simeq \frac{N}{c \ln N}.$$

Here $c = 2$ if a typical single pattern and $c = 4$ if all the patterns together are to be reproduced without error. This can be derived using tools from probability theory [We85, Mc87].

In order to obtain a complete description of the ability of a stochastic Hopfield network to recall memorized patterns one has to consider m as function of α as well as of T. One then obtain a complete phase diagram of the *"order parameter"* $m(T, \alpha)$. The phase boundary between "functioning memory" and total "confusion" or "amnesia" is given by a line in the (α–T) diagram, depicted in Fig. 4.6. For $T = 0$ this line begins at $\alpha_c = 0.138$ and ends, as we just have found, at $T = 1$ for $\alpha = 0$. At the phase boundary m falls discontinuously to zero, except for $T = 0$, where the change is continuous, but not differentiable.[4] This behavior is illustrated in Fig. 4.7.

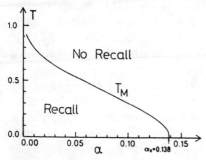

Fig. 4.6. Memory recall is possible only in a finite region of temperatures T and storage density $\alpha = p/N$.

Fig. 4.7. Recall quality m as a function of the temperature T for several storage densities α.

In addition to its usefulness in the formal treatment of neural-network properties the introduction of a stochastic neuron evolution law has important practical advantages. The thermal fluctuations reduce the probability that the network becomes caught in a spurious, undesired locally stable configuration. The statistical analysis [Am85b] reveals that configurations of the type (3.23) become unstable above the temperature $T_3 \approx 0.46$ (we shall explicitly prove this in Sect. 18.2). Extended numerical simulations of Hopfield networks have shown that their information-retrieval properties are substantially improved by thermal noise, with the best results obtained in the range $T_3 < T < 1$ [Pe89b]. Simulations [Bu87a] of neural networks, which were modeled in close analogy to their physiological counterparts, have also supported this result: a certain level of noise improves the memory efficiency of neural networks by eliminating spurious memory states.

---

[4] Note that these results strictly apply only in the thermodynamic limit $N \to \infty$, i.e. for systems with infinitely many neurons.

Use the program ASSO (see Chapt. 22) to learn and re-call the first five letters of the alphabet: A–E. Choose the following parameter values: $(5/5/0/1)$ and $(0/T/0/1;2)$, i.e. random updating, and threshold zero, and experi-ment with the value of the temperature T.

Use the program ASSO to learn and recall a certain number p of random patterns. Choose the following pa-rameter values: $(p/0/0/1)$ and $(0/T/0/1;2)$, i.e. random updating, and threshold zero, and experiment with the value of the temperature T. How many patterns can be recalled? Explore the boundary of working memory in the $T - p$ plane (see Fig. 4.6).

5. Cybernetic Networks

5.1 Layered Networks

The neural network models discussed so far were all aimed at storing and recalling given information. Memory is an important function of the brain, but far from the only one. Another important task of the central nervous system is to learn reactions and useful behavior that permit survival in an often hostile environment; probably this was the original purpose of the development of the brain during evolution.

In a highly simplified perspective one may identify this role of the brain with that of the supervising element in a control circuit. We shall therefore introduce the term *cybernetic* networks for neural networks that provide an optimal reaction or answer to an external stimulus. Such networks usually have a structure that fundamentally differs from that of memory networks. In particular, the synaptic connections are generally not symmetric; often they are maximally asymmetric, i.e. unidirectional. As a result, the theories of thermodynamic equilibrium systems have no direct application to cybernetic networks, and much less general insights are known into their behavior.

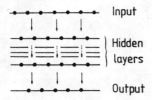

Fig. 5.1. Schematic view of a feed-forward, layered neural network (perceptron).

The best-studied class of cybernetic networks are the so-called *feed-forward, layered* neural networks, where information flows in one direction between several distinct layers of neurons, like along a one-way street, as illustrated in Fig. 5.1. At one end is the *input* layer composed of sensory neurons, which receive external stimuli, at the other end is an *output* layer often composed of motor neurons, which cause the desired reaction.[1] Following

---

[1] In complex networks (such as the brain) the output layer may feed into other subnetworks that further process the output.

Rosenblatt, who was the first to study such systems in detail, a feed-forward layered network is also called a *perceptron* [Bl62a, Bl62b].

In mathematical terminology, the input–output relation defines a *mapping*, and the feed-forward neural network provides a *representation* of this mapping. One therefore also speaks of mapping neural networks. The main questions that will occupy us in the following are: What mappings can a feed-forward neural network represent for a given complexity of architecture? How can the representation be established through a learning algorithm? Is the network able to represent the correct mapping also for such input–output pairs which are not contained in the training set (the ability to generalize)?

5.2 Simple Perceptrons

5.2.1 The Perceptron Learning Rule

In principle a perceptron can contain an arbitrary number of layers of neurons in addition to the input and output layers, but only rather simple cases have been studied in depth. Here we begin with the so-called *simple* perceptron, where the input layer feeds directly into the output layer without the intervention of inner, or "*hidden*", layers of neurons. We shall denote the states of the input neurons by σ_k, $(k = 1, \ldots, N_i)$; those of the output neurons are labeled by S_i, $(i = 1, \ldots, N_o)$. The activation of the output neurons by the input layer may be determined by the nonlinear function $f(x)$ as

$$S_i = f\left(\sum_k w_{ik}\sigma_k\right) . \tag{5.1}$$

The fact that the variables σ_k occur only on the right-hand side of this relation, while the S_i occur only on the left-hand side, is expression of the directedness of the networks: information is fed from the input layer into the neurons of the output layer, but not vice versa. The function $f(x)$ may be considered a stochastic law, where it determines the probability of the values $S_i = \pm 1$, or as a continuous function, if the neurons are assigned analog values. Here we concentrate on the latter case, for which a standard choice of function is

$$f(h) = \tanh(\beta h) . \tag{5.2}$$

The task now is to choose the synaptic connections w_{ik} in such a way that a certain input σ_k leads to the desired reaction, specified by the correct states of the neurons in the output layer, which we denote by $S_i = \zeta_i$. Of course, this condition must not only be satisfied for a single input, but for a number of cases indicated by the superscript μ:

$$S_i^\mu \equiv S_i[\sigma_k^\mu] = \zeta_i^\mu, \qquad \mu = 1, \ldots, p . \tag{5.3}$$

An explicit function allowing the calcuation of the w_{ik} from the input σ_k^μ and the output ζ_i^μ is not known. However, it is possible to construct iterative procedures which converge to the desired values of synaptic connections, if those exist in principle.[2]

We first consider the simplest case, i.e. the step function $f(x) = \text{sgn}(x)$, describing a deterministic network, and a single reaction that is to be learned. We start from some arbitrary configuration of weights w_{ik}. In this case we have either $S_i = \zeta_i$ or $S_i = -\zeta_i$, depending on whether the reaction of neuron i is correct or not. In the former case the synaptic connections at the ith output neuron are properly left unchanged, while they have to be modified in the latter case. Here we can again apply Hebb's rule, which is now interpreted to mean that the synaptic connection w_{ik} is incremented by a multiple of the product of the states of the kth input neuron and the ith output neuron. The formal expression of this rule is:

$$\delta w_{ik} = \epsilon(1 - S_i\zeta_i)\zeta_i\sigma_k = \epsilon(\zeta_i\sigma_k - S_i\sigma_k) \,. \tag{5.4}$$

The last representation allows for an alternative interpretation of this learning rule: the correct combination of stimulus and reaction is *learned*, the incorrect one is *unlearned*.[3] The parameter ϵ, which describes the intensity of this learning process, should be chosen in such a way that optimal convergence is achieved, when repeatedly applying (5.4). It is customary to choose values $\epsilon \ll 1$. If we have to satisfy several of these stimulus–reaction associations, we can simply sum over the required synaptic modifications:

$$\delta w_{ik} = \epsilon \sum_\mu (\zeta_i^\mu - S_i^\mu)\sigma_k^\mu \,. \tag{5.5}$$

This law of synaptic plasticity is known as the *perceptron learning rule*.

5.2.2 "Gradient" Learning

The perceptron learning rule (5.5) yields the desired result for simple perceptrons, but there is no obvious generalization to more complex neural networks with hidden layers of neurons. For this purpose it is more appropriate to start from the expression

$$D = \frac{1}{2} \sum_\mu \sum_i (S_i^\mu - \zeta_i^\mu)^2 = \sum_{\mu,i} (1 - S_i^\mu\zeta_i^\mu) \tag{5.6}$$

for the deviation between the actual and the correct output and to apply a method of minimization. This approach is conceptually simplified if we consider neurons with analog values that are directly determined by the nonlinear

---

[2] This condition is not always fulfilled in the case of simple perceptrons, as we shall see in Sect. 5.2.3.

[3] A related method for refreshing the memory of Hopfield-type networks will be introduced in Sect. 10.2.1; see (10.33).

response function $f(x)$. With the help of (5.1) we then obtain the deviation as an explicit function of the synaptic connections:

$$D[w_{ik}] = \frac{1}{2}\sum_{\mu}\sum_{i}\left[\zeta_i^\mu - f\left(\sum_k w_{ik}\sigma_k^\mu\right)\right]^2 \equiv \frac{1}{2}\sum_{\mu,i}[\zeta_i^\mu - f(h_i^\mu)]^2. \quad (5.7)$$

In order to lower the value of D, we now compute the gradient with respect to the synaptic couplings:

$$\frac{\partial D}{\partial w_{ik}} = -\sum_\mu [\zeta_i^\mu - f(h_i^\mu)]f'(h_i^\mu)\frac{\partial h_i^\mu}{\partial w_{ik}} = -\sum_\mu \Delta_i^\mu \sigma_k^\mu, \quad (5.8)$$

with the abbreviation

$$\Delta_i^\mu = [\zeta_i^\mu - f(h_i^\mu)]f'(h_i^\mu) = [\zeta_i^\mu - S_i^\mu]f'(h_i^\mu) . \quad (5.9)$$

Using the gradient expression, a small variation of the synapses of the form

$$\delta w_{ik} = -\epsilon\frac{\partial D}{\partial w_{ik}} = \epsilon\sum_\mu \Delta_i^\mu \sigma_k^\mu \quad (5.10)$$

decreases the deviation D. The method works because the gradient always points in the direction of strongest change of the function; hence the negative gradient follows the direction of steepest descent. The factor ϵ should be chosen sufficiently small to avoid overshooting the goal. In order to see formally why the value of D is lowered, we perform a linear Taylor expansion:

$$\delta\Big(D[w_{ik}]\Big) = \sum_{ik}\frac{\partial D}{\partial w_{ik}}\delta w_{ik} = -\epsilon\sum_{ik}\left(\frac{\partial D}{\partial w_{ik}}\right)^2 < 0. \quad (5.11)$$

Apart from the appearance of the derivative of the nonlinear function $f(x)$ in Δ_i^μ, the equation (5.10) governing synaptic plasticity has the same form as the simple perceptron learning rule (5.5), and hence can be considered its generalization. Evaluation of the derivative is particularly simple for the function (5.2), where it is given by

$$f'(x) = \beta\left[1 - f(x)^2\right] . \quad (5.12)$$

The disadvantage of gradient learning is that it cannot be directly applied to perceptrons built from deterministic, binary-valued neurons. Its great advantage is that it can be easily generalized to multilayered perceptrons. As we show in the next subsection, this is essential.

5.2.3 A Counterexample: The Exclusive-OR Gate

The existence of a simple but effective learning algorithm for perceptrons without hidden neurons makes these very attractive as neural-network models. Unfortunately, there exist elementary problems that cannot be handled by such a system [Mi69].

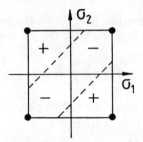

Fig. 5.2. Architecture of the XOR gate.

Fig. 5.3. Disconnected areas of equal output value in the space of input variables.

Probably the simplest example is provided by the task of representing the logical exclusive-OR("XOR") function $S = \text{XOR}(\sigma_1, \sigma_2)$, where S takes the value "true" if either σ_1 or σ_2 is true, but not both. When this function is to be represented by a deterministic perceptron with two input neurons and one output neuron, the network architecture (5.1) is explicitly given by

$$S = \text{sgn}(w_1\sigma_1 + w_2\sigma_2 - \vartheta) , \qquad (5.13)$$

where ϑ denotes the threshold polarization potential of the output neuron (see Fig. 5.2). The values "true" (T) and "false" (F) of the logical variables are denoted by the values $+1$ and -1. The XOR function then implies the following conditions for the argument of the sign function in (5.13):

σ_1	σ_2	ζ	condition
$+1$	$+1$	-1	$+w_1 + w_2 - \vartheta < 0$
$+1$	-1	$+1$	$+w_1 - w_2 - \vartheta > 0$
-1	$+1$	$+1$	$-w_1 + w_2 - \vartheta > 0$
-1	-1	-1	$-w_1 - w_2 - \vartheta < 0$

When we subtract the first two inequalities we obtain $2w_2 < 0$, while subtracting the last two inequalities we get $2w_2 > 0$. The two conditions are clearly contradictory and cannot be satisfied simultaneously, i.e. the desired task cannot be performed by the network for *any* choice of the synaptic parameters w_1, w_2 and ϑ. This result is not entirely surprising, because there are only three adjustable parameters but a total of four conditions. Although these are only inequalities, the example shows that the available freedom to choose the synapses may not be sufficient to solve the problem. The XOR problem is called *linearly inseparable*, since there is no way of dividing the space of input variables into regions of equal output by a single linear condition.

This condition has a simple geometrical interpretation. The space of input variables $\sigma_i = \pm 1$ consists of the corners of an N_i-dimensional hypercube.

A binary valued function is linearly separable if an $(N_i - 1)$-dimensional hyperplane can be found which separates the regions of function values $+1$ and -1. It is obvious from Fig. 5.3 that this condition is not met for the XOR-problem where two such separating hyperplanes (i.e. lines) are required. The class of linearly inseparable functions can not be represented by a perceptron.

This kind of criticism of the concept of simple perceptrons, which was particularly emphasized by the MIT school [Mi69], almost put an end to the study of neural networks in the late 1960s, at least concerning their possible use as general information processing machines. As will become clear in the next section, this was probably an exaggerated response. However, we have to consider multilayered perceptrons in order to avoid problems of nonrepresentability, and this requires that we first learn how to handle and "teach" networks with hidden neurons.

6. Multilayered Perceptrons

6.1 Solution of the XOR Problem

Before we enter into the discussion of how one can derive a learning rule for multilayered perceptrons, it is useful to consider a simple example. We choose the exclusive-OR (XOR) function, with which we are already familiar. In order to circumvent the "no-go" theorem derived for simple perceptrons in the previous section, we add a *hidden layer* containing two neurons which receive signals from the input neurons and feed the output neuron (see Fig. 6.1). We denote the states of the hidden neurons by the variables s_j, $(j = 1, \ldots, N_h)$.[1] The synaptic connections between the hidden neurons and the output neurons are denoted by w_{ij}; those between the input layer and the hidden layer by $\overline{w}_{jk}$. The threshold potentials of the output neurons are called ϑ_i; those of the hidden neurons are called $\overline{\vartheta}_j$.

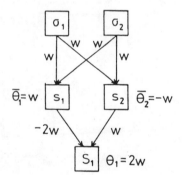

Fig. 6.1. Neural network with two hidden neurons representing the XOR function.

The state of our network with one hidden layer is then governed by the following equations:

$$s_j = \operatorname{sgn}(\overline{h}_j) = \operatorname{sgn}\left(\sum_k \overline{w}_{jk}\sigma_k - \overline{\vartheta}_j\right), \tag{6.1}$$

---

[1] Capital letters S will henceforth symbolize output neurons, while hidden neurons are denoted by lower-case letters s.

$$S_i = \text{sgn}(h_i) = \text{sgn}\left(\sum_j w_{ij}s_j - \vartheta_i\right). \tag{6.2}$$

In the case of our example, the XOR function, the indices j and k run from 1 to 2; the index i takes only the value 1. As the table below shows, the following choice of synaptic couplings and threshold potentials – indicated in Fig. 6.1 – provides a solution to our problem:

$$\overline{w}_{jk} = w, \qquad \overline{\vartheta}_1 = w, \qquad \overline{\vartheta}_2 = -w \;;$$

$$w_{11} = -2w, \qquad w_{12} = w, \qquad \vartheta_1 = 2w \;. \tag{6.3}$$

We check the validity of this solution by explicit evaluation of the XOR function for all four possible input combinations:

σ_1	σ_2	ζ_1	$\overline{h}_1$	$\overline{h}_2$	s_1	s_2	h_1	S_1
$+1$	$+1$	-1	w	$3w$	$+1$	$+1$	$-3w$	-1
$+1$	-1	$+1$	$-w$	w	-1	$+1$	w	$+1$
-1	$+1$	$+1$	$-w$	w	-1	$+1$	w	$+1$
-1	-1	-1	$-3w$	$-w$	-1	-1	$-w$	-1

As one can see, the hidden neuron s_1 plays the role of a logical element representing the AND function, while the other hidden neuron s_2 emulates a logical OR element. The combination of these two elements allows the generation of the XOR function. This is not an accident. We shall show in Sect. 6.3 that any Boolean function can be represented by a feed-forward network with one hidden layer and appropriate architecture.

The success of feed-forward networks with hidden layers, where simple perceptrons fail, can be traced back to the problem of linear separability of the input space into regions of positive and negative output (for a single output neuron). At the end of Chapt. 5 we argued that the XOR problem cannot be solved by a simple perceptrons, since for it the argument space is not linearly separable. For a perceptron with one hidden layer this condition is not required; it can solve all problems where the argument space is divided into two convex open or closed regions of arbitrary shape. In the case of two hidden layers the regions need not even be contiguous or simply connected; given sufficient complexity of the two hidden layers, the perceptron can handle any type of topological division of the input space into regions with positive and negative output [Li87].

6.2 Learning by Error Back-Propagation

To use multilayered networks efficiently, one needs a method to determine their synaptic efficacies and threshold potentials. A very successful method, usually called *error back-propagation*, was developed independently around 1985 by several research groups [Cu86, Pa85, Ru86a]. It is based on a generalization of the gradient method discussed in Sect. 5.2.2. Here we formulate this algorithm for a generic three-layered network with analog-valued neurons, schematically represented in Fig. 6.2. In analogy to (6.1, 6.2) the equations governing the state of the network are

$$S_i = f(h_i), \qquad h_i = \sum_j w_{ij} s_j - \vartheta_i , \tag{6.4}$$

$$s_j = f(\overline{h}_j), \qquad \overline{h}_j = \sum_k \overline{w}_{jk} \sigma_k - \overline{\vartheta}_j . \tag{6.5}$$

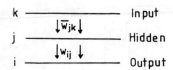

Fig. 6.2. General architecture of a feed-forward neural network with one hidden layer of neurons.

We demand that the synapses and threshold potentials be chosen such that the output deviation function

$$D\left[w_{ij}, \vartheta_i, \overline{w}_{jk}, \overline{\vartheta}_j\right] = \frac{1}{2} \sum_\mu \sum_i \left[\zeta_i^\mu - f(h_i^\mu)\right]^2 \tag{6.6}$$

becomes as small as possible. Thus we search for a minimum (which ideally should be a global minimum) of the error surface defined by (6.6). For this purpose we compute the gradient of D with respect to every parameter and then change the value of the parameters accordingly.[2] In the first step we consider only synaptic connections at the output neurons:

---

[2] The theory of numerical analysis holds in stock various methods of finding minima of a nonlinear function $g(\mathbf{x})$ in a space of high dimension. These are iterative algorithms which generate a sequence of approximations which (hopefully) converge to the position of a minimum $\mathbf{x}_1, \mathbf{x}_2, \dots \to \mathbf{x}_{\min}$. Some of them are discretized versions of the classical *method of Newton*, which in its original form calls for the inversion of the Hessian matrix $\mathcal{H}$ of second derivatives

$$\mathbf{x}_{n+1} = \mathbf{x}_n - \mathcal{H}^{-1} \nabla g(\mathbf{x}_n)$$

and is computationally expensive. A popular and more efficient alternative is the *conjugate gradient method* described in [Pr86]. However, for many applications it is entirely sufficient to employ the simple and naive *method of steepest descent*

$$\mathbf{x}_{n+1} = \mathbf{x}_n - \epsilon \nabla g(\mathbf{x}_n) ,$$

to be used in (6.7, 6.8).

$$\delta w_{ij} \;=\; -\epsilon \frac{\partial D}{\partial w_{ij}} = \epsilon \sum_{\mu} [\zeta_i^\mu - f(h_i^\mu)] f'(h_i^\mu) \frac{\partial h_i^\mu}{\partial w_{ij}} = \epsilon \sum_{\mu} \Delta_i^\mu s_j^\mu, \quad (6.7)$$

$$\delta \vartheta_i \;=\; -\epsilon \frac{\partial D}{\partial \vartheta_i} = \epsilon \sum_{\mu} [\zeta_i^\mu - f(h_i^\mu)] f'(h_i^\mu) \frac{\partial h_i^\mu}{\partial \vartheta_i} = -\epsilon \sum_{\mu} \Delta_i^\mu \quad (6.8)$$

with the same abbreviation as in (5.9):

$$\Delta_i^\mu = [\zeta_i^\mu - f(h_i^\mu)] f'(h_i^\mu) \, . \tag{6.9}$$

In the next step we consider the parameters associated with synaptic connections between the input and hidden layer. The procedure is exactly the same, except that we have to apply the substitution rule of differentiation once more:

$$\delta \overline{w}_{jk} \;=\; -\epsilon \frac{\partial D}{\partial \overline{w}_{jk}} = \epsilon \sum_{\mu, i} [\zeta_i^\mu - f(h_i^\mu)] f'(h_i^\mu) \frac{\partial h_i^\mu}{\partial s_j} \frac{\partial s_j}{\partial \overline{w}_{jk}}$$

$$= \; \epsilon \sum_{\mu, i} \Delta_i^\mu w_{ij} f'(\overline{h}_j^\mu) \frac{\partial \overline{h}_j}{\partial \overline{w}_{jk}} \equiv \epsilon \sum_{\mu} \overline{\Delta}_j^\mu \sigma_k^\mu, \tag{6.10}$$

$$\delta \overline{\vartheta}_j \;=\; -\epsilon \frac{\partial D}{\partial \overline{\vartheta}_j} = \epsilon \sum_{\mu, i} [\zeta_i^\mu - f(h_i^\mu)] f'(h_i^\mu) \frac{\partial h_i^\mu}{\partial s_j} \frac{\partial s_j}{\partial \overline{\vartheta}_j}$$

$$= \; \epsilon \sum_{\mu, i} \Delta_i^\mu w_{ij} f'(\overline{h}_j^\mu) \frac{\partial \overline{h}_j}{\partial \overline{\vartheta}_j} \equiv -\epsilon \sum_{\mu} \overline{\Delta}_j^\mu, \tag{6.11}$$

with the new abbreviation

$$\overline{\Delta}_j^\mu = \left(\sum_i \Delta_i^\mu w_{ij} \right) f'(\overline{h}_j^\mu) \, . \tag{6.12}$$

One should note that the equations (6.10, 6.11) determining the synaptic adjustments have the same form as the synaptic equations (6.7, 6.8) derived earlier. Only the expression for $\overline{\Delta}_j^\mu$ differs from that for Δ_i^μ, from which it may be obtained recursively.

One interesting aspect of this recursion relation is that it resembles (6.5), which determines the state of the two final layers of the neural network. In a sense, therefore, the error-correction scheme works by propagating the information about the deviation from the desired output "backward" through the network, against the direction of synaptic connections. It is doubtful, though not entirely impossible, whether a related procedure can be realized in biological neural networks[3] . What is certain is that the algorithm of error

---

[3] Even if error back-propagation may be biologically implausible, its use as a computational algorithm can be helpful in neurobiological studies. For example [Lo89] have been able to understand the function of interneurons found in the nervous system of a leech by comparison with a simulated neural network trained with back-propagation.

back-propagation is well suited for representation on electronic computers, either in hardware or software realizations.

The method is easily generalized to neural networks with more than one hidden layer of neurons. An equation of the form (6.12) always expresses the parameters Δ in terms of those obtained for the previous layer (in the backward propagating sense); the synaptic modifications are determined by equations such as (6.10) and (6.11). As an exercise we explicitly write down the full set of equations for a feed-forward network with two layers of hidden neurons. The variables pertaining to the additional layer, here taken to be the one directly connected to the input layer, are denoted by an additional "bar", e.g. $\bar{s}_k$, $\bar{\bar{w}}_{jk}$, $\bar{\bar{\vartheta}}_k$, and $\bar{\bar{h}}_k$ (see Fig. 6.3).

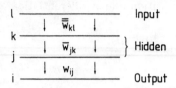

Fig. 6.3. General architecture of a feed-forward neural network with two hidden layers of neurons.

The states of the neurons in the various layers are then determined from the three equations

$$S_i = f(h_i), \qquad h_i = \sum_j w_{ij}s_j - \vartheta_i , \tag{6.13}$$

$$s_j = f(\bar{h}_j), \qquad \bar{h}_j = \sum_k \bar{w}_{jk}\bar{s}_k - \bar{\vartheta}_j , \tag{6.14}$$

$$\bar{s}_k = f(\bar{\bar{h}}_k), \qquad \bar{\bar{h}}_k = \sum_\ell \bar{\bar{w}}_{k\ell}\sigma_\ell - \bar{\bar{\vartheta}}_k . \tag{6.15}$$

The equations defining the adjustment of synapses are

$$\delta w_{ij} = \epsilon \sum_\mu \Delta_i^\mu s_j^\mu, \qquad \delta\vartheta_i = -\epsilon \sum_\mu \Delta_i^\mu, \qquad \Delta_i^\mu = [\zeta_i^\mu - S_i^\mu]f'(h_i^\mu); \tag{6.16}$$

$$\delta\bar{w}_{jk} = \epsilon \sum_\mu \bar{\Delta}_j^\mu \bar{s}_k^\mu, \qquad \delta\bar{\vartheta}_j = -\epsilon \sum_\mu \bar{\Delta}_j^\mu, \qquad \bar{\Delta}_j^\mu = f'(\bar{h}_j^\mu)\sum_i \Delta_i^\mu w_{ij}; \tag{6.17}$$

$$\delta\bar{\bar{w}}_{k\ell} = \epsilon \sum_\mu \bar{\bar{\Delta}}_k^\mu \sigma_\ell^\mu, \qquad \delta\bar{\bar{\vartheta}}_k = -\epsilon \sum_\mu \bar{\bar{\Delta}}_k^\mu, \qquad \bar{\bar{\Delta}}_k^\mu = f'(\bar{\bar{h}}_k^\mu)\sum_j \bar{\Delta}_j^\mu \bar{w}_{jk}. \tag{6.18}$$

Learning in networks with hidden layers is, therefore, easily implemented. However, no general convergence theorem for the learning process is known for such networks.

6.3 Boolean Functions

Deterministic neural networks with binary neurons are ideally suited to represent logical, or *Boolean*, functions. These are functions whose arguments and function values are logical variables taking only the two values "true" and "false" ("T" and "F"), denoted by "1" and "0", or "+1" and "−1" in the Boolean representation. The range of definition of a Boolean function with N arguments covers 2^N elements, with a choice between two function values for each element. Hence there exist 2^{2^N} different Boolean functions with N arguments, a vast number even for moderately large N. Nonetheless, every Boolean function can be represented by a feed-forward neural network with a single, albeit very large, hidden layer [De87b].

We take an input layer of N binary neurons $\sigma_k (k = 1, \ldots, N)$ to represent the function arguments and a single output neuron S to represent the function value: $S = F(\sigma_1, \ldots, \sigma_N)$. The logical truth values "T" and "F" are represented by the neuron states $+1$ and -1, and 2^N neurons denoted by $s_j, j = 0, \ldots, (2^N - 1)$ form the hidden layer. It will become clear in a moment why the index j runs from 0 to $(2^N - 1)$ instead from 1 to 2^N. We assume that the network is fully connected in the forward direction.

We fix the strengths of the synaptic connections between the input and hidden neurons uniformly, except for their sign: $|\overline{w}_{jk}| = 1$. The sign is determined as follows. Take any neuron from the hidden layer and write its index j as a binary number. Owing to the above choice of counting j from 0, this binary number has at most N digits, or exactly N digits, if leading zeros are added:

$$j^{\mathrm{bin}} = (\alpha_1, \ldots, \alpha_N) \tag{6.19}$$

with α_ν equal to 0 or 1. We now set $\overline{w}_{jk} = +w$ when $\alpha_k = 1$, and $\overline{w}_{jk} = -w$ when $\alpha_k = 0$. In closed notation this definitions reads

$$\overline{w}_{jk} = (2\alpha_k - 1)w . \tag{6.20}$$

We also choose the threshold potentials of all hidden neurons uniformly, namely $\overline{\vartheta} = (N - 1)w$.

In order to recognize the meaning of these assignments, we compute the total synaptic potential at the jth hidden neuron:

$$\overline{h}_j = \sum_k \overline{w}_{jk}\sigma_k - \overline{\vartheta}_j = w\left[\left(\sum_k (2\alpha_k - 1)\sigma_k\right) - (N - 1)\right] . \tag{6.21}$$

All N terms in this sum over k are of modulus 1; hence their sum is only equal to N if all terms are equal to $+1$. Otherwise the value of the sum is at most $(N - 2)$. This implies that the total synaptic potential is only positive if $(2\alpha_k - 1) = \sigma_k$ for all k, i.e. only a single neuron in the hidden layer will and can become active for a given input. This neuron is given by $(\sigma_k + 1)/2$ in binary representation:

$$s_j = \left\{ \begin{array}{ll} +1, & j = j_0 \\ -1, & j \neq j_0 \end{array} \right\} \quad \text{with } j_0^{\mathrm{bin}} = \left(\frac{\sigma_1 + 1}{2}, \ldots, \frac{\sigma_N + 1}{2}\right) . \tag{6.22}$$

Each argument of the Boolean function thus activates exactly one hidden neuron, which is specific to that argument.[4] All we still have to do is find out how the specific state of the hidden layer of neurons can be utilized to generate the correct output value S.

For this purpose we must determine the appropriate strengths w_j of the synaptic connections between the hidden neurons and the output neuron, and its activation threshold ϑ. [Here we need only a single index at the synaptic couplings, because there is only one output neuron.] Since each hidden neuron is mapped onto a unique input, i.e. a unique function argument $(\sigma_1, \ldots, \sigma_N)$, we may set:

$$w_j = \begin{cases} +1, & \text{if } F(\sigma_1, \ldots, \sigma_N) = \text{``T''} \\ -1, & \text{if } F(\sigma_1, \ldots, \sigma_N) = \text{``F''} \end{cases} . \tag{6.23}$$

Assigning the value $\vartheta = -\sum_j w_j$ to the activation threshold, the effective polarization potential at the output neuron is given according to (6.22) by

$$h = \sum_j w_j s_j - \vartheta = -\sum_j w_j + 2w_{j_0} + \sum_j w_j = 2w_{j_0} , \tag{6.24}$$

where j_0 is the binary number denoting the function argument. In view of (6.23) the correct sign of the desired output value is obtained from the deterministic neural evolution equation $S = \text{sgn}(h)$. Because we have not made any restricting assumptions as to the nature of the Boolean function, this completes the proof that every Boolean function can be represented in this manner.

As is the case with many mathematical proofs, it only states that the representation is possible in principle. The scheme on which the proof is based may be totally useless in practice, because the number of required hidden neurons (2^N) is much too large. However, one must realize that things can hardly be different, since we did not make any assumptions concerning the function $S = F(\sigma_1, \ldots, \sigma_N)$. The representation of arbitrary Boolean functions is a computationally "hard" problem.[5] Here this finds its expression in the fact that an exponentially large number of hidden neurons are generally required.

In special cases it is possible to succeed with far fewer neurons in the hidden layer. Boolean functions, which can be represented by a number of neurons that is a low power of the number of arguments N, have been called NERFs ("Network Efficiently Representable Functions"). We have already studied one example of such a function in Sect. 6.1, namely the exclusive-OR (XOR) function. The XOR function has two logical arguments, but can be represented on a network containing only 2, instead of 4, hidden neurons.

---

[4] A neuron with this property is often referred to as the "grandmother neuron". This term derives from the (probably incorrect) notion that a specific neuron is activated in the brain, when one thinks about a certain concept, such as one's grandmother.

[5] It belongs to the class of NP-complete problems (see Footnote 1 in Sect. 11.1).

This most economical representation, shown in Fig. 6.1, is compared in Fig. 6.4 to the standard representation involving four hidden neurons.

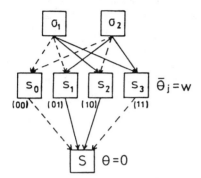

Fig. 6.4. Standard representation of the XOR function by a three-layer perceptron with four hidden neurons.

An entirely different question is whether a Boolean function of N arguments can be "learned" by a neural network in an amount of time that only grows polynomially with N. This question has been studied by Valiant [Va84] in a somewhat more general context, where the learning protocol allowed for two different levels: (1) Examples of argument vectors for which the function is TRUE are given at random ("positive" examples); (2) for any chosen argument vector the function value is provided (the "oracle"). Valiant showed that logical functions of the form (conjunctive normal form)

$$F(\sigma_i) = \wedge(\vee\sigma_i) \tag{6.25}$$

can be learned from a polynomial number of positive examples alone, whereas also the second learning level (the "oracle") is required to learn logical functions of the disjunctive normal form

$$F(\sigma_i) = \vee(\wedge\sigma_i) \tag{6.26}$$

in polynomial time. A related question of practical interest is whether, and under what conditions, the function can be learned from a few examples of the mapping: argument $\rightarrow$ function value. This is commonly called the problem of *generalization*. Ideally, the network should be able to generalize correctly from input–output relations that were included in the learning set to those that were not. We shall return to this interesting question in Sect. 9.2.

Experiment with the program PERBOOL learning Boolean functions with 2–5 logical arguments by error back-propagation (see Chapt. 24). Begin with the XOR function; try to learn it with 1, 2, and 4 hidden neurons. The first choice fails, since it is equivalent to no hidden neuron; the second case corresponds to the network shown in Fig. 6.1; the last case corresponds to Fig. 6.4. Compare the weights obtained by training the network with those obtained from formal analysis (see the description of the figures).

6.4 Representation of Continuous Functions

Another practically interesting application of multilayered neural networks is the prediction of functions which are known only at a certain number of discrete points. Problems of this type occur almost everywhere, e.g. in the prediction of economic or social data, in weather forecasting, in signal processing, and so on. Of course, there already exist a variety of useful methods of time-series analysis and forecasting. Neural networks are intrinsically nonlinear systems, and it is worthwhile investigating their forecasting capabilities. Extensive studies have been performed by Lapedes and Farber [La87b], who were able to show that two hidden layers suffice for the representation of arbitrary, "reasonable" functions of any number of continuous arguments.[6]

The nonrigorous but constructive proof [La88] is based on the fact that all continuous functions can be expressed as superpositions of simple, localized "bump" functions. Typical examples are the squares or trapezoids used in the derivation of the Riemann integral to approximate a function; more sophisticated examples are the *basis-spline* functions widely used in numerical analysis. We shall show by construction that a feed-forward neural network with a single hidden layer can generate such a function in one dimension, and that two hidden layers are sufficient in any higher dimension. (As noted above, this is not the minimal number of required hidden layers, but this is not essential since the back-propagation algorithm works for any number of hidden layers.)

We assume that the hidden layers of the network are constructed from analog-valued neurons with output values between 0 and 1, and we use the

---

[6] It can actually be shown that even a single layer of hidden neurons has sufficient flexibility to represent any continuous function [Fi80, He87b, Ho89d, Cy89]. The proof, which is based on a general theorem of Kolmogorov concerning the representation of functions of several variables [Ko57], is rather formal and does not guarantee the existence of a reasonable representation in practice, in contrast to the constructive proof that can be given for networks with two hidden layers.

Fermi function $f(x)$ (4.3) to describe post-synaptic response. In order to be able to represent arbitrary continuous functions we allow the output neurons to assume any real value and use the linear function $f(x) = x$ for the response of the neurons in the output layer. The network is accordingly described by (6.13–6.15), except for the first part of (6.13), which is replaced by a linear relation:

$$S_i = h_i, \qquad h_i = \sum_j w_{ij} s_j - \vartheta_i , \qquad (6.27)$$

$$s_j = f\left(\bar{h}_j\right), \qquad \bar{h}_j = \sum_k \bar{w}_{jk} \bar{s}_k - \bar{\vartheta}_j , \qquad (6.28)$$

$$\bar{s}_k = f\left(\bar{\bar{h}}_k\right), \qquad \bar{\bar{h}}_k = \sum_\ell \bar{\bar{w}}_{k\ell} \sigma_\ell - \bar{\bar{\vartheta}}_k . \qquad (6.29)$$

We begin with functions of a one-dimensional variable x. The function

$$g(x) = f(x) - f(x - c), \qquad c > 0 , \qquad (6.30)$$

where $f(x)$ is the Fermi function, describes a "bump" of width c and normalized height 1, which is localized at the point $x = c/2$. By means of a linear mapping of the variable x we can transform (6.30) into a bump of arbitrary width and height, and move it to any desired location on the x axis:

$$g(x) = \alpha[f(ax - c_1) - f(ax - c_2)] . \qquad (6.31)$$

This equation is just of the form obtainable by a combination of the two network equations (6.27) and (6.28).

A second hidden layer allows the generation of multidimensional bump functions. At first glance it is tempting to create a two-dimensional bump function by multiplying two one-dimensional functions in the form $g(x, y) = g_1(x)g_2(y)$. However, this operation exceeds the capability of our neural network, because the synapses are summation, not multiplication, devices. In other words, signals can only be added, but never multiplied, at a synapse. This problem may be resolved in the following way. In two dimensions the one-dimensional bump function represents an infinitely extended, straight "ridge". The sum of two such functions running in two different directions in the two-dimensional plane

$$g_1(x) + g_2(y) = \alpha[f(ax - b) - f(ax - c) + f(ay - d) - f(ay - e)] , (6.32)$$

reaches its highest value (about 2α) at the centre of the intersection between the two ridges. If we choose α as a very large number, and introduce an activation threshold between α and 2α at the neurons in the second hidden layer, the function $f(x)$, when applied to the right-hand side of (6.32) suppresses all parts of the intersecting bump lines except in the immediate vicinity of the centre of the intersection. For $\alpha \gg 1$ we hence obtain in

$$g(x, y) = f\Big(\alpha[f(ax - b) - f(ax - c) + f(ay - d) - f(ay - e) - 1.5]\Big)(6.33)$$

a suitable bump function for the representation of functions in two dimensions (see Fig. 6.5). By adding further contributions to the argument of the outer Fermi function in (6.33) we can easily generalize the construction to any number of dimensions.

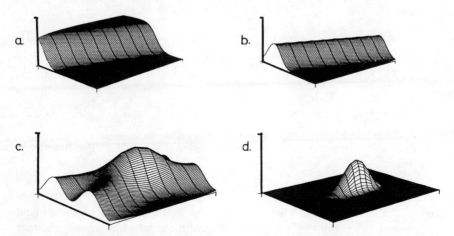

Fig. 6.5. Transition (a) from a sigmoidal threshold to (b) a ridge to (c) a pseudobump generated by two intersecting ridges to (d) the true bump.

Once again the question may be posed whether this representation of continuous functions in terms of localized bump functions can be learned by the gradient method with error back-propagation. It is also unclear whether it is the best way of representing a given function on the neural network. Irrespective of these uncertainties it must be noted that neural networks with two hidden layers have proved to be powerful instruments of forecasting and signal prediction [La87b, La87c], only to be matched by the recently developed local linear method [Fa88] (see also Sect. 7.1).

Experiment with the program PERFUNC, learning continuous functions with 2–5 arguments by error back-propagation (see Chapt. 25). Try various network topologies in order to study the ability of the network to represent straight lines, sinusoidal functions, and arbitrary powers. Refer to Chapt. 25 for advice.

7. Applications

Layered feed-forward neural networks learning by error back-propagation have been applied to a number of problems. We cannot discuss all of these sometimes quite impressive applications here in detail, but we shall discuss a few selected examples in the present and the next chapter.

7.1 Prediction of Time Series

Lapedes and Farber [La87b] studied the ability of layered feed-forward neural networks to forecast time series, which is an important problem in economics, meteorology, and many other areas.

7.1.1 The Logistic Map

Most difficult in this context is the prediction of those time series which exhibit chaotic behavior, i.e. for which an infinitesimal change in the initial value results in exponentially growing, dramatic deviations of values in the future [Sc84]. Remarkably simple functional relationships can show such a complete lack of long-term predictability, e.g. the so-called *logistic map*:

$$x(t+1) = F[x(t)] \qquad \text{with} \quad F[x] = 4x(1-x), \qquad 0 \leq x \leq 1 . \qquad (7.1)$$

For a neural network the prediction of $x(t+1)$ from $x(t)$ becomes simply the task of learning to represent the quadratic function $F(x)$. Of course, the network does not "know" that the function is so simple, it has to "find out" on its own on the basis of the available training data $(x^\mu, F[x^\mu]), \mu = 1, \ldots, p$.

Lapedes and Farber used a feed-forward neural network consisting of one input and one output neuron, and a single layer of five hidden neurons. There was also a direct synaptic connection between the input and output layers. With error back-propagation this network was trained with a set of 1000 pairs. The activation function of the hidden units was the Fermi function:

$$f(h) = \left(e^{2h} + 1\right)^{-1} = \frac{1}{2}\left(1 + \tanh(h)\right) ; \qquad (7.2)$$

the output neuron had linear response. The performance of the network was then tested with 500 randomly chosen additional points $x(t)$, where the network should predict the correct $x(t+1)$. The network was found to perform this task very well, with a mean square deviation between prediction and correct value of 1.4×10^{-4}. An analysis of the synaptic connections, in the spirit of our discussion in Sect. 6.4, revealed that the function (7.1) was represented in the form

$$
\begin{aligned}
F[x] \;=\; & -0.64 f(-1.11x - 0.26) - 1.3 f(2.22x - 1.71) \\
& -2.285 f(3.91x + 4.82) - 3.905 f(2.46x - 3.05) \\
& +5.99 f(1.68x + 0.60) + 0.31x - 2.04 \,.
\end{aligned}
\tag{7.3}
$$

As shown in Fig. 7.1, this provides a very good representation of the parabolic function (7.1) in the interval $0 \le x \le 1$.

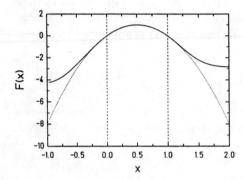

Fig. 7.1. Representation of the logistic map function $F[x]$ (*dashed line*) by a superposition of Fermi functions (*solid line*).

7.1.2 A Nonlinear Delayed Differential Equation

A more demanding test of predictive ability is provided by the time series generated by the nonlinear, delayed differential equation

$$
\frac{\mathrm{d}x}{\mathrm{d}t} = \frac{ax(t - \tau)}{1 + [x(t - \tau)]^{10}} - bx(t)
\tag{7.4}
$$

with $a = \tfrac{1}{5}$ and $b = \tfrac{1}{10}$, which was first investigated by Mackey and Glass [Ma77]. Lapedes and Farber studied (7.4) for the delay parameters $\tau = 17$ and $\tau = 30$, for which the time series $x(t + n\Delta)$ with fixed step length $\Delta = 6$ exhibits chaotic behavior. To represent the Mackey–Glass equation on a neural network, they chose a network architecture as illustrated in Fig. 7.2 with four input neurons, two hidden layers of ten neurons each, and a single output neuron, representing the value of $x(t + P)$, where P was a multiple of Δ.

There are two ways of obtaining predictions for different values of $P = n\Delta$ with $n > 1$ (see Fig. 7.3). In the first method (D), one actually *trains* the network for the specific value P, so that the synaptic strength depends on the chosen value. In the second method (E), the network is trained to predict

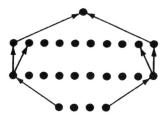

Fig. 7.2. Network architecture employed to predict the Mackey–Glass equation (7.4).

the value $P = \Delta$, and the *prediction* is iterated n times to yield an estimate for the value at $P = n\Delta$. The latter is, of course, the more severe test of the network's ability to predict the time series generated by the Mackey–Glass equation, because the errors from each iteration step enter into the next one, causing the danger of error magnification.[1]

The predictive ability of the feed-forward neural network trained by error back-propagation is compared in Fig. 7.3 with that obtained by three commonly used method of time series prediction, the *linear predictive method* (B), and the Gabor polynomial prediction method for polynomials of sixth order with direct (C) or iterated (A) prediction. Obviously the network provides by far the most reliable prediction of the time series (curves D and E), and surprisingly the iteration method (curve E) works best! Somehow the network must be able to capture the essence of the mapping described by the Mackey–Glass equation, with as few as 500 points of the time series used to train the network.

The recently developed *local linear method* of Farmer and Sidorowich [Fa88] predicts the time series with roughly comparable accuracy. This method requires much less time for adaptation of parameters than is needed for training the neural network (which took about one hour of CPU time on a CRAY X-MP), but the prediction must be based on many more points of the time series (more than 10^4). This is a clear disadvantage in the case of empirical time series, where the amount of known data is often rather limited. A comparison of various methods for nonlinear prediction can be found in [Ca89b].

7.1.3 Nonlinear Prediction of Noisy Time Series

The prediction of future values in a time series generated by a combination of deterministic and stochastic processes is frequently quite difficult, since it is not easy to distinguish between real noise and apparent irregularities which are the result of deterministic but chaotic behavior of the underlying mapping. The standard method is that of linear prediction, where a linear function is adjusted to the known past values of the time series by means of a least squares fit. This method often does not yield good predictive power

---

[1] Because of the chaotic nature of the time series, this second method is bound to fail for large values of n, however good the predictive quality may be.

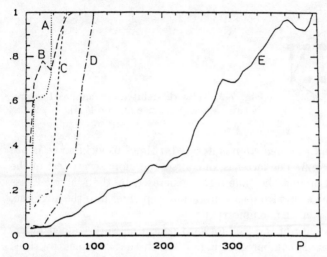

Fig. 7.3. Prediction of the Mackey–Glass time series for various prediction times P (for $\tau = 30$). The neural network (*curves D and E*) performs better than standard predictive methods (*curves A–C*). (From [La87])

if the underlying mapping is strongly nonlinear. Unfortunately, construction of the optimal nonlinear prediction method requires an a priori knowledge of the properties of underlying deterministic mapping, which is not available in many cases of practical interest.

H. Reininger [Re90] has studied the ability of simple, layered neural networks to generate almost optimal nonlinear predictors of noisy time series. The number of input neurons N_i corresponded to the number of data points on which the individual prediction was based (the *order* of the prediction); only $N_i = 1, 2$ were considered. The number of hidden neurons N_h was chosen reasonably small. Some of the considered time series were:

$$
\begin{align}
(1) \qquad x(t) &= \eta(t) + 0.9x(t-1) \,, \\
(2) \qquad x(t) &= \eta(t) + 0.21x(t-1)^2 \,, \\
(3) \qquad x(t) &= \eta(t) + 3\tanh[3x(t-1)] \,, \\
(4) \qquad x(t) &= \eta(t) + 3\tanh[3x(t-1)] + 3\sin[x(t-2)] \,, \qquad (7.5)
\end{align}
$$

where $\eta(t)$ is a normal distributed stochastic variable (here taken with variance $\langle \eta(t)^2 \rangle = 1$). A measure of the quality of the prediction is provided by the so-called *predictive gain* G, which is defined as[2]

$$
G = 10\log \frac{\langle x(t)^2 \rangle}{\langle [x(t) - \tilde{x}(t)]^2 \rangle} \,, \qquad (7.6)
$$

where $\tilde{x}(t)$ is the prediction for $x(t)$ and the brackets indicate an average over many predictions. The smaller the deviation between the predicted value

---

[2] The gain is here defined in units of dB (decibel).

$\tilde{x}(t)$ and the actual value $x(t)$, the higher is the predictive gain. Since the stochastic part $\eta(t)$ cannot be predicted by any method, the largest possible gain is given by

$$G_{\text{opt}} = 10 \log \frac{\langle x(t)^2 \rangle}{\langle \eta(t)^2 \rangle} \, . \tag{7.7}$$

The results of the study, in which the network was trained with data sets of 2000 elements with the error back-propagation algorithm, are shown in the following table:

Series	N_{i}	N_{h}	G_{opt}	G_{net}	G_{lin}
1	1	1	6.12	6.02	6.12
2	1	2	0.63	0.63	0.05
3	1	1	9.55	9.55	6.80
4	2	8	11.64	9.03	2.92

The column labeled G_{net} shows the gain produced by the layered neural network; that labeled by G_{lin} indicates the gain achieved by the standard linear prediction method. Except for the first series, which is based on a linear time series, the network fares considerably better: in some cases it even provides optimal prediction. These results show that nonlinear predictors can be effectively realized by means of multilayered perceptrons.

7.2 Learning to Play Backgammon

An amusing way of studying the ability of neural networks to acquire some form of "intelligence" is to teach them to play competitive games, such as backgammon. Success in playing backgammon, as opposed to deterministic games like chess, involves a combination of skill and chance, and no preconceived strategy can be perfect. Another related feature that makes backgammon well suited for a neural network is that moves are selected more on the basis of position-based judgements involving pattern recognition, than on look-ahead, tree-search computations, which are not very useful owing to the probabilistic element introduced by the dice.

The concept followed by Tesauro and Sejnowski [Te88] was to train the network to *select* moves (out of a predetermined set of allowed moves), rather than to *generate* them, thus avoiding the difficulties associated with teaching the intricacies of move legality. Equipped with a "preprocessor" that generated the legal moves, the network was trained to score all alternative moves on a relative level, scores ranging from -100 (worst possible move) to $+100$ (best possible move). The position and its change caused by the move were encoded by 459 input neurons (18 each for the 24 basic board locations, as well as some neurons for the black and white bar).[3] Various network architectures were tried; the best-performing network had two hidden layers of neurons with 24 elements each. However, the differences in performance, when fewer or even no hidden neurons were employed, were not very large. The network always had a single output neuron representing the score.

The data base for training contained slightly more than 3000 board positions (out of an estimated total of about 10^{20}), a particular dice roll for each position, and a set of legal moves. A small subset of these moves had been scored by a human expert; for the larger part of the moves, however, a score was chosen at random with a slight negative bias (from the interval $[-65, +35]$). This somewhat unsatisfactory technique had to be introduced in view of the large number of possible moves. The negative bias was introduced to account for the expert's bias to select good moves for scoring. The network was primarily trained on the explicitly scored moves, but it was also exposed to the randomly scored moves to avoid exaggerated bias in the training set.

The performance of the trained network was measured in various ways, the most interesting being its play against the commercial software program "GAMMONTOOL".[4] In its best configuration the network defeated GAMMONTOOL in almost 60% of its matches, which constitutes a fairly impressive achievement. Inclusion of the precomputed features in the input coding proved to be essential: without these the network could produce a winning rate of only 40%. A good performance (54% winning rate) was also achieved if the precomputed features were substituted by the individual terms of the position–move evaluation function used in the GAMMONTOOL program. Obviously, the network was able to combine these successfully into a scoring function, which was superior to that developed by the human software programmers. This ability may indicate other useful applications in similar cases, replacing painstaking trial-and-error tuning by hand.

Finally, the network was matched against a human backgammon expert (one of its creators, G. Tesauro), in a series of 20 games. Remarkably, the network won, 11 : 9 ! However, the human opponent claimed later on the basis of a detailed analysis that this surprising success was facilitated by pure luck in rolling the dice, since the network committed serious blunders in 9 of the games, of which it still managed to win 5 owing to favorable dice throws. The analysis of these games also showed that the network had acquired an "understanding" of many global concepts and strategies as well as important tactical elements of advanced backgammon play. The errors it committed were, at least in most cases, not random mistakes but instead fell into well-defined categories, which conceivably could be eliminated by adding appropriate cases to the training set.

---

[3] In addition, a few input neurons were used for the representation of some precomputed features of the position that are useful for the evaluation. This procedure, equivalent to giving the network useful "hints", greatly helped in producing a network capable of competent play.

[4] A product of Sun Microsystems, Inc.

7.3 Prediction of the Secondary Structure of Proteins

H. Bohr and collaborators [Bo88] trained a neural network to predict the so-called secondary structure[5] of proteins from their local sequence of amino acids with an accuracy of close to 70 percent. This task differs from the preceding ones in that the solution is not known in principle. The traditional approach, which involves consideration of the various forces between amino-acid residues and their interaction with the solving agent, only allows for correct prediction of the secondary structure in about one half of all cases on average [Ch78]. One therefore has to trust here in the ability of the network to extract from the training data set functional relations that have so far escaped human intelligence. As the results have shown, this pragmatic approach – originally suggested by Qian and Sejnowski [Qi88], see also [Ho89a] – is surprisingly successful in spite of its lack of physical transparency.

The network employed by Bohr et al. received as input the amino-acid coding of a certain fraction of the protein; a window size of 51 consecutive amino acids proved to be a useful choice.[6] The twenty different types of amino acids were coded by a sequence of 19 zeros and a single one, represented by 19 inactive and one active input neuron. Glycine, for example, was coded by 00000010000000000000, alanine by 00000000000000001000, and so on. The chosen window length therefore required $51 \times 20 = 1020$ input neurons. The input was then processed by a layer of 40 hidden neurons, while the output layer consisted only of two neurons, which indicated the presence or the absence of a certain type of secondary structure. Separate neural networks were trained to detect either α helix, β sheet, or random coil. The total number of synapses and thresholds amounted to about 40 000.

To train the network, the authors chose 56 proteins (comprising a total of about 10 000 amino acids) from the Brookhaven Protein Data Bank. The training was performed on a subset consisting of n of these proteins, which was extended in steps of eight until the training set contained all 56 proteins. During this process the remaining $(56 - n)$ proteins were used to monitor the performance of the network in predicting the presence of the specified category of secondary structure in unknown proteins. The results of this process are shown in Fig. 7.4 for the network assigned to detect the presence of the α-helix structure. After completion of the training, the score approached 73%.

---

[5] One distinguishes three levels of protein structure: the *primary* structure describes the sequence of amino acids, the *secondary* structure determines the local stereo-chemical form of the amino-acid chain (e.g. α helix, β sheet, β turn, or random coil), whereas the *tertiary* structure describes how the whole macromolecule is folded up in space.

[6] Window sizes between 7 and 91 were tried, with 51 giving the best results. This window length roughly corresponds to the range of hydrogen bonds between amino acid residues in the β-sheet structure.

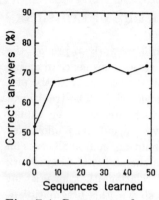

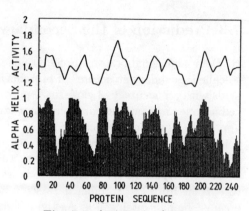

Fig. 7.4. Percentage of correct predictions of the presence of an α-helix structure versus the number of proteins in the training set (after [Bo88].

Fig. 7.5. Activity in the output neuron signalling α-helix structure versus amino-acid number in rhodopsin. *Histogram*: neural network [Bo88]. *Upper curve*: conventional method (shifted upward by 0.9).

The ability of the network to predict secondary structure was then compared with that of conventional algorithms on the protein *rhodopsin*, which plays a role in bacterial photosynthesis. The network's ability to detect the presence of an α-helix structure was clearly superior to the more traditional method, as shown in Fig. 7.5. The network actually confirmed older electron microscopical work on the secondary structure of rhodopsin, whose validity had been questioned by more recent studies utilizing other analytic techniques [Bo88].

7.4 NET-Talk: Learning to Pronounce English Text

One of the most promising fields of application of neural networks is speech recognition. One reason is that the representation of language by acoustical speech signal does not follow simple rules, and thus the conversion is difficult to implement on programmable electronic computers. The virtue of neural-network architectures is that they can be made to learn the representation simply from examples of spoken language without knowledge of the underlying rules. It does not matter if the training process takes very long, since it need be done only once. Once the representation has been learnt, the processing of a speech signal would be fast because of the parallel architecture of the network.

Versatile speech recognition by neural networks is still a great challenge. What has been achieved in practice is the (much simpler!) reverse process, i.e. of pronouncing written text. For that purpose, Sejnowski and Rosenberg [Se87] trained a multilayered feed-forward neural network to distinguish groups of letters in English words associated with individual phonemes. These could then be converted into real spoken language by a commercial speech generator, so that the text was read aloud.

The network looked at a text sequence of seven letters at a time, and therefore contained seven groups of 29 input neurons that each encoded one letter of the text fraction, for which the phonetic value was to be determined. There was a layer of 80 hidden neurons, and the 26 neurons in the output layer represented the various possible phonetic values, such as position in the mouth (labial, glottal, etc.), the phoneme type (voiced, nasal, etc.), vowel frequency, and punctuation. The network was trained to associate a phonetic value with a certain group of letters by the examples provided by 1024 words of the English language. After 50 training cycles the network had reached an accuracy of about 95% for the words contained in the training set. In a test of its ability to generalize, the network was then able to pronounce new text correctly with an accuracy of around 80%. In view of the complexity of the relationship between phonetic value and spelling in the English language, this is a quite remarkable achievement. However, in comparison with software programs designed to convert English text into phonemes based on rules derived from linguistic research the network still performed rather badly [Kl87]. It is fair to say that the design of NET-talk took only a fraction of the effort which went into the development of commercial text-to-speech conversion programs.

How does the network solve its task? Using a scaled down version of the NET-talk network, an attempt has been made to analyze the way in which the hidden neurons contribute to the recognition of the phonemes, by looking at the product of the activation of a hidden neuron and its synaptic connection strength to the output neuron representing the recognized phoneme, i.e. its contribution to the activation of this output neuron [Sa89]. This "contribution analysis" revealed, for example, that some hidden neurons were not relevant to the recognition of any phoneme, and thus could be eliminated from the network without loss of performance. However, many details of how the network implements the phonetic representation remain poorly understood.

8. More Applications of Neural Networks

Layered feed-forward neural networks have been applied to a great variety of practical tasks with varied success. As opposed to the representation of mathematical mappings, where general existence theorems have been proven, little is known about whether a perceptron network with hidden neurons will succeed given a specific task. In particular, this is true when no formal solution of the problem is known and the network is expected to discover the solution completely on its own. After all, neural networks of the low degree of complexity discussed here are surely *not* more intelligent than human beings.

For example, early applications of neural networks have been made in such diverse areas as:

- *Hyphenation.* A network was trained to predict the hyphenation of randomly chosen long words, which were presented through a six-letter window. The input layer consisted of 26 neurons for each of the six letters, one neuron for each letter in the alphabet. The hidden layer contained 20 neurons, and a single output neuron signaled whether hyphenation is possible in the middle of the six-letter window. After training the network on 17 228 words, hyphenation was predicted correctly with about 99-percent accuracy [Br88a].
- *Nuclear binding energies.* The binding energy B of atomic nuclei is a function of their proton and neutron numbers Z and N. Since this function is almost linear over a wide range, it is more meaningful to predict the separation energy of the last neutron: $S(Z, N) = B(Z, N) - B(Z, N - 1)$. A three-layered perceptron with 60 neurons in each layer (the representation of the numbers Z, N, and S was somewhat artificial) was trained to model the function $S(Z, N)$, using values from about 1000 nuclear isotopes. The network was then able to predict the separation energy within 3 percent for around 60 percent of other nuclei, but failed rather badly on 20 percent of the nuclei it had not seen in the training phase [Br88a]. Further applications of neural networks to the study of nuclear structure can be found in [Ge93]
- *Robotics.* The control of trajectories of robotic manipulators operating in production plants of under hostile environmental conditions is a task for which neural networks appear to be well suited. One example is the CMAC model (cerebellar model articulation controller) [Ma69], which is based on

an analysis of the cerebellum, the part of the brain of vertebrates which controls body movements and maintains static equilibrium. The CMAC is a feed-forward network which can be trained by supervised learning to generate a complex nonlinear mapping from a space of input signals, which include the actual position and orientation of the limbs and joints, to a set of output signals, which drive robotic actuators. More elaborate mechanisms for motor control in the primate brain are discussed in [Ec89].

A demanding task will be the construction of autonomous robots which interpret their environment from visual and other sensory data and respond in real-time with the correct motor-control reactions, e.g. avoiding collisions with objects which obstruct their path. A step in this direction is ALVINN, a neural network developed at Carnegie-Mellon University which is able to steer a car along a winding road [Po91, Po93] under a variety of environmental conditions. ALVINN has a surprisingly simple feed-forward architecture which uses only a coarse image of the road ahead as its input and is able to perform in real time. The training uses back-propagation based on the performance of a human driver.

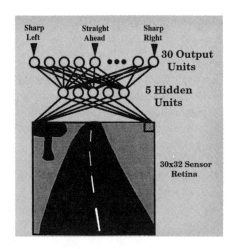

Fig. 8.1. The architecture of ALVINN, a feed-forward neural network with five hidden units which generates the steering signal for a vehicle. The input is a 30×32 pixel image of the road ahead.

– *Biomedicine.* Neural networks can classify patterns even in cases where it is difficult to formulate simple rules to distinguish the desired categories. This ability has led to a potentially rewarding application in medical diagnostics. The method of cytodiagnosis developed by G.N. Papanicolaou in the 1920s allows the detection of certain types of cancer in an early stage by microscopically screening mucous smears for malignant cells. Human experts are able to spot and classify such abnormal cells but it is desirable to automate the screening procedure. A single microscopic slide of a 'Pap smear' test for cervical cancer contains typically 100 000 cells, only a few of which may be abnormal. Fatigue of the human expert can lead to the danger of a falsely negative rating.

The system PAPNET created by Neuromedical Systems Inc. employs a neural network to automatically search for abnormal cells. Digitized images of unusual cells identified by a conventional classification system are presented to a neural network which was trained to recognize the shapes of malignant cells. In this way for each microscopic slide a collection of the most suspect cases is assembled for inspection by a pathologist who has the final say. In this way the danger of falsely negative test reports is virtually eliminated.

- *Commercial applications.* These include such diverse subjects as prediction of stock and commodity prices, signature verification on bank checks, and risk estimation in loan underwriting. Such applications will be discussed below.

In the following sections we will provide some more details concerning the application of neural networks to various real-world problems. Hopefully, this small sampling can give the reader insight into the broad applicability of neural networks.

8.1 Neural Networks in High Energy Physics

Pattern recognition has always been an important part of high-energy physics data analysis. Until the advent of hardware neural processors, these tasks were better suited to the capabilities of the human brain than that of a electronic computer. In recent years there have been a number of applications of neural networks to nuclear and high-energy physics for heavy-quark tagging and event classification [Be93, Ab92, Bo91, Be91], jet identification [Lo90, Lo91], triggering [De90a], track finding [De87a, De90b], mass reconstruction [Lo92], cluster detection [De88a], and nuclear spectroscopy [Ol92] .

8.1.1 The Search for Heavy Quarks

The use of artificial neural networks in the search for heavy quarks has become practical recently. The ability of networks to construct an internal representation of the training data without being hardwired makes them ideal candidates for this application. In a recent paper by K. Becks et al. [Be93] the application of neural networks to the identification of decays of the Z^0 particle into a bottom quark (b) and an anti-bottom quark ($\bar{b}$) was studied[1].

---

[1] The Z^0 particle is a neutral vector boson (charge q = 0, spin J = 1, isospin I = 1, and isospin projection I_3 = 0) which is one of the mediators of the weak interaction along with the W^+ and W^- bosons. These particles are produced in high-energy electron-positron collisions. The bottom quark is one of the six types, or flavors, of quarks: up, down, strange, charm, top, and bottom

In addition to designing a neural decay classifier Becks et al. also performed a comparison between neural methods and traditional multivariate statistical methods.

The process of identifying $b\bar{b}$ events is called *tagging*. Tagging allows experimentalists to separate heavy quark (**bottom**, **charm**, **top**) events from light quark (**up**, **down**, **strange**) events. In this experiment it is necessary to tag all $b\bar{b}$ pairs in order to prepare the purest possible subsample of events for studying b physics. These b-quark events provide a direct measure of the $Z^0 \to b\bar{b}$ partial decay width, or the decay lifetime for $Z^0 \to b\bar{b}$. This partial decay lifetime is very sensitive to the mass of the top quark and so its precise determination is a major goal of electron-positron colliders.

For their studies Becks et al. used the JETSET 7.2 Monte Carlo event generator [Sj89] to simulate Z^0 decay events. The events were then passed through the full detector simulation package DELSIM [DE89, DE91]. In order to facilitate future experimental application the events were also passed through the same kinematical cuts which would be present in the actual experiment.

To compare the two methods the following measures are introduced:

$$\text{b finding efficiency}: \epsilon_b \equiv \frac{\text{number of correctly tagged b's}}{\text{total number of b's}}, \tag{8.1}$$

and

$$\text{b purity}: p_b \equiv \frac{\text{number of correctly tagged b's}}{\text{total number of tagged (correct + incorrect) b's}}. \tag{8.2}$$

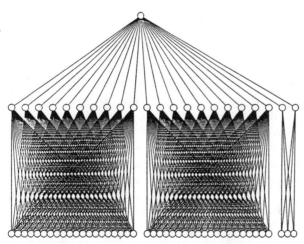

Fig. 8.2. Neural tagger network architecture. Benchmarks were computed with this topology.

The neural network consisted of 43 or 45 input neurons (depending on whether muon tagging information was used or not), and one partially connected hidden layer as shown in Fig. 8.2. The training method used was standard back-propagation as implemented in the JETNET neural network simulation package [Lo94]. The training data set included 10 000 events of which 5000 were b-quark events and 5000 were other flavor events.

Not suprisingly, when track information was presented in cartesian (x, y, z) form the neural network did not perform well. Therefore, in order to facilitate pattern recognition the cluster algorithm LUCLUS [Sj82, Sj87] was applied to events to identify *jets*. Jets are collections of particles with approximately collinear trajectories. A typical 2-jet event is shown in Fig. 8.3.

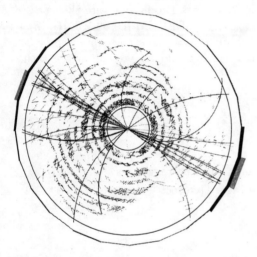

Fig. 8.3. A typical two-jet event generated in a high-energy collision.

As input only the 10 fastest particles from the two fastest jets were chosen. For each of these chosen particles the absolute values of the momentum and the transverse momentum with respect to the particle's jet axis were determined giving 40 variables. Of the remaining five variables three were chosen to be the angles between the three fastest jets. The remaining two variables were the average energy per tagged muon and the total number of reconstruced muons as reconstructed by a muon event tagger [DE92]. These last two variables were used to provide simple information about candidates for semi-leptonic decays into muons, since this information was shown to enhance the performance of both the neural and classical taggers.

Three statistical multivariate methods were chosen for comparison:

1. Log-Likelihood Method (Maximum Entropy Method)
2. Mahalanobis Distances
3. Fisher's Discriminant Analysis

For more information about multivariate statistical methods we suggest the books of Manly [Ma94b] and Srivastava and Carter [Sr83]. Applications of these methods to b-quark tagging can be found in [Va89, Pr90].

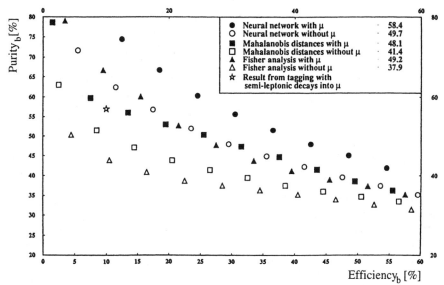

Fig. 8.4. Comparison of neural and classical event taggers. Neural network results are indicated by circles, Manhalanobis distances by squares, and Fisher discriminant analysis by triangles. Filled symbols indicate results with muon information included. The star indicates the efficiency and purity of the muon tagger.

Figure 8.4 shows a plot of the tagging purity versus efficiency for the neural network and classical taggers. In addition to the performances of each model with and without muon information, the efficiency and purity of the semi-leptonic decay tagger is indicated. The performance of the log-likelihood method was so poor that it was not included in the figure.

As Fig. 8.4 shows, the neural tagger was superior to the other classical methods given the same amount of information. The neural tagger gave an approximately 10% increase in purity when compared to the classical methods. In addition, studies of joint classification – events are classified according to the combined results of two methods – showed that the neural tagger appears to find all of the relevant information contained in the inputs. This is to be contrasted with the combination of classical methods for which significant performance gains were achieved. It may be that in its tested configuration the neural tagger has extracted all of the information possible from the inputs, at least there was no information presented by the classical taggers that was not detected by the neural tagger.

In another collaboration by P. Abreu et al. [Ab92] a neural tagger was trained to separate Z^0 decays into three classes: b events, c events, others (u, d, and s). This allowed them to measure both the b and c partial decay widths. Their neural network had 19 input nodes, one hidden layer of 25 nodes, and a three-node output layer. The components of the output vector were assigned to one of the three event classes and the network was trained with standard back-propagation from the JETSET package.

It was their conclusion that the neural network was stable against a wide range of systematic uncertainties, and was able to consistently generalize the data. Therefore, it seems that neural networks can also be used to reliably assign events with a probability of coming from b, c, or uds events.

8.1.2 Triggering

A trigger is a device that decides whether an event is "interesting" or "not-interesting" given a subset of the event data. With large numbers of particles produced at very high event rates it is important to identify which events merit further analysis. Otherwise, the necessary data collection rates and storage capacity required become unmanageable, if not impossible. Traditional triggering algorithms rely on approximating the triggering function as a linear combination of known functions, and vary the parameters of this function to maximize the signal to background ratio. Two possible representations are linear combinations of polynomials in the input variables and Fourier series. However, neural networks can also reproduce any function and the hardware needed for neural triggers is considerably less complicated than traditional fitting hardware.

An example of the application of a neural trigger has been provided by L. Lönnblad et al. [Lo90]. In this paper the authors study the application of a neural event classifier to separate jets formed from gluons from those formed by quarks. Being able to distinguish the origin of a jet of hadrons is very important since this provides experimental tests of the confinement mechanism[2] and evidence for the existence or non-existence of a three-gluon coupling in e^+e^- annihilation.

As above, the inputs were preprocessed with the LUCLUS algorithm and the network was trained with standard back-propagation. In the test runs presented to the neural trigger the network correctly identified 85%–90% of the jets. The success rate of assignment based on traditional energy based methods was on the order of 65%. Therefore, at least for this example neural networks can provide significant improvements in triggering applications. Further extension to jet identification in hadron-hadron collisions is expected.

---

[2] Quarks cannot propagate by themselves, but are *confined* to the hadrons (protons, neutrons, pions, etc.) they compose.

8.1.3 Mass Reconstruction

In another paper Lönnblad et al. studied the application of neural networks to reconstructing the invariant mass of hadronic jets [Lo92]. By using the fact that $E = mc^2$ it is possible to reconstruct the total energy of a collision event, the invariant mass, by measuring the energies and momenta of all collision products. To illustrate their approach they studied the decay of W vector bosons: $W \rightarrow q\bar{q} \rightarrow$ hadrons produced in proton–anti-proton annihilation. If the signals were clean the invariant mass, M_W, of the W-boson could be determined by adding the energies of the particles in each of the quark jets. Unfortunately, the hadrons produced in the collision are not restricted to the q and $\bar{q}$ jets; there are also remnants from projectile hadrons and hadrons created by other mechanisms. In addition, there is always the problem of noise in the detector.

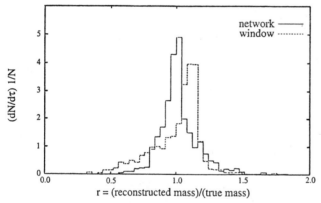

Fig. 8.5. Comparison of neural and conventional mass reconstruction.

The conventional method for mass reconstruction calls for computing the invariant mass by restricting trajectories included to a cone surrounding the jet trajectory. This will eliminate some of the background signal but is still somewhat crude. Figure 8.5 shows a plot of the ratio of the reconstructed mass and the true mass for the neural and conventional mass reconstructions. Lönnblad's studies indicate that the performance of the neural algorithm is better than conventional techniques. This can be seen in the fact that the reconstructed mass is more sharply peaked about the correct value and the resulting distribution is more symmetric. The authors cited the ability of the network to capture information about the tails of the gluon bremsstrahlung distribution as one of the reasons for enhanced performance.

8.1.4 The JETNET Code

As mentioned above, there is a software package called JETNET that is designed to provide general-purpose neural simulation for high-energy physics applications. Although it has been designed with high-energy physics in mind, the package is versatile enough for general application. JETNET is now in version 3.0 and can be downloaded from an ftp site at `thep.lu.se` in the directory `pub/Jetnet/` or at `freehep.scri.fsu.edu` in the directory `freehep/analysis/jetnet`. If you are familiar with the World Wide Web you can access the web page at `http://www.thep.lu.se`. This page maintains a JETNET program summary and an html version of the JETNET documentation. JETNET is designed to run on many different platforms: DEC Alpha, DECstation, SUN, Apollo, VAX, IBM, Hewlett-Packard, and others with an F77 compiler.

JETNET 3.0 implements a back-propagation and self-organizing map (see Sect. 15.2.3) algorithm on feed-forward multilayer networks. The package allows the user to change the network topology, learning parameters, activations functions, and updating schemes. The amount of user control over these parameters makes this a very useful tool.

8.2 Pattern Recognition

Neural networks have many applications in the area of pattern recognition. The high-energy physics examples given above are just specific instances of this type of application. To this day, there continues to be research into using neural networks for pattern recognition. For example there have been recent papers on the application of neural networks to handwritten text recognition and segmentation [Ba93, Su93, Sr93, Gi93, Vs93, Ki94], 3-D object classification using visual, thermal, sonar, and radar signals [Ba92, Te93, St94, Ha94a, Na94, Dr95], robotic place learning [Ba94a], speech recognition [Wa89b, Le90, Te90, Is90, De94a, Mo95], cartography [Kr94], speech translation [Wa91, Wo94], part-of-speech tagging [Sc94], and face recognition [Va94]. There also exist some recent articles concerning the theoretical applicability of neural networks to pattern recognition [Li93, Th94]. As with the high-energy physics applications, the literature on neural networks applied to pattern recognition is extensive (entire journal volumes have been dedicated to this topic), so instead of trying to provide a comprehensive picture of recent developments, we will review the results of a few representative applications.

8.2.1 Handwriting and Text Recognition

The post office alone has considerable interest in developing automated address readers to further optimize their mail sorting capabilities. In addition, banks and other companies are interested in automated processing of checks and handwritten documents. The postal services of most first- and second-world countries already use some form of automated text recognition; however, the success rates for systems currently in the field are not acceptable when applied to handwritten text. For instance, the French postal service, La Poste, uses an address recognition system developed by a French company, CGA-HBS, which is able to read *typed* addresses correctly 90% of the time, including the zip code and distribution information, with less than a 1% confusion rate. Unfortunately, for handwritten addresses the results are not as good, with about 67% correctly identified.

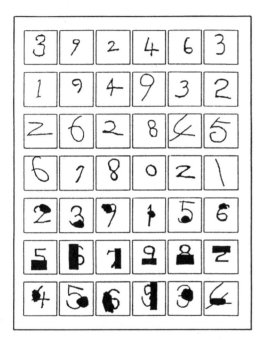

Fig. 8.6. Examples of test patterns. Figure taken from [Ki94].

The models used in the past for text recognition relied on custom engineered practices, such as manually designed rule-based systems. However, researchers have realized that these systems appear to be "engineering dead ends" [Ba93]. Another approach to the problem does not attempt to master the specific details of a given recognition task, but instead encourages the use of algorithms that rely as little as possible upon task-specific information. A neural network classifier is one type of recognition tool that falls into this latter class. However, recent evidence indicates that in most cases a hybrid of linear regression, rule-based systems, and neural networks perform better than any of the three alone [Gi93, Su93, Ki94].

Table 8.1 lists a performance comparison of different types of handwritten digit recognition. An analysis of the accuracy of each of the models indicates that classifier M (mixed statistical/structural) is the most accurate.[3] However, for overall correctness, the neural network models perform better than the others.

Table 8.1. A comparison of handwritten digit recognition performance. There were 18 468 digits in the training set and 2711 digits in the test set. Data taken from [Sr93].

Recognizer	Features	Classification	Correct
Polynomial discriminant [Sc78]	Pixel pairs (1241,binary)	Linear discriminant	93.6%
Mixed classfier [DA82]	Topological: arcs, caves, holes Histogram-based: leftmost pixel posit. (26, integers and real)	Mixed	89.4%
Stroke-based [Ku91]	Strokes, holes, contour profiles (6)	Rule based	86%
Contour chains [Du80]	arc, bay, curl, inlet, null, spur, stub, wedge (8)	Rule-based (>130 rules) Decision Tree	83%
Combination P+M+S+C	Classifier decision confidences (40, real)	Neural network (80 hidden units)	96%
Contour chains [Sr93]	Contour chain code (464, integers)	Neural network (20 hidden units)	96%
Histogram [Sr93]	Histogram (72, real $\in [0,1]$)	Neural network (40 hidden units)	95%
Gradient [Sr93]	Gradient (192, binary)	Neural network (40 hidden units)	96.4%
Gabor [Sr93]	Gabor coefficients (72, real)	Neural network (40 hidden units)	94%
Morphology [Sr93]	Morphology (85, real $\in [0,1]$)	Neural network (35 hidden units)	96%

Handwritten character recognition is made difficult by the increased number of symbols, and the different types of character shapes for print and

---

[3] Accuracy is defined as a weighted sum of the number of identification errors plus rejects: Accuracy = α Errors + Rejects, where α measures the number of rejects needed to match the cost of making one identification error.

cursive writing. The success rates for handwritten character recognition performance are on the order of 93% for uppercase letters, 85% for lowercase, and 63% for cursive writing [Sr93]. H.J. Kim and H.Y. Yang have recently published a paper [Ki94] in which they use a hybrid system called AINET. This system consists of a simple recursive network which is initialized and trained using expert knowledge for feature extraction. Table 8.2 lists a comparison between the performance of a standard multi-layer perceptron and the AINET initialized with and without the use of rule-based weight initialization (indicated as AINET + and AINET – respectively). Although the overall performance of this model is not as good as some of the neural network recognizers listed above, we can see from this data that training with expert knowledge greatly enhances the performance of a standard neural network recognizer.

Table 8.2. AINET handwritten digit recognition rates. From [Ki94].

Model	Typical handwritten digits Recognition rate	Partially distorted digits Recognition rate
Multi-Layer Perceptron	52.0%	33.0%
AINET –	69.5%	57.0%
AINET +	81.0%	78.0%

8.2.2 Speech Recognition

There are two reasons why layered neural networks appear to be ideal tools for implementing mapping functions applicable to speech recognition. Firstly, one hidden layer is sufficient to approximate arbitrarily well any continuous function. Secondly, it has been shown that long-term temporal correlations in speech data cannot be captured with conventional linear or nonlinear predictive models. On the other hand, neural networks which combine linear and nonlinear terms are shown, analytically and by simulation, to be able to capture these long-term temporal correlations. In a recent paper by Deng et al. [De94a], the application of three-layered feed-forward neural networks to speech recognition was studied. In this model the neural network actually forms the heart of a Markov state-dependent nonlinear auto-regressive time series model which gives rise to a nonlinear hidden Markov model (HMM) neural predictor unlike the standard feed-forward implementations. A HMM can be viewed as a type of recurrent network which, when presented with a segment of an image, passes through a probabilistic state sequence to reconstruct the full image. The papers by [Ba72, Po88, Vs93] provide more information on HMM theory.

Deng et al. tested the neural HMM using a limited vocabulary of "CV" syllables (C stands for consonant and V for vowel) obtained from six male speakers. They further restricted the data by requiring C to be one of the six stop consonants (/p/, /t/, /k/, /b/, /d/, /g/) and by using only the vowel /i/. This set was chosen because recognition of syllables with stop consonants is known to be a difficult task [De91, De90c] and because these syllables exhibit the temporal correlations mentioned above. Table 8.3 compares the performance of standard HMM recognizers with that of the neural HMM recognizers. The data show that for every speaker, an architecture with mixed linear and nonlinear hidden units performs better than the linear, nonlinear, or standard HMM predictors alone. The conclusion of the paper was that high-accuracy speech recognition using neural networks is possible.

Table 8.3. Comparison of recognition of CV syllable for HMM recognizers using standard HMM and various combinations of neural HMMs. From [De94a].

Speaker	Standard HMM	Linear Predictive Neural HMM	Nonlinear Predictive Neural HMM	Mixed Predictive Neural HMM
1	85.7%	84.5%	92.8%	97.6%
2	91.7%	91.7%	96.4%	100.0%
3	80.9%	88.1%	89.3%	89.3%
4	90.8%	78.6%	86.9%	89.3%
5	94.0%	96.4%	94.0%	96.4%
6	71.4%	70.2%	69.0%	84.5%
Average	84.1%	84.9%	88.1%	92.9%

8.2.3 3D Target Classification and Reconstruction

The military also has some applications of neural networks in mind. One area in which neural networks are being used is in the identification or reconstruction of three-dimensional objects using visual, thermal, sonar, and radar information. Recently, B. Bai and N. Farhat presented some results from the application of neural networks to the reconstruction and identification of possibly "noncooperative targets" [Ba92]. For testing, Bai and Farhat constructed three scale models: a B-52 airplane, a Boeing 747, and a NASA space shuttle. They then tested the performance of a standard feed-forward neural network in two applications: (a) reconstructing a two-dimensional image of the object using microwave radar waves with a frequency band of approximately 2 GHz to 26.5 GHz and (b) identifying a target using only a limited number of frequency response echos.

The neural networks performed well in both applications. In the reconstruction test the neural networks exhibited a very good tolerance to noise in the signals. This is an important fact since in real world applications radar signals are never noise-free. In the simpler task of object identification, the neural networks performed astonishingly well, with a 100% correct identification of the targets when given three or more successive views of the objects.

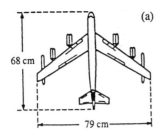

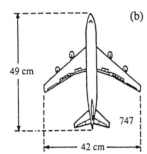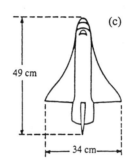

Fig. 8.7. The three targets used: (a) a B-52 airplane; (b) a Boeing 747 airplane; (c) a NASA type space shuttle. From [Ba92].

The network used in the identification task was trained by presenting the network with Fourier transforms of radar echos from the targets. Each target was positioned in 100 different aspect views equally spaced over a 20° range. Figure 8.8 shows the recognition rates for the B-52, Boeing 747, and space shuttle using radar echos from three randomly chosen views for testing. Final identification was made by recording the identification of the network for each aspect and using the prediction of the majority (2 out of 3 views in this case).

These results are encouraging. The neural network is shown to be robust, in that noisy versions of the training data are correctly identified; however, the network has no way of returning a response of "I don't know what that object is". Further research should include additional outputs for giving an uncertainty factor. This could possibly be implemented by collecting network identifications from multiple views and constructing statistics from these.

8.3 Neural Networks in Biomedicine

Another area in which neural networks are finding application is biomedicine. The ability of neural networks in pattern recognition, and the fact that they can be trained by example, makes them appealing to medical researchers. Frequently, researchers do not fully understand the detailed dynamics of complicated biological systems, but with neural networks they can train systems on

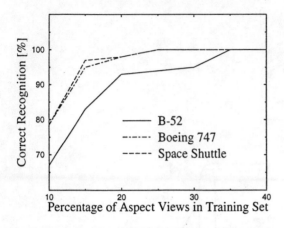

Fig. 8.8. Correct recognition versus the training set size. Three randomly selected views were presented during testing. From [Ba92].

actual data with little prior information. Neural networks are being applied in the areas of biological signal processing [Mi92], medical image processing [Mi92], and medical decision-support [Hr90, Ka94]. Although neural networks are finding successful application in some medical tasks, there are applications for which neural networks do not seem to perform well. Again, the choice of architecture and data preprocessing strongly effects the results of neural network solutions.

A review of neural network applications in medical imaging and signal processing has been published by A.S. Miller et al. [Mi92]. The most promising application of neural networks appears to be in the presentation of magnetic resonance (MR) images. An MR image consists of an array of two-dimensional slices of a three-dimensional image. Neural networks can aid in the interpretation of these images by classifying different tissue types and identifying features, like blood vessels or tumors. It has been shown that images segmented with a neural classifier have better delineation of tissue regions with fewer isolated points than conventional Bayesian classifiers.

Neural networks are also being applied to the analysis of X-ray radiographs, lung scintigrams, ultrasound, computer tomography, and ECG, EEG and EMG signals. One example of an EEG application is in the detection of epileptiform discharges (EDs) [We94b]. The health of epileptic patients depends on the medical staff's response, and any advanced warning of a full epileptic seizure is important.

Medical decision-support is another area where neural networks are being applied. Two reviews are currently available on the subject [Hr90, Ka94]. The subfield of computer-aided diagnosis is reviewed by D.F. Stubbs [St90]. The benefits of neural networks in medical decision-support and diagnosis are that they can be trained by example, as contrasted to expert systems which must have their rules laboriously constructed, and that neural networks are better suited for probabilistic and sometimes ill-defined diagnosis problems since they can have some degree of generalization.

Indications are that neural networks can give major advances in these areas; however, most applications are still in their early stages. The problems which must be addressed are: collection and classification of training and testing data, input preprocessing, and network architecture.

8.4 Neural Networks in Economics

The application of neural networks to economics has become a "hot" topic within the last five years, with at least four books currently available on the subject [Mi90, Tu92, Ba94b, De94b]. This trend began approximately five years ago with the publication of several papers including articles on the application of neural networks to stock price prediction [Sc90], and stock selection [Wo90].

In these original papers, the researchers concluded that neural networks could be applied to economics and finance, but found limited success with their original models. Research in this area continued over the next years through specific applications to financial time series prediction [Ta91, Ch92, Co92, Wo92, We94a, Le94, Le94, Mc95a] and the general theory of using neural networks for time series prediction [Ma91, Ke91, Da93, Wa93a, We94a, Wa94, Ta95]. In addition, there are currently at least ten commercial trading applications available which use neural networks and artificial intelligence [De94a].

Unfortunately, there is inadequate benchmarking of the neural network methods and traditional regression techniques due to the competitive nature of this application. A paper by Refenes et al. does offer a comparision [Re94], but because of the proprietary nature of their system, definite conclusions about the relative performances are difficult to draw from this work. This seems to be a major problem for scientists approaching this application, because if there were a system that *really* worked, chances are they would be the last to know.

There have been a number of papers which apply neural networks to the prediction of economic time series. F. Wong and P.Y. Tan have applied neural networks to the prediction of Singapore's gross domestic product (GDP), and currency exchange rates [Wo92]. Their model uses feed-forward networks trained with an optimized back-propagation algorithm called FastProp. Both linear and nonlinear hidden units along with direct connections from inputs to outputs are used. Wong and Tan also investigated using a genetic algorithm to determine the network topology (See Chapter 16). Figure 8.9 shows the prediction of Wong's NeuroForecaster® for Singapore's GDP. It has been found [Ta91, Sh90] that neural networks perform better than conventional methods for series with short memory, but for time series with long memory, neural networks and the traditional Box-Jenkins model produce comparable results.

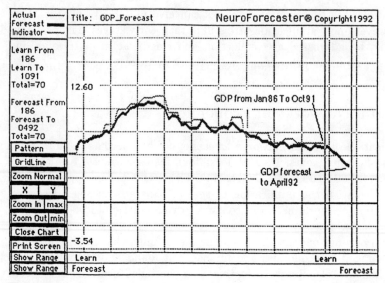

Fig. 8.9. A 6-month ahead forecast of Singapore's national GDP. The thin line is the actual quarterly GDP, and the thick line is the forecast. The network was trained with a 6-month window to perform a 6-month-ahead forecast. From [Wo92].

In another application P.J. McCann and B.L. Kalman have applied neural networks to the prediction of the gold market [Mc95a]. Instead of using the standard mean square deviation of the market prediction and the actual next day's value, McCann and Kalman use the profitability of the network as the performance measure. As shown in Fig. 8.10, the network consisted of 10 inputs, 5 hidden units, and 1 feedback or context unit. The feedback unit was constructed by copying one of the hidden units from the previous feed-forward pass into the input layer. The output was a two-dimensional vector trained on the possible outputs: $(+1, -1)$ buy gold, $(-1, +1)$ sell gold, $(0, 0)$ no decision. The inputs to the network were an oil index, an aggregate commodities index, the Standard & Poor's 500 index, a dollar index, a bond index, the 30-year and 10-year bond yields, the Sterling currency index, and the gold mining index for the period between September 1, 1988, through January 24, 1994.

Figure 8.11 shows a plot of the gold market and network trades during the test period day 901 to day 1369. Trades indicated by a "b" are from a network with (0,0) outputs subtracted from the error function, and those indicated by "c" were trained on an extended set from day 1 to day 1169. Network b made six trades before its performance degraded, for a profit of 11% in about 3 months. Network c made four trades during the test period, for a profit of 21%. The authors point out that if these trades were also made in the gold mining company stock index, the paper profits were 50% and 64% respectively!

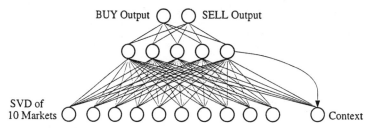

BUY Output   SELL Output

SVD of
10 Markets   Context

Fig. 8.10. A network for predicting the gold market. Direct connections between input and output layers are not shown. From [Mc95a].

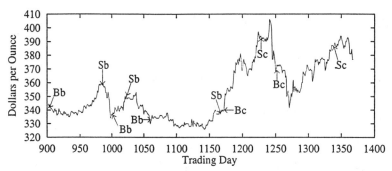

Fig. 8.11. Trades on the gold market which were executed by the neural network depicted in Fig. 8.10 (B = buy, S = sell). From [Mc95a].

As mentioned above, network b's performance degraded after about 3 months of trading. The authors claim that this degradation can be attributed to the evolution of world financial markets over time [Mc95a]. This is not necessarily convincing, nonetheless, neural networks seem to be able to make accurate predictions about gold market turning points within a limited prediction window. The decision by the authors to move away from an exact numerical prediction of the gold price, to trying to train networks to recognize patterns in the input data, like turning points, is essential. In fact, it seems that the most successful financial applications of neural networks do just this.

This claim is further evidenced by the research of B. LeBaron and A.S. Weigend into neural network prediction of stock market volumes from December 3, 1962, through September 16, 1987 [Le94]. In this paper LeBaron and Weigend discuss a method for determining the relative performances of neural and linear predictors. Previous work on time series predicition has shown that the performance of neural and linear predictors depend on the way in which the data is segmented into training and testing sets. In some cases authors would find that the neural models performed better, while others found the opposite. LeBaron and Weigend suggest that the proper way to compare two methods is to average over all possible segmentations presented to the models, in addition to averaging of the model initialization. Their goal

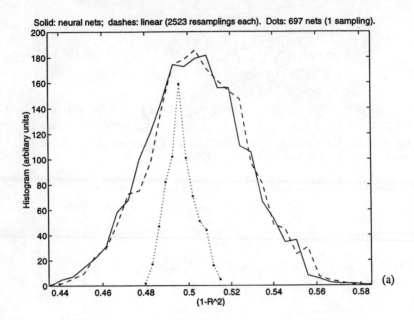

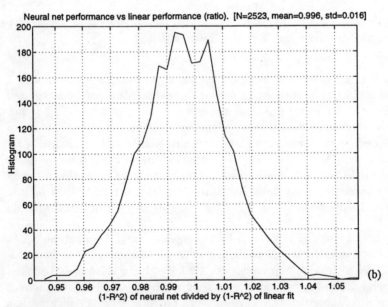

Fig. 8.12. (a) Histogram of $(1 - R^2)$ forecast performance. Neural network results are indicated by a solid line and linear predictor results by a dashed line. The dotted line is from a subset given only one segmentation with random network initialization. (b) Histogram of the ratio of $(1 - R^2)$ network performance divided by the $(1 - R^2)$ linear predictor performance. From [Le94].

was to show that on average the neural networks performed better than linear predictors in predicting the numerical value of the stock market volume.

Figure 8.12 shows a histogram of the performances of the neural and linear predictors and the ratio of $(1 - R^2)$ performances for each model.[4] From this figure we see that the neural network and linear predictor performances are nearly identical. Figure 8.12b suggests that the linear model is even slightly better than the neural model for this task, when averaged over segmentations and initialization. As Fig. 8.12a shows it is possible to have better performance with a neural model for a given data segmentation, but in general linear and neural models perform similarly in the numerical prediction of the stock market volume.

Another advantage that LeBaron and Weigend's technique allows is the determination of a probability distribution for the prediction rather than simply the next day's expected value. This distribution is constructed by keeping track of the predictions of the many differently initialized/segmented networks. Figure 8.13 shows the prediction distributions during the period between September 17, 1987, through October 19, 1987. This set is special since a stock market crash occurred on October 19 and 20. Although the model's performance is not very good, we can see how the model's predictions spread out as the crash is approached.

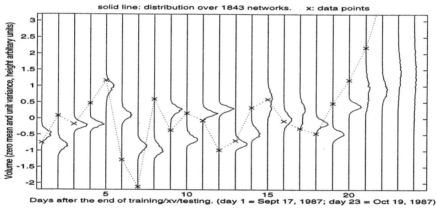

Fig. 8.13. Prediction distributions for the stock market volume during the period leading up to the October 1987 crash. The crash occurred on day 23 of this figure. From [Le94].

---

[4] R is the correlation coefficient between the forecast and the target.

Modify the code of program PERFUNC (see Chapt. 25) to generate a feed-forward neural network for your own favorite application. Consult Chapts. 19 and 23 for useful hints (and don't forget to save the source code PERFUNC.C before you begin). Your only limits are your imagination and the speed of your computer!

9. Network Architecture and Generalization

9.1 Building the Network

9.1.1 Dynamic Node Creation

While the gradient-learning algorithm with error back-propagation is a practical method of properly choosing the synaptic weights and thresholds of neurons, it provides no insight into the problem of how to choose the network architecture that is appropriate for the solution of a given problem. How many hidden layers are needed and how many neurons should be contained in each layer? If the number of hidden neurons is too small, no choice of the synapses may yield the accurate mapping between input and output, and the network will fail in the learning stage. If the number is too large, many different solutions will exist, most of which will not result in the ability to generalize correctly for new input data, and the network will usually fail in the operational stage. Instead of learning salient features of the underlying input–output relationship, the network simply learns to distinguish somehow between the various input patterns of the training set and to associate them with the correct output.

In general there are two approaches to finding the optimal network architecture. The first one involves starting from a larger than necessary topology which is trained to learn the desired mapping. Then individual synapses or entire neurons are eliminated if they are not actively used or carry little weight. By this process of clipping and pruning, the network is eventually reduced in size. The shortcomings of this approach are that one first has to deal with an unnecessarily large network, which is computationally wasteful, and that the pruning process may get stuck in an intermediate-size solution, which cannot be smoothly deformed into the optimal network architecture.

The second approach follows an opposite line, starting with a small network and *growing* additional neurons until a solution can be found. If performed in a sufficiently careful manner, this method is guaranteed to find the smallest possible network that solves the task, at least for architectures involving only a single layer of hidden neurons. However, it is necessary to retrain the complete network after the addition of each single new neuron, in order to make sure that a further increase in size occurs only if convergence of the learning procedure cannot yet be achieved. The simplified version of this approach, where only the newly added neuron and its synapses are trained and all old parameters remain frozen, does not, in general, find the optimal solution.

The second approach involving full retraining of the network was studied by Ash [As89], who called it the *dynamic-node-creation* method. The decision to add a new hidden neuron and its synaptic connections with the input and output layers is based on the following consideration. When the network is trained (by the back-propagation algorithm), the mean squared deviation D_t between actual and desired output (6.6) decreases with time t. However, when no solution can be found, the rate of decrease slows down drastically before D_t has reached the desired small value; a new neuron should be added then. More precisely, the trigger condition for creation of an additional neuron was

$$\frac{|D_t - D_{t-\delta}|}{D_{t_0}} < \Delta_{\mathrm{T}}, \qquad (t \geq t_0 + \delta), \tag{9.1}$$

where t_0 is the time of creation of the previous new neuron, δ denotes the interval over which the slope of the error curve D_t is calculated, and Δ_{T} is the trigger slope. The last condition in (9.1) ensures that at least δ training steps are taken before a new neuron is added. The whole procedure is stopped when D_t becomes sufficiently small or some other measure of convergence is satisfied.

The method was tested on various problems taken from [PDP86], where the minimal network architecture is known. In most cases a solution with this optimal topology was found; in some cases the obtained solutions were slightly larger. Figure 9.1 shows how the method fared for the task of adding two binary three-digit numbers (called ADD3 in [PDP86]). Here seven hidden neurons were required, but it is not known whether that is the minimal number for networks with a single hidden layer. One clearly sees how the error curve D_t always levels off, but drops again sharply after a new neuron is added to the hidden layer. Convergence is best observed for the quantity M_t, which represents the largest squared error at time t for any output neuron in any pattern contained in the training set. M_t drops precipitously after the addition of the seventh neuron.

One remarkable result of this study [As89] was that the total computational effort was never more than 40% larger, and sometimes considerably smaller, when neurons were dynamically generated, than in the case where one started right away with the optimal network architecture. Apparently the effort spent in training the network when it still is too small is not lost, but prepares the ground for training the network of optimal size. In some cases, neuron generation during training was even the *only* way to get convergence in back-propagation learning. An interesting, but open question is whether the network architectures and the specific solutions found by the dynamic-node-creation method also yield optimal properties concerning generalization to input data not contained in the training set.

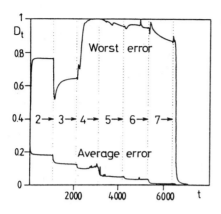

Fig. 9.1. Development of the squared error functions D_t (*lower curve*) and M_t (*upper curve*) for the ADD3 task. At each vertical line a new neuron is added to the network.

9.1.2 Learning by Adding Neurons

The freedom to choose the network architecture at will can be exploited to devise a strategy that permits the network to *learn* to represent a function mapping without the need to apply the error back-propagation algorithm, if one allows for the addition of many layers of hidden neurons [Me89c]. The idea behind this strategy (called the "tiling" algorithm) is surprisingly simple. Let us assume we want to represent the mapping of a binary (Boolean) function with N arguments. We then start from the simple perceptron architecture, i.e. an input layer containing N neurons and a single neuron S_1 in the output layer. In general, as we discussed in Sect. 5.2.3, this network architecture will not permit a faithful representation of the function. For some input vectors the output neuron will be able to yield the correct function value, but for some others it will not. Nonetheless, one can optimize the solution, so that the correct function value is represented in as many cases as possible [Ga87].

If the output neuron S_1 takes the correct value $\zeta^{(\mu)}$ for each of the function arguments ($\mu = 1, \ldots, p$), we are finished; if not, we add more neurons to the output layer. For that purpose we divide the input vectors, i.e. the function arguments in the training set, into classes yielding the same state of the output neuron(s). We start with all input vectors yielding $S_1 = +1$. Those for which the desired function value is also $\zeta = +1$ have been learned correctly, those for which $\zeta = -1$ have not. We now add a second neuron S_2 to the output layer and try to learn the representation $S_2 = \zeta$ only for the class of inputs with $S_1 = +1$, using the perceptron learning algorithm. Then we do the same for the second class of inputs, i.e. those with $S_1 = -1$. If this procedure leads to the result that two input vectors (function arguments) with different function values ζ also correspond to different states of the output neurons (S_1, S_2), we have obtained a "faithful" representation of the functional mapping, although not yet the desired one.

Otherwise, we have to add a third neuron, and attempt to learn the correct function values $S_3 = \zeta$ for the class of those function arguments that produce the same output state (S_1, S_2), and so on, until the mapping of the input vectors on the output states $(S_1, \ldots, S_q)$ is faithful in the above sense, that different values of ζ are represented by different output states. We then start a new output layer with the neuron S_1' and consider the previous layer of neurons $(S_1, \ldots, S_q)$ as the new input layer. Utilizing the perceptron algorithm, we try to learn the mapping $(S_1, \ldots, S_q) \longrightarrow S_1' = \zeta$. Again, if this is not successful, we add more neurons to the new output layer and proceed in the same way as before, until the mapping $(S_1, \ldots, S_q) \longrightarrow (S_1', \ldots, S_{q'}')$ is faithful in the sense explained above. Then we add another output layer, and continue this procedure, until a single neuron in the last layer allows the function value ζ to be represented correctly for all arguments. The procedure is guaranteed to converge, because one can show that subsequent layers decrease in size, i.e. $q' < q$, and so on [Me89c].

Mézard and Nadal tested this algorithm for random Boolean functions with $N = 8$ arguments. They found that on the average 7 layers were needed to represent the function with a total of about 55 hidden neurons. This is not so bad, since the formal proof given in Sect. 6.3 requires a single hidden layer of $2^N = 256$ neurons. The rather unusual complexity with regard to the number of layers is compensated by the conceptual simplicity of the learning algorithm, which is based on the elementary Hebb rule rather than on error back-propagation. This allows the use of deterministic binary neurons during the training phase, whereas the gradient-descent method employed in the back-propagation technique requires training with continuous-valued or at least stochastic neurons. It is also not unreasonable to imagine that a related mechanism could be at work in biological neural networks. An initially small assembly of neurons might be able to "recruit" neighboring neurons, until the network has become large enough to learn a particular task with the help of a simple Hebbian mechanism of synaptic plasticity.

9.2 Can Neural Networks Generalize?

9.2.1 General Aspects

Up to now we have mostly discussed how a feed-forward, layered neural network (perceptron) can learn to represent given input–output relations, such as the optimal reaction to an external stimulus, by a suitable choice of synapses and activation thresholds. If this were all there is, the neural network would only act as a convenient storage and recall device for known information, similar to the associative memory networks discussed in Chapt. 3. The deeper intention is, of course, to use the network, after completion of the training phase, to process also inputs that were never learned. In other words, one would like to know whether the neural network can *generalize* the acquired knowledge in a meaningful way. Experience has shown that networks often succeed in this task, but not always, and sometimes only to a certain degree. This has various reasons and conditions:

1. The new and previously unknown input must not deviate too much from the examples used in the learning phase. Consider the task of predicting linear functions, where the network may be trained to extrapolate straight lines given by two points. After appropriate training the neural network will be able to forecast other straight lines, if their slope lies within the range of the learned cases, but it will fail miserably when asked to predict a straight line whose slope lies far outside this range.

2. The generalization of a given set of examples is never unique. Whenever it appears to be so to common sense, this judgement is based on additional implicit assumptions. In many cases one is only interested in the *simplest* possible generalization, but again the concept of simplicity is often strongly biased. The difficulties encountered in generalization are easily illustrated by the example of the Boolean functions. For N logical arguments there are, as discussed above, 2^{2^N} different such functions. By giving the function values for p different arguments many of these are excluded, but a multitude of $2^{2^N - p}$ possible generalizations remain. Any further reduction requires additional assumptions as to which generalizations are acceptable.

3. The information required for the correct generalization must not be covered by other properties of the examples in the training set. Anyone knows the technique, often applied in crime stories, of hiding the essential clue among many other, much more obvious details of information.[1] A neural network will doubtlessly fall victim to similar tricks.

4. The information required for the correct generalization must not lie hidden too deep in the form of the input. An obvious example is provided by the task of finding the number of prime factors contained in a given binary number. This is a notoriously hard problem, whose difficulty is exploited in constructing break-safe codes. No neural network will be able to solve this task even after seeing an extended training set. One also recognizes that the manner of presentation of a given task can make a big difference to how fast it can be learned; for instance, if the number were presented in its prime factor decomposition, instead of in binary form, the network would easily solve the problem of counting the number of factors!

It should have become clear from these rather cursory remarks that generally valid statements concerning the ability of neural networks to "generalize" are difficult. Exemplary studies, e.g. those by Scalettar and Zee [Sc88, Sc87] concerning the ability of a neural network to differentiate left and right, or large and small, have shown that a limited capability for generalization certainly exists. Whether one is impressed by the results obtained in these and related studies (see also [Pa87b, An88b, Zi88]), greatly depends on one's expectations. One should note, however, that the human ability for general-

---

[1] This technique was exploited by Edgar Allen Poe in "The Purloined Letter".

ization and abstraction must be patiently and often painfully learned. How far it is fair to expect such talents from artificial neural networks, whose architecture is very much simpler than that of the human brain, remains unclear.

9.2.2 Generalization and Information Theory

Is there anything mystical about the apparent ability for generalization observed in properly trained neural networks, or is it a natural consequence of the information fed into the network during the training? This intriguing question was studied by Anshelevich et al. [An89b], who applied the tools of information theory to analyze the ability of neural networks to perform generalization. In particular, we are interested in the question how many examples p^* the network has to learn until it can successfully generalize by induction. For that purpose, let us define the *capacity* C of a neural network as the number of different input–output mappings (algorithms) it can represent by appropriate selection of its synaptic efficacies. Neglecting the possibility that two different sets of synaptic weights can represent the same mapping, C is equal to the total number of different synaptic configurations of the network. If all algorithms occur with equal probability (this is, of course, an unproven assumption), the amount of information that must be fed into the network to teach it one particular algorithm is given by $\log_2 C$.

On the other hand, as a result of learning the correct response to p input patterns the network acquires $-pN_o \log_2 q$ bits of information, where q is the probability of correct response in any one of the N_o output neurons. We shall assume that the output neurons are binary, so that $q = \frac{1}{2}$. Hence, in order to provide the network with sufficient information to learn any particular algorithm it is capable of performing, the network must be trained with at least

$$p^* = \frac{1}{N_o} \log_2 C \tag{9.2}$$

input patterns. Since this number is derived on the basis of a probabilistic argument, there is no guarantee that some particular set of p^* training patterns uniquely fixes the synaptic connections of the network and enables it to generalize correctly. However, the argument should be valid for an average training set. If p^* is equal to the total number of possible input patterns, i.e. $P = 2^{N_i}$ for binary input neurons, the network can learn the algorithm, but it has no ability to generalize owing to a lack of further input cases. A measure of the network's ability to perform generalization is therefore given by the ratio p^*/P.

Anshelevich et al. [An89b] tested their argument by simulations with a feed-forward network with one hidden layer containing N_h binary neurons. The network operated according to the familiar equations

$$s_j = \theta\left(\sum_k \overline{w}_{jk}\sigma_k - \overline{\vartheta}_j\right) , \qquad S_i = \theta\left(\sum_j w_{ij}s_j\right) , \tag{9.3}$$

where σ_k, s_j, and S_i take the values 0 and 1, and the synaptic weights were also restricted to discrete values ($\overline{w}_{jk} = -1, 0, 1; w_{ij} = 0, 1$) to facilitate the counting of possible synaptic configurations. The thresholds of the hidden neurons were fixed by $\overline{\vartheta}_j = \sum_k \theta(\overline{w}_{jk}) - 1$, i.e. the number of excitatory synapses at the hidden neuron j less one. The capacity of the network was thus

$$C = 3^{N_i N_h} 2^{N_h N_o} (N_h!)^{-1} , \tag{9.4}$$

since the sequence of the hidden neurons in a fully interconnected layered network is irrelevant. Using Stirling's formula to approximate the factorial, the relative number of training patterns required before the network is able to perform generalization is then predicted to be

$$\frac{p^*}{P} = \frac{N_h}{N_o 2^{N_i}} \left(N_i \log_2 3 + N_o - \log_2(N_h/e)\right) . \tag{9.5}$$

To see how this prediction compares with real simulations, a network was trained to perform bitwise addition (defined as the XOR function) of two binary numbers, so that $N_i = 2N_o$. The network was trained with a variable number p of input patterns, and then its average error on all possible P input patterns was measured. The error can be defined by the usual expression (6.6) of mean square deviation between real output $S_i^{(\mu)}$ and the desired output ζ_i^μ:

$$D[p/P] = \frac{1}{pN_o} \sum_{\mu=1}^{P} \sum_{i=1}^{N_o} \left(\zeta_i^\mu - S_i^{(\mu)}\right)^2 . \tag{9.6}$$

$(1 - D)$ is a measure of the efficiency of generalization of the network. If the prediction (9.5) is correct, the average value $\langle D \rangle$ for a random choice of p input patterns should (approximately) vanish when p exceeds p^*. This is indeed borne out by the simulations, the results of which are represented in Fig. 9.2. The dashed line represents a linear extrapolation to the point p^*/P, where generalization should be perfect according to the analytical prediction (9.5). The different curves in Fig. 9.2 correspond to the network parameters (a) $N_i = N_h = 4$, (b) $N_i = N_h = 6$, and (c) $N_i = N_h = 8$.

Intuitively it is clear that the ability of a network to generalize from learned examples must decrease with a rising number of hidden neurons. The smallest network that can perform a certain task must also have the greatest ability to generalize. This expectation is supported by (9.5), which says that the required number of patterns in the training set grows almost linearly with N_h, and it is also proved by the results of the simulation, shown in Fig. 9.3. Here the network with $N_i = 8$ was studied for various sizes of the hidden layer. The full error curves are shown for $N_h = 8, 16, 32$; for higher values of

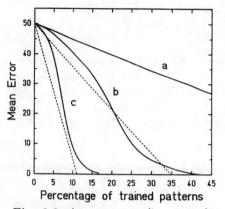

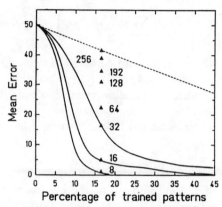

Fig. 9.2. Average error (in percent) as function of the relative number of learned patterns. The *dashed lines* indicate the analytical result.

Fig. 9.3. Dependence of the efficiency of generalization on the capacity of the network, determined by the number of hidden neurons $N_{\rm h}$.

$N_{\rm h}$ only the error at $p/P = 0.164$ is given. The dashed line represents the error function $\langle D \rangle = \frac{1}{2}(1 - p/P)$ expected in the absence of any ability to generalize.

Almost identical results were (earlier) obtained by Carnevali and Patarnello [Ca87b]. They studied the learning curve of Boolean networks constructed from elementary units with two inputs and one output that realize one of the $2^{2^2} = 16$ possible Boolean functions of two variables. (These units cannot be represented by individual neurons but, as discussed in Sect. 6.3, by ensembles of seven neurons in three layers.) The units are numbered from 1 to $N_{\rm G}$. Each unit i can have any of its inputs connected to the output of a unit $k < i$, or an input gate may represent one of the N primary inputs of the network. The network is completely described by specifying, for each unit, the Boolean function it implements and where its two inputs are connected.

Carnevali and Patarnello considered networks with $N = 4$ and $N_{\rm G} = 2, 3, 4$, realizing Boolean functions of four variables. There are $2^{2^4} = 65536$ such functions, but only a fraction of these can be realized by the networks with the specific type of architecture considered here, although the total number of architecturally different networks, $W(N_{\rm G})$, is much larger (see Table below). Apparently many different networks represent the same Boolean function $F(\sigma_1, \ldots, \sigma_4)$; we call this number $H(F)$. The fact that the majority of functions cannot be realized at all implies that $H(F) = 0$ in most cases.

The following table shows the number of different networks, W, and number of realizable functions, $N_{\rm F}$, for the Boolean structure considered by Carnevali and Patarnello [Ca87b].

N_G	W	N_F
2	102 400	526
3	58 982 400	3000
4	46 242 201 600	13624

It is now not difficult to surmise that the magnitude of $H(F)/W$ is a measure of the ease with which the function F can be learned by the network. Using simulated annealing (see Sect. 10.2) to train the network, Carnevali and Patarnello probed this conjecture by checking how many examples N_E for the function argument $\rightarrow$ value mapping were needed to teach the network a particular Boolean function. Since a Boolean function of four arguments is completely specified by $2^4 = 16$ function values, the largest possible value for N_E is 16. The analysis showed that N_E was very closely given by

$$\Delta S = -\log_2[H(F)/W] . \tag{9.7}$$

ΔS, which describes the amount of information gained by knowing that the network represents the function F, can be considered as the *entropy* of the implementations of the function F on the network. Its value turned out to be virtually independent of N_G, i.e. of the complexity of the network.

Again we are confronted with the result that the ability to generalize has nothing mysterious, but is a direct consequence of the laws of statistical physics: for a given amount of information fed into the network it will (most likely) choose the generalization that has the greatest probability of being realized, or the greatest entropy. Learning and generalizing from a subset of examples can only occur because of the wide variability of the number of representations $H(F)$. There is a very restricted set of "simple" functions (about 500 out of a total of 65536), for which $\Delta S < 16$; all other functions are "beyond the grasp" of the network.[2] The statistical approach to learning and generalization in neural networks has also been studied in detail by S. Solla and collaborators, who have developed an explanation of the rather strange behavior of learning curves (the error often drops suddenly to zero after a long period of slow progress) on the basis of thermodynamical concepts [So89, Ti89].

In recent years the tools of statistical physics have been applied to the problems of optimal learning and generalization with great success. Readers interested in these developments should consult the extensive review article [Wa93b] which also contains many references to the literature.

---

[2] Details of the network structure have a strong influence on which problems are "simple" and which are "hard" to learn. It would be very helpful to have a way of estimating $H(F)$ for various functions F on a given network without explicitly counting all realizations, but such a method is not known. In many cases, problems intuitively considered "simple" are also simple in the technical sense defined here, but this rule is not generally valid.

9.3 Invariant Pattern Recognition

9.3.1 Higher-Order Synapses

In the forms discussed up to now neural networks can learn to perform a variety of basic identification tasks, but they are not yet able to solve all the complex problems arising in the recognition of optical patterns. An example is provided by a perceptron composed of an input layer with N^2 neurons arranged in a square, and an output layer with two neurons. This simple perceptron is now confronted with the task of differentiating between simple patterns containing either a single vertical or a single horizontal line of N active input neurons, on the criterion whether the bar is horizontal or vertical (see Fig. 9.4). Experience shows that the network necessarily fails, because two different vertical lines are not sufficiently closely correlated to pull one of the two output neurons into their common range of attraction.[3]

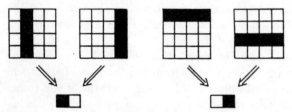

Fig. 9.4. A perceptron fails to differentiate between horizontal and vertical bars independently of their position.

Although the failure appears disastrous at first, it is not as bad as it seems. In the brain, also, any single neuron is not responsible for the detection of all vertical lines, independent of their location within the field of vision; rather it addresses this task only for patterns observed in a small region of the retina. Nevertheless, the example raises an important question, namely that of *invariant* perception or pattern recognition. Ideally, we would like the network to recognize a stored pattern independent of its precise location, orientation, or size. Indeed, our brain can (mostly) recognize an object independent of its distance, angle of vision, or speed with which it moves. We can even recognize three-dimensional objects from their two-dimensional picture. Although we may be fooled sometimes by an involuntary misapplication of this ability, such *trompe-l'œil* situations are rare.

Neural networks of the type discussed up to now are generally not capable of invariant pattern recognition. The reason why they fail is the linearity of the equation describing how the various signals arriving at different synapses

---

[3] In the sense of the scalar product, the two input patterns are even orthogonal, since they have no active neuron in common.

on the same neuron combine to a single post-synaptic polarization potential:

$$h_i = \sum_k w_{ik} s_k + \vartheta_i \ .$$

(9.8)

Invariant pattern recognition requires the computation of essentially non-linear characteristics of the stored patterns. This is well known from the algebraic theory of invariants. For example, the invariant properties of a symmetric matrix $A = (a_{ik})$ can be represented in the form

$$I_n = \text{Tr}(A^n) \ .$$

(9.9)

Biological neurons are not really linear summing devices of synaptic inputs, in contrast to the oversimplified "neurons" of McCulloch–Pitts neural network models. The synaptic response may be a complicated and nonuniversal function of the presynaptic signal, and the decision whether a neuron fires may not be expressible as a linear combination of post-synaptic potentials. It is well possible that the nonlinear features of biological neurons could play a decisive role in the capacity of the brain for invariant pattern recognition.

Probably the simplest way of introducing nonlinear synaptic connections consists in allowing several presynaptic neurons to send their input signal into the same synapse. Presynaptic inhibition is a special case of this phenomenon, which is known to occur in biological nerve nets (see Sect. 1.1). In order to emphasize the geometrical aspects of invariance properties, we consider the input neurons to be arranged in a one- or two-dimensional array, labeled by their position x: $s(x)$. The output neurons will be labeled by a discrete index i: S_i. The most general form the post-synaptic potential can take is then

$$
\begin{aligned}
h_i\big[s(x)\big] \ &= \ w_i^{(0)} + \sum_x w_i^{(1)}(x)s(x) + \sum_{x,y} w_i^{(2)}(x,y)s(x)s(y) \\
&\quad + \sum_{x,y,z} w_i^{(3)}(x,y,z)s(x)s(y)s(z) + \cdots \\
&= \ \sum_m \sum_{x_1,\ldots,x_m} w_i^{(m)}(x_1,\ldots,x_m)s(x_1)\cdots s(x_m) \ ,
\end{aligned}
$$

(9.10)

combined with the activation condition $S_i = f(h_i)$. These higher-order neurons, called *sigma-pi units* by Rumelhart et al. [Ru86b], can be employed to define the conditions of invariant perception [Gi88]. The necessary condition that the perceptron reacts in the same way to the original pattern $s(x)$ and any one of a class of transformed patterns $s^T(x)$ is that [Pi47]

$$h_i\big[s(x)\big] = h_i\big[s^T(x)\big] \ .$$

(9.11)

If the transformation is purely of a geometrical nature, the transformed pattern corresponds to a transformation among the input neurons: $x \to \hat{T}x$. To be specific, we consider *translations* for which the transformation law reads

$$s^T(x) = s(\hat{T}x) = s(x + a) \ .$$

(9.12)

Since the expression (9.10) involves summations over all input neurons, their order can be rearranged so that the transformation is absorbed into the arguments of the synaptic weights $w_i^{(m)}$:

$$
\begin{aligned}
h_i\big[s^T(x)\big] &= \sum_m \sum_{x_1,\ldots,x_m} w_i^{(m)}(x_1,\ldots,x_m) s(\hat{T}x_1) \cdots s(\hat{T}x_m) \\
&= \sum_m \sum_{x_1,\ldots,x_m} w_i^{(m)}(\hat{T}^{-1}x_1,\ldots,\hat{T}^{-1}x_m) s(x_1) \cdots s(x_m) \\
&= \sum_m \sum_{x_1,\ldots,x_m} w_i^{(m)}(x_1 - a,\ldots,x_m - a) s(x_1) \cdots s(x_m) ,
\end{aligned}
$$

(9.13)

where $\hat{T}^{-1}$ is the inverse transformation of the input pattern. If the network is to be capable of invariant pattern recognition, condition (9.11) with the expressions (9.10) and (9.13) must hold for all possible input patterns, in turn requiring the synaptic weights to be invariant functions:

$$
w_i^{(m)}(x_1 - a,\ldots,x_m - a) = w_i^{(m)}(x_1,\ldots,x_m) . \tag{9.14}
$$

The only way to satisfy this constraint is that the synaptic strengths depend only on the separation between two input neurons. We thus obtain the following result:

$$
w_i^{(1)}(x_1) = w_i^{(1)}; \qquad w_i^{(2)}(x_1,x_2) = w_i^{(2)}(x_1 - x_2); \qquad \text{etc.} \tag{9.15}
$$

The additional freedom gained by introducing higher-order synapses is thus reduced again, at the benefit of obtaining independence of translation. Similar conditions are derived, if one requires invariance under rotation or scale invariance. Even invariance under some combined tranformations can be imposed, such as rotational and scale invariance (but not translation invariance).

The action of the higher-order synapses can be understood, from a more general point of view, as the evaluation of nonlocal correlations between input patterns. Bialek and Zee [Bi87a] have argued, in the context of statistical mechanics, that such nonlocal operations are an essential requisite of invariant perception. There is no doubt that human vision allows for a very large class of invariances, achieving close to optimum performance [Ba80], but it is not known to what extent the brain relies on nonlocal information processing for that purpose [Ze88]. The question remains how such higher-order networks perform in practice. The networks can be trained with the help of a properly adapted error back-propagation algorithm [Ru86b], so this poses no problem. Giles et al. [Gi88] have trained a small network to recognize the letters "T" and "C" independent of their size, obtaining satisfactory performance over a difference in scale of a factor of four. A similar performance was achieved for the case of translation invariance.

Bischoff et al. [Bi92] have studied autoassociative recurrent networks with higher-order connectivity (up to fifth order) and found an increase in storage capacity ([Ho88b], see also page 243) and robustness. Encoding translation invariance in the manner sketched above led to a decreased performance. A comparison with randomly diluted networks of the same size showed that this can be explained by the reduction of the number of degrees of freedom in weight space caused by the constraint (9.15).

9.3.2 Preprocessing the Input Patterns

A second approach to invariant pattern recognition deals with the problem by encoding the input patterns in a clever way. In fact, folklore has it that a neural network can be made to solve any task, given an appropriate preprocessor.[4] There are strong indications that just this type of signal preprocessing occurs in the visual systems of man and other animals [Fi73, Sc77]. We shall see in Sect. 15.2.3 how neural networks can learn to distort input patterns geometrically in very complex ways. In fact, preprocessing mechanisms such as the one we shall discuss next have been suggested to be operative in the human brain [Br82, Re84].

The preprocessing required to transform the two-dimensional input pattern into a form that is simultaneously invariant against translation, rotation, and scaling consists of three steps. First, the picture $s(x, y)$ is Fourier transformed:

$$\tilde{s}(k_x, k_y) = \int \mathrm{d}x \int \mathrm{d}y \, \mathrm{e}^{\mathrm{i}k_x x + \mathrm{i}k_y y} s(x, y) . \tag{9.16}$$

A translation of the input patterns corresponds to a multiplication of its Fourier transform by a constant phase, hence $|\tilde{s}(k_x, k_y)|$ is invariant against translations in the plane. In the second step one goes over to polar coordinates in the **k**-plane, writing

$$(k_x, k_y) = (\mathrm{e}^q \cos \phi, \mathrm{e}^q \sin \phi) . \tag{9.17}$$

A rotation of the original picture now corresponds to a shift in the polar angle ($\phi \to \phi + \phi_0$), while a scale transformation by the factor r corresponds to the shift ($q \to q - \ln r$). Rotations and rescalings have thus been converted into translations in a new coordinate space (q, ϕ). Finally, applying a second Fourier transformation

$$\hat{s}(u, v) = \int \mathrm{d}q \int \mathrm{d}\phi \, \mathrm{e}^{\mathrm{i}uq + \mathrm{i}v\phi} |\tilde{s}(\mathrm{e}^q \cos \phi, \mathrm{e}^q \sin \phi)| , \tag{9.18}$$

one finds that the thrice-transformed pattern $|\hat{s}(u, v)|$ is invariant against all three types of transformation.

---

[4] A trivial but not very elegant preprocessor for translationally invariant pattern recognition would simply shift the input pattern slowly around until it "locks in" with one of the stored patterns [Do88].

Fuchs and Haken [Fu88b] have studied the performance of a pattern-recognizing system equipped with such a preprocessor.[5] Figure 9.5 shows how the method works. The upper and the lower part show the successive stages of processing the original and the shifted, rotated, and scaled picture of a human face. In their final form (the right-most of the four stages) the two patterns are perceived to be virtually identical even by the naked eye. At least in principle, we conclude, invariant pattern recognition by a neural network is feasible with the help of this preprocessor. Since (finite) Fourier transformations can be performed very rapidly by parallel-processing computers, this approach may even be practical on a large scale.

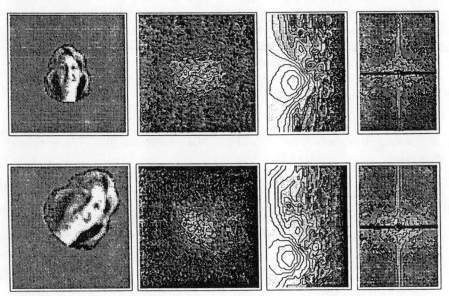

Fig. 9.5. Three stages of processing an image (*left*) into a translation-, rotation-, and scale-invariant form (*right*) (from [Fu88b]).

The technique discussed above is not easily generalized to provide invariance against other nonlinear transformations of the pattern, in particular genuine distortions of the figure. Here one has to rely on techniques extracting topological features, such as nodes, from the input pattern and compare these with those of the memorized patterns. Such techniques, e.g. *graph matching*, form computationally hard (possibly NP-complete) tasks. Since neural networks have been proposed to be able to provide practical solutions to such

---

[5] The system studied by Fuchs and Haken was not a neural network, but a content-addressable memory built from nonlinearly coupled synergetic units [Ha87]. One can expect, however, that the preprocessor coupled to a Hopfield-type neural network would perform similarly.

problems,[6] it is reasonable to attempt to implement such topological-feature detectors on neural networks, see [Ma87, Bi87b, Kr88].

---

[6] See Chapt. 11 on combinatorial optimization.

10. Associative Memory:
Advanced Learning Strategies

10.1 Storing Correlated Patterns

As we discussed in Sects. 3.1 and 4.3, the ability to recall memories correctly breaks down if the number p of stored patterns exceeds a certain limit. When the synaptic connections are determined according to Hebb's rule (3.12), this happens at the storage density $\alpha = p/N = 0.138$. The reason for this behavior was the influence of the other stored patterns as expressed by the fluctuating noise term in (3.13). As we already pointed out at the end of Sect. 3.1, this influence vanishes exactly if the patterns are orthogonal to each other as defined in (3.16). On the other hand, the power of recollection deteriorates even earlier if the stored patterns are strongly correlated. Unfortunately, this happens in many practical examples. Just think of the graphical representation of roman letters, where "E" closely resembles "F" and "C" resembles "G", or of a typical list of names from the telephone book, which are probably highly correlated.

Use the program ASSO (see Chapt. 22) to learn and recall the 26 letters of the alphabet: A–Z. Choose the following parameter values: (26/26/0/1) and (1/0/0/1;2), i.e. sequential updating, temperature and threshold zero, and experiment with the permissible amount of noise. Is any letter stable? Repeat the exercise with the first six letters of the alphabet and study the network's ability to recall the similar letters "E" and "F".

The nature of the problem is not so different from that encountered in the previous section in connection with layered feed-forward networks. The perceptron learning rule provides the perfect learning strategy for simple perceptrons without hidden layers of neurons. However, these devices are not particularly useful in practice, since they fail to solve even some very simple tasks. This is the reason why the perceptron concept fell out of grace for almost twenty years, although multilayered perceptrons with hidden neurons do not suffer from such ailments. But without a practical learning algorithm, which became first available with error back-propagation, they did not provide a practical alternative.

Similarly, some of the practical difficulties encountered with associative-memory networks are not of a fundamental nature, but rather a consequence of the inadequacy of the elementary form (3.12) of Hebb's learning rule. As we discussed in Sect. 3.1, Hebb's rule is based on the concept of the Hamming measure (3.1) of distance between different patterns. In mathematical terms, this distance measure or *metric* is called the *Euclidean* metric in the space of patterns. More general measures of the distance between different patterns are conceivable and may be more useful if the patterns are correlated. For example, in the case of the letters "E" and "F" a distance measure based solely on the bottom part of the letter would easily discriminate between the patterns. However, as the letters "K" and "R" show, this simple choice does not yet provide a general solution, not even for the alphabet.

10.1.1 The Projection Rule

Nonetheless, it turns out that the problem of discriminating between correlated patterns has a remarkably simple solution, which even permits the storage of $p = N$ arbitrarily correlated patterns, as long as they are linearly independent. To see how it works, we form the matrix of scalar products between all pairs of patterns ($\sigma_i^\mu = \pm 1$):

$$Q_{\mu\nu} = \frac{1}{N} \sum_i \sigma_i^\mu \sigma_i^\nu \qquad (1 \le \mu, \nu \le p) . \tag{10.1}$$

For linearly independent patterns the matrix $Q_{\mu\nu}$ is invertible, and we can define the following improved synaptic coupling strengths [Ko84, Pe86b]:

$$\tilde{w}_{ij} = \frac{1}{N} \sum_{\mu,\nu} \sigma_i^\mu (Q^{-1})_{\mu\nu} \sigma_j^\nu . \tag{10.2}$$

Mathematically, (10.2) corresponds to a projection technique that eliminates the existing correlations between patterns, hence this learning rule is often called the *projection rule*. With this choice the interaction of the stored patterns due to the fluctuating term in (3.13) vanishes exactly, as can be easily seen by computing the post-synaptic potentials in the presence of one of the memorized patterns σ_i^λ:

$$\begin{aligned}
\tilde{h}_i = \sum_j \tilde{w}_{ij}\sigma_j^\lambda &= \frac{1}{N} \sum_{\mu,\nu} \sigma_i^\mu (Q^{-1})_{\mu\nu} \sum_j \sigma_j^\nu \sigma_j^\lambda \\
&= \sum_{\mu,\nu} \sigma_i^\mu (Q^{-1})_{\mu\nu} Q_{\nu\lambda} = \sum_\mu \sigma_i^\mu \delta_{\mu\lambda} = \sigma_i^\lambda.
\end{aligned} \tag{10.3}$$

We conclude that every stored pattern represents a stable network configuration, independent of correlations among the patterns. Of course, the condition $p \le N$ continues to limit the memory capacity, since at most N linearly independent patterns can be formed from N units of information.

As it stands this statement is not entirely correct because some patterns may be effectively memorized without being explicitly represented in the synaptic couplings. Such a phenomenon is not new to to us; we have seen in Sect. 3.3 (3.23) that linear combinations of stored patterns may also be stable memory states. As Opper has shown [Op88] this occurs for the *iterative learning algorithm* of Krauth and Mézard [Kr87a] (see also Sect. 10.1.3), which permits the storage of up to $2N$ different patterns in an optimal way. Only N of these patterns are stored according to the projection rule, the others are memorized without being explicitly stored. However, for $p > N$ the learning process converges very slowly. (Techniques to accelerate convergence were suggested in [Ab89, An89a].) The optimal storage capacity of a neural network will be discussed in detail in Chapt. 20.

The prescription (10.2) for the coupling strengths is also called the *pseudoinverse solution*. Essentially it performs an inversion of the set of pattern vectors σ_i^μ which can be viewed as a matrix with N columns and p rows. The role of the pseudoinverse is most easily understood if we look at (10.3) at a fixed site, dropping the index i. Then we have $\tau^\mu = \sum_j \sigma^\mu{}_j \, w_j$ or in matrix notation $\tau = \underline{\sigma} \, \mathbf{w}$. Here τ^μ stands for the output which is to be evoked when the input pattern σ_j^μ is presented to the network. We have to solve a system of p linear equations for the N-dimensional weight vector $\mathbf{w}$. Since in general the matrix $\underline{\sigma}$ is not square ($p < N$) we cannot invert it directly. However, one can introduce the pseudoinverse matrix [Ko84]

$$\underline{\sigma}^{\mathrm{pi}} = \underline{\sigma}^T \left(\underline{\sigma}\,\underline{\sigma}^T\right)^{-1} \tag{10.4}$$

so that $\mathbf{w} = \underline{\sigma}^{\mathrm{pi}} \, \tau$, (cf. (10.2)), solves the problem:

$$\underline{\sigma}\mathbf{w} = \underline{\sigma}\,\underline{\sigma}^{\mathrm{pi}}\tau = \underline{\sigma}\,\underline{\sigma}^T \left(\underline{\sigma}\,\underline{\sigma}^T\right)^{-1} \tau = \tau \, . \tag{10.5}$$

In an autoassociative memory τ happens to coincide with σ but this is not essential, the pseudoinverse solution also works for heteroassociation and for perceptrons without hidden layers.

The quality of pattern recall deteriorates with growing temperature T and memory utilization $\alpha = p/N$ [Ka87].[1] As in the case of Hebb's rule the quality of recollection is described by the parameter m defined in (4.15). $m = 1$ denotes perfect memory recall, whereas $m = 0$ indicates total amnesia. The regions of working and confused memory are shown, together with the value of m at the phase boundary, in Fig. 10.1. The *radius of attraction R* of the stored patterns is shown in Fig. 10.2 as function of storage density α for the models of Kanter and Sompolinsky [Ka87] (curve a) and Personnaz et al. [Pe86b] (curve b). Here the radius of attraction is defined as $R = 1 - m_0$, where m_0 is the smallest overlap a pattern s_i can have with a stored pattern σ_i to be recognized with certainty by the network. As one sees, the elimination

---

[1] Note that the diagonal couplings $\tilde{w}_{ii}$ are set to zero in [Ka87]. If the diagonal terms are retained [Pe86b], the critical memory capacity remains $\alpha_c = 1$ at $T = 0$, but in the presence of tiny fluctuations the stored patterns cannot be recalled above $\alpha = 0.5$. (One says that the *radius of attraction* of the stored patterns is zero.)

of the diagonal couplings w_{ii} in the model of [Ka87], curve (a), has a very beneficial effect.

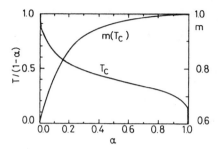 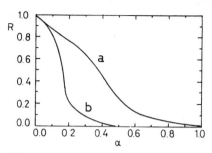

Fig. 10.1. Regions of working memory and total confusion, and recall quality m at the phase boundary labeled "T_c".

Fig. 10.2. Average radius of attraction R as function of storage density in the models [Ka87] (*curve a*) and [Pe86b] (*curve b*).

10.1.2 An Iterative Learning Scheme

The practical application of the projection learning rule for large, memory-saturated networks suffers from the need to invert the $(p \times p)$-matrix $Q_{\mu\nu}$, which poses a formidable numerical problem. Fortunately, the matrix inversion need be performed only once, when the patterns are stored into the network. The ingrained memory can then be recalled as often as desired without additional effort. A practical method of implementing the projection rule is based on an iterative scheme, where the "correct" synaptic connections are strengthened in order to stabilize the correlated patterns against each other [Di87]. For the sake of simplicity, we demonstrate this method only for the deterministic network ($T = 0$).

Because of the neuron evolution law $s_i(t+1) = \text{sgn}[h_i(t)]$ any pattern σ_i^μ represents a stable network configuration, if h_i has the same sign as σ_i, i.e.

$$\sigma_i^\mu h_i = \sum_j w_{ij} \sigma_i^\mu \sigma_j^\mu > 0 \tag{10.6}$$

for every neuron i. If the expression (10.6) is only slightly positive, any small perturbation, i.e. $s_i \neq \sigma_i^\mu$ for a few neurons i, can change its sign. In order to achieve greater stability of the desired memory patterns, we demand that the expression (10.6) be not only positive but also greater than a certain threshold $\kappa > 0$. For a single pattern, Hebb's rule (3.7) yields $h_i = \sigma_i$. As a consequence, the condition $\sigma_i h_i = 1$ is always satisfied in this case. It appears natural to take the stability threshold at $\kappa = 1$ also in the general case of several stored patterns, and to demand that the synaptic connections be chosen such that

$$\sigma_i^\mu h_i = \sum_j w_{ij} \sigma_i^\mu \sigma_j^\mu = 1 \qquad (10.7)$$

for all neurons i.

An obvious method of achieving the desired result begins with choosing the synaptic connections initially according to Hebb's rule:

$$w_{ij} = \frac{1}{N} \sum_\mu \sigma_i^\mu \sigma_j^\mu . \qquad (10.8)$$

In the next step we check, one after the other for all stored patterns, whether the condition (10.7) is fulfilled. If this is not the case, we modify the synapses according to the prescription

$$w_{ij} \rightarrow w_{ij}' = w_{ij} + \delta w_{ij} \qquad (10.9)$$

with

$$\delta w_{ij} = \frac{1}{N} \left(1 - \sigma_i^\mu h_i\right) \sigma_i^\mu \sigma_j^\mu , \qquad (10.10)$$

where μ denotes the pattern just under consideration.[2] With these modified synaptic connections we obtain for the same pattern μ

$$
\begin{aligned}
\sigma_i^\mu h_i' &= \sigma_i^\mu h_i + \sum_j \delta w_{ij} \sigma_i^\mu \sigma_j^\mu \\
&= \sigma_i^\mu h_i + \frac{1}{N} \sum_j \left(\sigma_i^\mu\right)^2 \left(\sigma_j^\mu\right)^2 \left(1 - \sigma_i^\mu h_i\right) = 1 ,
\end{aligned} \qquad (10.11)
$$

since $\left(\sigma_i^\mu\right)^2 = 1$. Thus, after updating all synapses, the threshold stability condition (10.7) is satisfied for the considered pattern. When we proceed to the next pattern $(\mu + 1)$, the synaptic couplings will be modified again, so that (10.7) becomes valid for the pattern now under consideration. However, (10.7) may cease to be satisfied for the previous pattern μ. After a full cycle over all stored patterns the condition is therefore only fulfilled with certainty for the last pattern, $\mu = p$, but not necessarily for all other patterns. The crucial question is whether this updating process converges, or whether it may continue indefinitely without reaching a stationary state, in which the threshold condition is satisfied by all patterns.[3]

In order to study this question, it is useful to introduce some abbreviations. We assume here that the synapses are adjusted sequentially, i.e. after inspection of the performance of the network for each single pattern. We define the deviation from the threshold in the ℓth updating cycle, and the sum of all deviations encountered up to that point, as

---

[2] In principle, we can do without the initial Hebbian choice of synapses. If we start with a completely disconnected network $(w_{ij} = 0)$, or *tabula rasa*, the first application of the modification law (10.9) results in synaptic connections with precisely the values assigned by Hebb's rule!

[3] The proof of convergence follows closely that of the perceptron convergence theorem given, e.g., in [Bl62a, Mi69].

$$\delta x_i^\mu(\ell) = 1 - \sigma_i^\mu h_i, \qquad x_i^\mu = \sum_{\ell'=1}^{\ell} \delta x_i^\mu(\ell') \, . \tag{10.12}$$

The synaptic modifications according to (10.10) for the νth pattern in the ℓth cycle can then be written in the form

$$\delta w_{ij} \equiv \frac{1}{N}\left(1 - \sum_k w_{ik}\sigma_i^\nu\sigma_k^\nu\right)\sigma_i^\nu\sigma_j^\nu = \frac{1}{N}\delta x_i^\nu(\ell)\sigma_i^\nu\sigma_j^\nu \, . \tag{10.13}$$

Thus, after the completion of the ℓth updating cycle the synaptic connections can be expressed as

$$w_{ij} = \frac{1}{N}\sum_\nu x_i^\nu(\ell)\sigma_i^\nu\sigma_j^\nu \, . \tag{10.14}$$

Consider now what happens in the $(\ell+1)$th iteration cycle. If we have just reached the pattern μ, all previous patterns have contributed $(\ell+1)$ times to the synaptic modification process, whereas all others (including pattern μ) have made only ℓ contributions. The modification for the μth pattern is therefore given by

$$\begin{aligned}
\delta x_i^\mu &= x_i^\mu(\ell+1) - x_i^\mu(\ell) = 1 - \sum_k w_{ik}\sigma_i^\mu\sigma_k^\mu \\
&= 1 - \frac{1}{N}\sum_k\left[\sum_{\nu<\mu}x_i^\nu(\ell+1)\sigma_i^\nu\sigma_k^\nu\sigma_i^\mu\sigma_k^\mu + \sum_{\nu\geq\mu}x_i^\nu(\ell)\sigma_i^\nu\sigma_k^\nu\sigma_i^\mu\sigma_k^\mu\right].
\end{aligned}$$
$$\tag{10.15}$$

We now introduce the N matrices of dimension $(p \times p)$

$$B_i^{\mu\nu} = \frac{1}{N}\sum_k \sigma_i^\nu\sigma_k^\nu\sigma_i^\mu\sigma_k^\mu = \sigma_i^\nu\sigma_i^\mu Q_{\mu\nu}, \qquad (i = 1, \dots, N), \tag{10.16}$$

where $Q_{\mu\nu}$ is the symmetric overlap matrix defined in (10.1). This allows us to put (10.15) into the simple form

$$x_i^\mu(\ell+1) - x_i^\mu(\ell) = 1 - \sum_{\nu<\mu}B_i^{\mu\nu}x_i^\nu(\ell+1) - \sum_{\nu\geq\mu}B_i^{\mu\nu}x_i^\nu(\ell) \, . \tag{10.17}$$

Assuming that the iteration procedure converges, i.e.

$$\lim_{\ell\to\infty} x_i^\mu(\ell) = y_i^\mu \, , \tag{10.18}$$

the limiting values must satisfy

$$\sum_\nu B_i^{\mu\nu}y_i^\nu = 1 \tag{10.19}$$

for all values of i and μ. This is a linear system of Np equations for the quantities y_i^ν. The iteration procedure (10.16) is just the well-known Gauss–Seidel method for the iterative solution of a system of linear equations, here the equations (10.19). It can be shown that this method always converges if the matrix $B_i^{\mu\nu}$ has only positive eigenvalues, i.e. if $\sum_{\mu,\nu}B_i^{\mu\nu}z^\mu z^\nu$ is a positive semidefinite quadratic form [St80]. This condition is certainly satisfied in our case, since on account of the definition (10.16):

$$\sum_{\mu,\nu} B_i^{\mu\nu} z^\mu z^\nu = \frac{1}{N} \sum_{k} \left(\sum_{\nu} \sigma_i^\nu \sigma_k^\nu z^\nu \right)^2 \geq 0 . \tag{10.20}$$

We conclude that the iteration process is guaranteed to converge, yielding the synaptic connections

$$w_{ij} \to \overline{w}_{ij} = \frac{1}{N} \sum_\nu y_i^\nu \sigma_i^\nu \sigma_j^\nu . \tag{10.21}$$

Owing to the relation (10.16) between the matrices $B_i^{\mu\nu}$ and $Q_{\mu\nu}$ we can write the equation (10.19) for y_i^ν also in the form

$$\sum_\nu Q_{\mu\nu} \sigma_i^\nu y_i^\nu = \sigma_i^\mu , \tag{10.22}$$

where we have multiplied by σ_i^μ and made use of the property $(\sigma_i^\mu)^2 = 1$. Multiplying with the inverse of the matrix $Q_{\mu\nu}$ and utilizing the same relation we find

$$y_i^\nu = \sigma_i^\nu \sum_\mu (Q^{-1})_{\nu\mu} \sigma_i^\mu . \tag{10.23}$$

Upon inserting this into (10.21), which describes the synaptic strengths at the end of the iteration process, we obtain the result

$$\overline{w}_{ij} = \frac{1}{N} \sum_{\mu,\nu} (Q^{-1})_{\mu\nu} \sigma_i^\mu \sigma_j^\nu \equiv \tilde{w}_{ij} . \tag{10.24}$$

These are precisely the synaptic connections (10.2), $\tilde{w}_{ij}$, of the projection rule discussed at the beginning of this section, which solve the problem of storing correlated patterns.

10.1.3 Repeated Hebbian Learning

For most practical purposes it is not necessary to use precisely the optimal synaptic couplings $\tilde{w}_{ij}$, or, in other words, it is not essential to render the left-hand side of (10.7) exactly equal to one. We recall that the starting point of our considerations was the desire to make the expression $\sigma_i^\mu h_i$ significantly greater than the critical-stability threshold zero. This condition is also satisfied if we modify the synaptic connection in such a way that the left-hand side of (10.7) is *greater than* or *equal to* a given threshold κ, which may or may not be taken equal to 1. This condition has the important advantage that the iteration process is guaranteed to come to an end after a finite number of steps, and it yields a maximal memory capacity $\alpha_c = 2$.

The procedure then works exactly as described above, except that we strengthen all synapses of a "subcritical" neuron by a fixed amount $1/N$, i.e. we replace the expression (10.10) by [Di87, Kr87a, Ga88a, Fo88b]

$$\delta w_{ij} = \frac{1}{N} \sigma_i^\mu \sigma_j^\mu (1 - \delta_{ij}) \theta (\kappa - \gamma_i^\mu) , \tag{10.25}$$

with the step function $\theta(x)$ and the normalized stability measures[4]

$$\gamma_i^\mu = \sigma_i^\mu h_i / \|w_i\|, \qquad \|w_i\| = \left(\sum_j^{j \neq i} w_{ij}^2\right)^{1/2}. \tag{10.26}$$

Here we have explicitly dropped the synaptic self-couplings w_{ii}, which leads to a better performance of the memory, as discussed in the previous subsection. The same calculation as in (10.11) then yields

$$\gamma_i^{\mu\prime} = \gamma_i^\mu + \theta(\kappa - \gamma_i^\mu). \tag{10.27}$$

When this expression is larger than κ, the iteration has converged; otherwise the synaptic reinforcement must be repeated.

In a sense this procedure can be understood as repeated learning according to Hebb's rule, where the synapses are increased by the amount $\frac{1}{N}\sigma_i^\mu \sigma_j^\mu$ as often as necessary to obtain the required stability for all stored patterns. This procedure reminds one of the experience of learning new words of a foreign language, where it is usually necessary to repeat those words several times until they have entered the long-term memory. As everyone knows, this method works for sure – if only after an annoyingly large number of repetitions!

It is therefore important to optimize the learning rate as much as possible. For this aim Abbott and Kepler [Ab89] have modified (10.25) by introducing a new function $f(\gamma)$ that modulates the magnitude of synaptic change according to the remaining deviation from the desired stability goal:

$$\delta w_{ij} = \frac{1}{N}\sigma_i^\mu \sigma_j^\mu \left(1 - \delta_{ij}\right) f\left(\gamma_i^\mu\right) \|w_i\| \theta\left(\kappa - \gamma_i^\mu\right). \tag{10.28}$$

Two choices of this function were considered, namely the quasilinear function

$$f_{\mathrm{L}}(\gamma) = (\kappa + \delta - \gamma)\theta(\kappa + \delta - \gamma) - 2\gamma\theta(-\kappa - \delta - \gamma) \tag{10.29}$$

and the nonlinear function

$$f_{\mathrm{NL}}(\gamma) = (\kappa + \delta - \gamma) + \sqrt{(\kappa + \delta - \gamma)^2 - \delta^2}. \tag{10.30}$$

Here $\delta \ll 1$ is a parameter that controls the speed of learning. For the quasilinear function (10.29) the algorithm converges after less than $2N/\delta^2$ iterations.

The rate of convergence for a typical simulation with the parameters $\kappa = 0.43$, $\delta = 0.01$, and $N = 100$ is shown in Fig. 10.3. The storage density was $\alpha = 0.75$, i.e. 75 patterns were to be stored by the network. Curve (a) refers to the standard algorithm (10.25), while curves (b) and (c) refer to the optimized algorithm (10.28) with the functions f_{L} and f_{NL}, respectively. The advantage of the modified algorithm is obvious. An analytic expression for

---

[4] The various references given above deviate slightly in their definition of the stability measure. The interested reader is urged to consult the original literature for details. We also refer to Chapt. 22 where some additional information on the learning rules can be found.

the rate of convergence of the standard algorithm (10.25) has been derived by Opper [Op88], which shows that the convergence slows down dramatically when the critical memory density is approached.

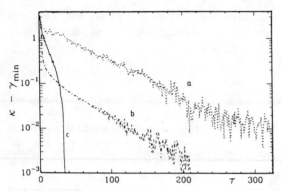

Fig. 10.3. Rate of convergence for the iterative learning rules (10.25) (*curve a*) and (10.28), (*curves b and c*). τ counts the number of iterations (from [Ab89]).

How large should the stability threshold κ be chosen? If κ is taken too large, no solution of the stability condition $\gamma_i^\mu > \kappa$ may exist. Then the algorithm will not converge. If κ is taken too small, the algorithm (10.25) converges rapidly and the stored patterns are stable, but the basins of attraction are small and the neural network will not necessarily recognize a slightly perturbed pattern. A certain amount of experience is required to find the optimal set of learning parameters. The dependence of the average radius of attraction of the stored patterns on the choice of κ was studied by Kepler and Abbott [Ke88], who found that it drops to $R = 0.1$ at $\kappa = 1$ for a saturated network.

Use the program ASSO (see Chapt. 22) to learn all 26 letters of the alphabet with any of the improved learning schemes 2–5. Compare their speed of convergence and the stability of the learned patterns against noise and thermal fluctuations.

Repeat the exercise of Sect. 3.4 with the program ASSCOUNT (cf. Chapt. 23) using the Diederich–Opper learning protocol on up to 10 numbers. Experiment with the parameters governing time delay.

10.2 Special Learning Rules

10.2.1 Forgetting Improves the Memory!

As we discussed in Sect. 3.3, the standard learning rule (Hebb's rule) leads to the emergence of undesirable local minima in the "energy" functional $E[s]$. In practice, this means that the evolution of the network can be caught in spurious, locally stable configurations, such as those in (3.23). Large networks usually contain a vast number of such spuriously stable states, many of which are not even linear combinations of the desired stability points. The learning rule for correlated patterns discussed in Sect. 10.1 does not guard against this problem. Thermal fluctuations do help to destabilize the spurious configurations, but at the expense of storage capacity. Moreover, the disappearance of all the spurious states at some finite T is not ensured.

A much better strategy is to eliminate the undesired stable configurations by appropriate modifications of the synaptic connections. Hopfield et al. [Ho83] have proposed to make use of the fact that the spurious minima of the energy functional $E[s]$ are usually much shallower than the minima that correspond to the learned patterns. Borrowing ideas developed in the study of human dream sleep, and discussed in Sect. 2.3, they suggested tracking these states by starting the network in some randomly chosen initial configuration and running it until it ends up in a stable equilibrium state s_i^∞. This may be one of the regular learned patterns, or one of the many spurious states. Whatever the resulting state is, the synapses are partially *weakened* according to Hebb's rule:

$$w_{ij} \to w_{ij} - \frac{\lambda}{N} s_i^\infty s_j^\infty \, , \tag{10.31}$$

where $\lambda \ll 1$ is chosen. This procedure of unlearning has two favorable effects. Most spurious equilibrium states of the network are "forgotten", since they are already destabilized by small changes in the synaptic connections w_{ij}. Moreover, the different regions of stability of the stored patterns become more homogeneous in size, since those with a larger range of stability occur more often as final configurations and are therefore weakened more than others.

The effect of this intentional forgetting is especially apparent in the sizes of the basins of attraction. This term denotes the set of all states, from which the network dynamics leads to a particular pattern. The change of the size of the basin of attraction of a given stored pattern, as the total memory load is increased, is illustrated in Fig. 10.4. The two axes labeled H_k and H_{N-k} in these figures represent a crude measure of the distance of an initial trial state s_i from the considered memory state σ_i^μ. They denote the partial Hamming distances between the trial state and the memory state, evaluated for the first k and the last $(N-k)$ of all $N = 200$ neurons, respectively.

$$H_k = \frac{1}{4} \sum_{i=1}^{k} (s_i - \sigma_i^\mu)^2, \qquad H_{N-k} = \frac{1}{4} \sum_{i=k}^{N-k} (s_i - \sigma_i^\mu)^2. \qquad (10.32)$$

In this specific case $k = N/2$ was taken, i.e. the axes represent the Hamming distance for the first and the last half of the neurons of the network. If the trial state s_i developed into the stored pattern μ, a black dot was plotted. For a single memory state (Fig. 10.4a), half of the trial states are found to evolve into the stored pattern σ_i, the other half ends up in the complementary pattern $(-\sigma_i)$. The basin of attraction thus represents a black triangle. For more memory states this region shrinks rapidly and takes on a highly ragged shape in the vicinity of α_c, as shown in Fig. 10.4b,c for 28 and 32 uncorrelated memory states, respectively. Figure 10.4d shows the result of applying the forgetting algorithm (10.31) 1000 times to the network loaded with 32 patterns [Ke87]. (The unlearning strength was $\lambda = 0.01$.) The basin of attraction grows strongly (by a factor of ten or more) and also takes on a more regular shape. The probability of retrieving the stored patterns is much improved.

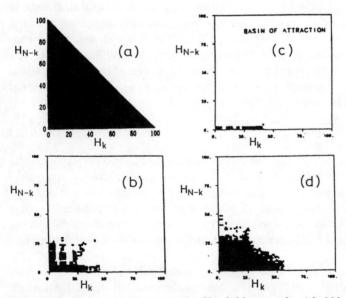

Fig. 10.4. Basins of attraction in a Hopfield network with 200 neurons for (a) 1, (b) 28, (c) 32 memory states. After deliberate forgetting the basin expands strongly (d). (From [Ke87]).

In a somewhat modified version of this method [Pö87] the network is allowed to develop from the stored patterns, deteriorated by random noise. One then not only weakens the synaptic connections by unlearning the final state s_i^∞, but also simultaneously relearns the correct starting pattern ν:

$$w_{ij} \rightarrow w_{ij} - \frac{\lambda}{N}\left(s_i^\infty s_j^\infty - \sigma_i^\nu \sigma_j^\nu\right) . \tag{10.33}$$

If the pattern was recalled without fault, the synapses remain unchanged according to this prescription. With this method the storage capacity can be increased to $\alpha = 1$, and the storage of strongly correlated patterns becomes possible.

Unfortunately, these methods do not eliminate the spurious stable states corresponding to linear combinations of stored patterns, such as (3.23). The total synaptic modification for all eight of these states taken together vanishes exactly, since the sum of the changes (10.31) adds to zero. However, a slightly different procedure works successfully [Ki87], where forgetting is controlled by the rule

$$w_{ij} \rightarrow w_{ij} - \frac{\eta}{N}\left(\sigma_i^\mu \sigma_i^\nu \sigma_i^\lambda\right)\left(\sigma_j^\mu \sigma_j^\nu \sigma_j^\lambda\right) . \tag{10.34}$$

Here μ, ν, and λ denote any triple of stored patterns. For $\eta > 1/3$ one finds that all eight spurious states (3.23) already become unstable at $T = 0$. At a finite value of the temperature parameter a smaller value of λ suffices for destabilization, and above $T = 0.46$ these spurious configurations become unstable because of the action of thermal fluctuations alone.

10.2.2 Nonlinear Learning Rules

An essential disadvantage of Hebb's rule (3.12) for p patterns

$$w_{ij} = \frac{1}{N}\sum_{\mu=1}^{p} \sigma_i^\mu \sigma_j^\mu \tag{10.35}$$

for many applications is that the synaptic strengths can vary over a wide range $-p/N \le w_{ij} \le +p/N$. This is particularly disturbing in hardware implementations of neural networks, since it requires electronic switching elements with a wide dynamical range. It is therefore natural to ask whether the range of allowed values of the w_{ij} can be limited with impunity. The extreme case would be to distinguish only between excitatory ($w_{ij} > 0$) and inhibitory ($w_{ij} < 0$) synapses, which are assigned the same absolute strength, but different signs:

$$w_{ij} = \frac{\sqrt{p}}{N}\,\mathrm{sgn}\left(\sum_{\mu=1}^{p} \frac{1}{\sqrt{p}}\sigma_i^\mu \sigma_j^\mu\right) = \pm\frac{\sqrt{p}}{N} . \tag{10.36}$$

This procedure, called *clipping* of synapses, is a special case of the nonlinear Hebb rule [He86, He87c, So86b], (see also Sect. 19.3)

$$w_{ij} = \frac{\sqrt{p}}{N}\,\Phi\left(\frac{1}{\sqrt{p}}\sum_{\mu=1}^{p} \sigma_i^\mu \sigma_j^\mu\right) , \tag{10.37}$$

where $\Phi(x)$ is an arbitrary, monotonously increasing function. The choice $\Phi(x) = x$ leads back to the standard (linear) Hebb rule (3.12), while $\Phi(x) = \text{sgn}(x)$ describes clipped synapses.

Clipping of synapses has a surprisingly small influence on the storage capacity of a network and on its ability to recall stored patterns. Compared to Hebb's linear rule the learning rule (10.36) acts as additional noise, causing a reduction of the critical storage density $\alpha_c = p_{\max}/N$ at $T = 0$ from $\alpha_c^{\text{Hebb}} = 0.138$ to $\alpha_c = 0.102$. It can be shown that in general the memory capacity is always less than in the linear Hebb case, $\alpha_c^{\Phi} \leq \alpha_c^{\text{Hebb}}$, for arbitrary synapses [He87c, He88a]. At low memory density, i.e. for small values of α, the error rate $(1 - m)/2$ in pattern recall is insignificantly larger for clipped synapses than for Hebbian ones. These deteriorations are compensated by the important simplifications resulting for the storage of the synaptic connections w_{ij} in hardware realizations, as well as in software simulations of the neural network on conventional digital computers.

Of particular interest are bounded synaptic strength functions $|\Phi(x)| \leq \Phi_0$. Here it is useful to modify the learning rule (10.37), in order to allow for the addition of more and more patterns to the memory: if the synapses after storing $(\mu - 1)$ patterns are denoted by $w_{ij}^{(\mu-1)}$, those obtained after adding the next pattern to the memory are defined as

$$w_{ij}^{(\mu)} = \Phi\left(\epsilon\sigma_i^{\mu}\sigma_j^{\mu} + w_{ij}^{(\mu-1)}\right) . \tag{10.38}$$

Whereas in the case of the linear Hebb rule the continued addition of more memory states eventually leads to the complete breakdown of the ability of the network to retrieve any stored pattern, memory degradation proceeds in a much gentler way for the nonlinear learning law (10.38) with bounded synaptic strengths. As more and more patterns are stored, the network approaches the limit of its storage capacity, α_c. However, instead of entering a state of total confusion, the network then experiences a gradual "blurring" of the older memory states, which are slowly replaced by the freshly learned patterns.

Because the whole memory can eventually be viewed as a sequence of clearly retrievable fresh patterns superimposed on increasingly deteriorated older patterns, such a memory structure is often called a *palimpsest*.[5] Memories of this type can serve as efficient *short-term memories*, because they permit the network to function as an information storage device continuously without encountering its capacity limit. Besides overwhelming physiological evidence that short-term-memory structures exist in the brain, such memories have important applications in electronic information processing as so-called *cache* memories.

---

[5] The term *palimpsest* derives from a practice used in the Middle Ages, when parchment was so precious that it was written upon several times, the earlier writing having been wholly or partially erased to make room for the next.

The nonlinear learning rule (10.38) was studied by Parisi [Pa86b] for the bounded function

$$\Phi(x) = \left\{ \begin{array}{ll} x & \text{for } |x| \leq 1 \\ \text{sgn}(x) & \text{for } |x| > 1 \end{array} \right. \tag{10.39}$$

shown in Fig. 10.5. As expected, the recall quality m of a stored state deteriorates after the addition of many other patterns to the memory, as depicted in Fig. 10.6. Although the total memory capacity remains bounded at any one time (Parisi found $\alpha_{max} = p_{max}/N \approx 0.04$), the network never loses its ability to learn new patterns. Similar models were studied by Nadal et al. [Na86] and various other authors [Ge87, He88c, Ge89]. A general discussion of their properties can be found in [Mo88].

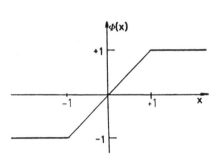

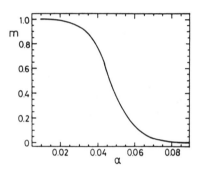

Fig. 10.5. Bounded synaptic-strength function used for palimpsest memory networks.

Fig. 10.6. The recall quality m of an older memory state fades after the addition of $p = \alpha N$ new states (from [Pa86b]).

Use the program ASSO (see Chapt. 22) to learn at least ten letters of the alphabet with any of the improved learning schemes 2–5, using standard values for the other parameters. Then choose option "m" in the search menu to limit the synaptic strength, or to allow only for binary synapses.

Use the program ASSO to learn at least ten letters of the alphabet with any of the improved learning schemes 2–5, using standard values for the other parameters, except selecting only positive (excitatory) or only negative (inhibitory) synapses on the first screen.

10.2.3 Dilution of Synapses

The assumption that all neurons are interconnected is not very realistic, especially if one is interested in modeling biological neural nets. One may ask whether a neural network can still act as an efficient associative memory if a large fraction of the synaptic connections are severed or *diluted*. The synaptic connectivity is usually diluted by eliminating connections at random, keeping only a fraction $d < 1$ of all synapses [So86b]. If this is done while the symmetry of the synaptic matrix w_{ij} is preserved, the memory capacity α_c of a network trained with Hebb's rule drops almost linearly with d, as illustrated in Fig. 10.7, where α_c and the critical recall quality $m_c = m(\alpha_c)$ are shown as functions of the fraction of destroyed synaptic connections.[6] This result clearly exhibits the *error resistivity* of neural networks. Even after elimination of a large fraction of synapses the network continues to operate quite reliably. Another approach is to dilute the synapses asymmetrically, i.e. to set $w_{ij} = 0$ but not necessarily also require $w_{ji} = 0$. This case can be treated analytically in the limit $d \rightarrow 0$ [De87c, Cr86]. One finds that the memory capacity per remaining synapse is about four times as large, $\alpha_c = 2d/\pi \approx 0.64d$, as for a fully connected network.

Virasoro has pointed out that the random destruction of synapses can lead to interesting effects when the stored patterns have a hierarchical similarity structure [Vi88]. By this one means that the patterns fall into several distinct classes or categories, the patterns belonging to a common class being strongly correlated [Pa86a, Do86]. A set of patterns with this property can be generated in the following way. First choose Q uncorrelated class patterns $\xi_i^\alpha, (i = 1, \ldots, N; \alpha = 1, \ldots, Q)$, where the $\xi_i^\alpha = \pm 1$ with equal probability. For each category α one now generates p_α correlated patterns $\sigma_i^{\alpha\mu}, (i = 1, \ldots, N; \mu = 1, \ldots, p_\alpha)$, taking the value $\sigma_i^{\alpha\mu} = \pm\xi_i^\alpha$ with probability $(1 \pm m)/2$. In the limit $m \rightarrow 1$ the patterns within the same category become more and more similar.

When the memory capabilities of the network deteriorate, e.g. because of overloading, the presence of thermal noise, or synaptic dilution, the network may reach a stage at which an individual pattern $\sigma_i^{\alpha\mu}$ can no longer be retrieved, but the recall of the corresponding class ξ_i^α is still possible.[7] Such a behavior can be of great interest, because the network can then perform the task of categorization, i.e. identify the class to which a given pattern that is

---

[6] This implies that the storage efficiency of the network does not really decrease, since fewer synapses are required in proportion. One has to keep in mind here that the complexity of the neural system is not described by the number N of neurons but by the total number of synaptic connections, $\frac{1}{2}dN(N-1)$, so that the true memory efficiency is given by α_c/d.

[7] The visual inability to recognize the difference between similar objects, e.g. human faces, is called *prosopagnosia* in clinical psychology. It is distinguished from the syndrome of *agnosia*, where the whole act of visual recognition is impaired, and which has been vividly described by Sacks [Sa87].

presented belongs.[8] That Hopfield–Little neural networks are capable of this task is related to the particular topology of the local minima of spin glasses, which form a hierarchical structure: broad, shallow minima contain narrower but deeper local minima which in turn contain even narrower and even deeper minima, and so on. This type of structure, which is called *ultrametric* [Ra86], may be the fundamental reason why the brain has the enormous power of classification and categorization of knowledge, which forms the basis of a large part of scientific thinking.[9]

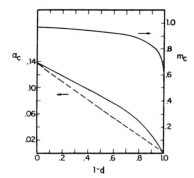

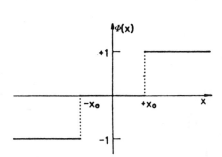

Fig. 10.7. Critical storage density α_c and recall quality m_c versus the fraction of destroyed synapses.

Fig. 10.8. Three-level synaptic response function.

A completely different concept of dilution suggests severing all connections, whose (absolute) strength falls below a certain threshold. This procedure, which might be appropriately termed "death by exhaustion", has been successfully applied in combination with general synaptic clipping [He87c, Mo87]. It corresponds to the choice (see Fig. 10.8)

$$\Phi(x) = \mathrm{sgn}(x)\theta(|x| - x_0) \tag{10.40}$$

of the nonlinear learning function in (10.37). A surprising result is that this systematic method of synaptic dilution actually enhances the efficiency of the

---

[8] Whether two such distinct transitions, first into a state of partial amnesia and later into total confusion, exist, can depend on the learning rule. Under gradual destruction of synapses the Hebb rule will not lead to two separate transitions, but the more general learning rule of Feigelman and Ioffe [Fe87] will.

[9] While the ultrametric structure of memory states follows naturally in a fully connected neural network with spin-glass character, a hierarchical structure of a more obvious nature seems to be realized in the brain. There is evidence that the neurons of the cortex on a larger scale are not interconnected in a completely disordered manner. Rather they appear to be grouped in a hierarchy of nested cortical subunits with a cluster size of several 100 or 1000 neurons at the lowest level. A generalization of the Hopfield–Hebb model to such a hierarchical-cluster structure has been discussed in [Su88].

network (compared with pure clipping, which gives $\alpha_c = 0.10$) for small values of x_0, with best results for the value $x_0 = 0.6$, yielding $\alpha_c = 0.12$. Of course, the reduction of synaptic strength to only three discrete values $(-1, 0, +1)$ is very convenient for hardware implementations, where the synapses can now be constructed from three-level switching elements.

Use the program ASSO (see Chapt. 22) to learn at least ten letters of the alphabet with any of the improved learning schemes 2–5, using standard values for the other parameters. Then choose option "m" in the search menu to dilute the synaptic connectivity. What fraction of dead synapses is tolerable?

10.2.4 Networks with a Low Level of Activity

Neurological experience has shown that neurons spend most of their time in the resting state and become active only once in a while. This contrasts with the assumption, made so far, that the two states $s_i = \pm 1$ occur with equal probability. The capacity of neural networks which store patterns with a low level of activity[10] was first studied by Willshaw [Wi69] and Palm [Pa81]. If the average value of neural activity for the stored patterns is denoted by

$$\bar{\sigma} = \frac{1}{pN} \sum_{i=1}^{N} \sum_{\mu=1}^{p} \sigma_i^\mu , \tag{10.41}$$

it is advisable to modify Hebb's rule in the more general case $\bar{\sigma} \neq 0$ according to [Am87b, Fe87]

$$w_{ij} = \frac{1}{N} \sum_{\mu=1}^{p} \left(\sigma_i^\mu - \bar{\sigma} \right)\left(\sigma_j^\mu - \bar{\sigma} \right) . \tag{10.42}$$

$\bar{\nu} = (1 + \bar{\sigma})/2$ denotes the probability that a neuron is active at any given moment. For $\bar{\nu} \ll 1$ it is useful to return to the original variables (3.2) n_i and ν_i , i.e. to describe the neurons in terms of McCulloch–Pitts variables instead of Ising variables. The synaptic excitation then is given by [Ts88a, Ts88b]

$$h_i = \sum_{j}^{j \neq i} w_{ij} n_j = \frac{1}{2} \left(\sum_{j}^{j \neq i} w_{ij} s_j + \sum_{j}^{j \neq i} w_{ij} \right) . \tag{10.43}$$

This corresponds to setting an excitation threshold $\vartheta_i = -\sum_{j \neq i} w_{ij}$ in the spin representation of the Hopfield network. It represents a special case of a

---

[10] In technical applications the use of patterns consisting mainly of 0s is known as *sparse coding*.

general rule saying that the activation threshold should be adjusted when the mean activity associated with the stored patterns, $\bar{\sigma}$, differs from zero. One can show that the memory capacity is improved substantially by this pattern-dependent choice of the threshold [Pe89c]. In the limit $\bar{\nu} = \frac{1}{2}(\bar{\sigma} + 1) \ll 1$ one finds $\alpha_c(\bar{\nu}) \to (2\bar{\nu}|\ln\bar{\nu}|)^{-1}$, which is the theoretical limit for the storage density [Ga88a] with an optimal choice of couplings, see Sect. 20.2. As Horner has demonstrated [Ho89c] the network is much less efficient if the network dynamics is based on the Ising-spin variables s_i. This is quite plausible since the distortion of the local fields h_i caused by the competing patterns will be much smaller if the (dominant) inactive neurons are represented by $n_i = 0$ instead of $s_i = -1$. This shows that the appropriate choice of the representation of the neurons may depend on details of the problem, such as the activity value.

11. Combinatorial Optimization

11.1 NP-complete Problems

The large class of often important numerical problems, for which the computational effort grows exponentially with size, is called *NP-complete*.[1] Simple examples are optimization problems, where the variables can take only discrete values. An important subclass is formed by the so-called combinatorial optimization problems. Here a *cost function E* has to be minimized, which depends on the order of a finite number of objects. The number of arrangements of N objects, and therefore the effort to find the minimum of E, grows exponentially with N.

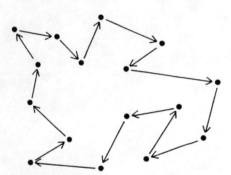

Fig. 11.1. A solution of the traveling-salesman problem for 16 cities.

Probably the most famous problem of this kind is that of the *traveling salesman*, who has to visit a number of cities and wants to find the shortest route connecting all locations. Here the cost function is given by the sum of all distances traveled between the cities, as illustrated in Fig. 11.1. There is no other way to find the best solution than to check out all possible orders of arrangement of the N cities and then select the shortest route. One easily sees that there are precisely $(N-1)!$ such arrangements: starting from some

---

[1] "NP" stands for "nondeterministic polynomial", implying that no algorithm is known that would allow a solution on a deterministic computer with an effort that increases as a power of the size parameter. A proof for the nonexistence of such an algorithm is not known either (see e.g. [Ga79]).

city, the salesman can travel to $(N-1)$ different locations; $(N-2)$ choices remain for the next stop, and so on, until he return to the city of origin. According to Stirling's rule we can write

$$(N-1)! \approx \sqrt{2\pi N}e^{N\ln N - N} , \qquad (11.1)$$

i.e. the number of possible choices grows faster than any power law. For just moderate values of N this exceeds the capacity of even the fastest available computers.

This line of argument is strictly correct only if one wants to find the very best solution. In practice it is often sufficient to know a good solution which differs only marginally from the best one. A procedure for finding such an almost optimal solution is therefore also of great interest. The classical methods of solving combinatorial optimization problems usually try to improve on an intelligent guess by exchanging the order of a few of the N objects. In this manner one arrives at a local minimum of the cost function, which is hopefully not very far from the absolute minimum. Of course, there is no guarantee that this is indeed so.

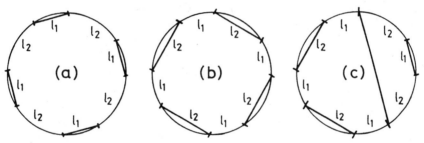

Fig. 11.2. Pairwise connection of points on a circle: (a) optimal solution, (b) suboptimal solution, (c) local modification of (b).

It is easy to see with an example why optimizing strategies that rely on local perturbations of the trial solution are bound to fail in certain cases. Consider $2N$ points arranged on a circle with perimeter L in such a way that the distance between nearest neighbors alternates between $\ell_1 = (L-a)/2N$ and $\ell_2 = (L+a)/2N$. The task is to connect these points pairwise such that the combined length of all connections is as small as possible (weighted matching problem). The optimal solution (see Fig. 11.2a) is obviously obtained if all pairs of points with separation ℓ_1 are connected; the total length is then $E_1 = N\ell_1 = (L-a)/2$. For large N the conjugate configuration (see Fig. 11.2b), where all pairs of points with separation ℓ_2 are connected, represents also a local minimum of the cost function. This minimum, which is stable against all local rearrangements of connections, is higher than the true minimum by an amount $\text{Delta}E = N(\ell_2 - \ell_1) = a$. Depending on the magnitude of a, this can be a considerable difference. However, the simplest local rearrangement, where the connections between two adjacent pairs of points are exchanged (see Fig. 11.2c), increases the cost function again by a significant amount, $[\ell_1 + (2\ell_2 + \ell_1)] - 2\ell_2 = 2\ell_1$. It is easy to see by trial and error that any rearrangement of connections between nonneighboring points leads to an even bigger increase, e.g. if the connections are made to run across the circle rather than along its perimeter. Any improvement in the total connection length requires the simultaneous rearrangement of (almost) all connections between pairs.

11.2 Optimization by Simulated Annealing

A different strategy, which has been applied to the solution of combinatorial optimization problems with great success, is the gradual *simulated annealing* of less than optimal combinations [Ki83, La87a]. The concept of annealing is derived from materials science, where it is used to describe the process of eliminating lattice defects in crystals by a procedure of heating, followed by slow cooling to room temperature. In the present context a lattice defect corresponds to the incorrect combination of two objects, e.g. the connection of the wrong pair of cities in the traveling-salesman problem.

If some material is cooled rapidly from the molten phase, its atoms are often captured in energetically unfavorable locations in the lattice. Once the temperature has dropped far below the melting point, these defects survive forever, since any local rearrangement of atoms costs more energy than is available in thermal fluctuations. The atomic lattice thus remains caught in a local energy minimum.[2] However, the thermal fluctuations can be enhanced if the material is reheated until energy-consuming local rearrangements occur at a reasonable rate. The lattice imperfections then start to move and annihilate, until the atomic lattice is free of defects – except for those caused by thermal fluctuations. These can be gradually reduced if the temperature is lowered so slowly that *thermal equilibrium is maintained at all times* during the cooling process. How much time must be allowed for the cooling depends on the specific situation, and a great deal of experience is required to perform the annealing in an optimal way: if the temperature is lowered too fast, some thermal fluctuations are frozen in; if one proceeds too slowly, the process never comes to an end.

The solution of an optimization problem proceeds in much the same way. One considers an ensemble of arrangements weighted by the Boltzmann factor $\exp(-E/T)$, where E is the cost function and T is a parameter that plays the role of a temperature. On lowering the value of T, the unfavorable arrangements become less and less likely, until only the optimal solution remains at $T = 0$. Strictly speaking this requires that the thermal-equilibrium distribution is conserved at all times during the cooling process, which would

---

[2] Rapid cooling is commonly used to increase the strength of metals, such as steel.

take an infinite amount of time. At any finite cooling rate some deviations from the exact Boltzmann distribution occur, with the possible consequence that a less than optimal solution is obtained at the end. In practical applications of this method one tries to avoid such problems by adding an extended period of constant temperature after each cooling step, so that equilibrium can be reestablished.

Numerical realizations of this procedure are usually based on the *Metropolis algorithm* [Me53] in combination with appropriate strategies for the creation of local rearrangements. The Metropolis algorithm generates an ensemble in thermal equilibrium by judicious selection of combinations according to their individual cost. The selection is based on the following principle: starting from some initial configuration, one chooses new, locally rearranged combinations by means of some appropriate random-selection mechanism. If the new state has a lower value E of the cost function ("energy") than the old one, it is selected as the new basis for the continuing process. If its energy is higher by an amount ΔE, the new combination is accepted only with probability $\exp(-\Delta E/T)$. Otherwise one retains the old, more favorable combination and continues with the choice of another "new", locally rearranged state, which is examined in the same way. The collection of accepted states approaches a thermal ensemble if it is continued indefinitely.

Modern solutions of the traveling-salesman problem are usually based on the Lin–Kernighan method for the generation of new arrangements of cities [Li73, Pr86]. This method employs two procedures in order to determine alternate tours: either one reverses the order of stations on a randomly selected part of the tour, or one excises a part of the tour and reinserts it unchanged at some other place. The numerical effort for this method grows like a moderate power of the number of cities N, so that problems with quite large values of N can be treated, as illustrated in Fig. 11.3.[3] The method of simulated annealing has also been successfully employed to find the optimal arrangement of integrated electronic circuits on semiconductor chips, where the number of elements lies in the range $N = 10^5$.

 The demonstration TSANNEAL solves the traveling-salesman problem by simulated annealing using the Lin–Kernighan prescription for the generation of moves (see Sect. 26.3).

---

[3] The search for the "best" tour among 50 cities takes a few minutes on a modern PC; searching through all possible 49! combinations would take about 10^{53} years.

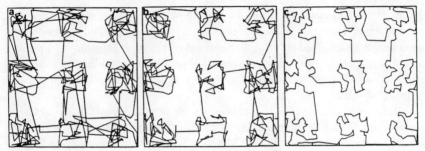

Fig. 11.3. Solution of the TSP for 400 cities by simulated annealing. The temperature parameters are: **(a)** $T = 1.2$, **(b)** $T = 0.8$, **(c)** $T = 0$ (from [Ki83]).

11.3 Realization on a Network

11.3.1 The Traveling-Salesman Problem

Hopfield and Tank [Ho85] have suggested that neural networks are well suited for the solution of combinatorial optimization problems. Let us illustrate the basic idea with the example of the traveling-salesman problem, which demands that one find the best order among the N cities to be visited. Expressed slightly differently, one would like to know the location α_i of the ith city in the tour, where none of the cities can occur twice. Introducing a square matrix containing $(N \times N)$ binary elements (neurons), the solution can be represented by entering a "1" into position α_i of the ith row, which indicates when the ith city is visited (see Fig. 11.3). Clearly, the matrix corresponds to an allowed solution if and only if each row and column contains exactly one entry "1" and $(N - 1)$ zeros.

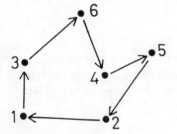

		Station					
		1	2	3	4	5	6
C	1	1	0	0	0	0	0
i	2	0	0	0	0	0	1
t	3	0	1	0	0	0	0
y	4	0	0	0	1	0	0
	5	0	0	0	0	1	0
	6	0	0	1	0	0	0

Fig. 11.4. Representation of a tour by an $(N \times N)$ table. *Rows*: number of city. *Columns*: number of station on the tour.

If the synaptic connections w_{ij} are chosen such that the energy functional $E[s]$ describes the total length of the path connecting all cities, the network should develop in the limit $T \to 0$ into the configuration corresponding to the shortest path. Of course, it is necessary to ensure that the arrangement of cities is acceptable in the sense discussed above. We are thus dealing with a special case of a very general class of problems, namely, to find the minimum of a function in the presence of constraints. The standard method of solution is by introducing the constraint via Lagrange multipliers.

We denote the elements of our neuron matrix by $n_{i\alpha}$, where the first index $i = 1, \ldots, N$ stands for the city and the second index $\alpha = 1, \ldots, N$ indicates the city's location in the travel route. An appropriate choice for the energy functional with constraints is

$$
\begin{aligned}
E[n] \;=\;& \frac{1}{2} \sum_{i,k}^{i\neq k} \sum_{\alpha,\beta}^{\alpha\neq\beta} w_{i\alpha,k\beta} n_{i\alpha} n_{k\beta} \\
=\;& \frac{1}{2} \sum_{i,k,\alpha}^{i\neq k} d_{ik} n_{i\alpha} (n_{k,\alpha-1} + n_{k,\alpha+1}) + \frac{A}{2} \sum_{i,\alpha,\beta}^{\alpha\neq\beta} n_{i\alpha} n_{i\beta} \\
& + \frac{B}{2} \sum_{i,k,\alpha}^{i\neq k} n_{i\alpha} n_{k\alpha} + \frac{C}{2} \left(\sum_{i,\alpha} n_{i\alpha} - N \right)^2 ,
\end{aligned}
\tag{11.2}
$$

where A, B, C are the Lagrange multipliers. The distance between cities i and k is denoted by d_{ik}. The first term measures the total length of the tour, which is to be minimized. The second term vanishes if at most one neuron $n_{i\alpha}$ is active for each city i ($n_{i\alpha} = 1$), i.e. if each city occurs at most once on the tour. The third term vanishes only if the station α is not occupied by two or more cities at once. The last term has its minimum when exactly N neurons are active, i.e. it makes sure that all cities are visited and all stations occur during the trip. Obviously, the function E takes its global minimum for the best tour. One easily checks that the energy functional (11.2) corresponds to the following choice of synaptic connections:

$$
\begin{aligned}
w_{i\alpha,k\beta} \;=\;& d_{ik}(1 - \delta_{ik})(\delta_{\alpha-1,\beta} + \delta_{\alpha+1,\beta}) \\
& + A(1 - \delta_{\alpha\beta})\delta_{ik} + B(1 - \delta_{ik})\delta_{\alpha\beta} + C ,
\end{aligned}
\tag{11.3}
$$

and the constant excitation threshold $\mu_{i\alpha} = CN$. The coupling matrix is symmetric and sparse, i.e. only a few of all N^4 elements are different from zero.

Does the dynamic evolution of the network defined by (2.1) lead to the minimum of the function E, if one starts from an arbitrary initial configuration? The calculations performed by Hopfield and Tank indicated that this would be the case [Ho85]. However, extensive studies by other authors have shown that the neural network does *not* evolve into the minimum of E in most cases [Wi88]. Studying various examples with only 10 cities they found that even after 1000 iterations of the network an acceptable tour was obtained in only 8% of all cases, and that these were rarely among the shortest tours. Typical examples for the solutions found by the network are shown in Fig. 11.5. This disappointing behavior was not significantly improved by variations of the Lagrange multipliers A, B, C, or by slight modifications of the synaptic connections.[4]

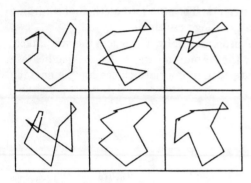

Fig. 11.5. Some acceptable but not optimal solutions found with the Hopfield–Tank method for 10 cities (after [Wi88]).

 Play with the demonstration program TSPHOP (see Sect. 26.1) and vary the network parameters. You should *not* feel frustrated if you cannot bring the network to find acceptable solutions of the traveling-salesman problem.

This leads to the more fundamental question, whether neural networks are better suited to solving NP-complete problems than standard digital computers. More recent studies have cast serious doubts on this claim. E.g., making use of the techniques of complexity theory, it has been shown that no network of polynomial size in N can exist which solves the traveling-salesman problem for N cities to a desired accuracy [Br88b] (unless all NP-complete problems are actually of polynomial complexity, which is believed not to be true). This result is of great practical interest, since present techniques allow the construction of neural networks that operate fast, but have a rather limited number of neurons. However, neural network models based on multivalued neurons (so-called *Potts models* in the terminology of the magnetic analogy) have been recently suggested to provide effective practical solutions of the traveling-salesman problem [Pe89d] (see Sect. 26.2).

---

[4] Our own attempts at solving the traveling-salesman problem with help of the Hopfield–Tank method also had little success. (Misquoting Arthur Miller, one may be tempted to speak of "the death of the traveling salesman".)

Use the demonstration program TSPOTTS (see Sect. 26.2) to solve the traveling-salesman problem.

11.3.2 Optical Image Processing

Neural networks have also been invoked for the solution of numerous other optimization problems. Another combinatorial problem is the so-called weighted matching problem mentioned in Sect. 11.1, where a number of points must be pairwise connected such that the sum of the lengths of all connections is as short as possible. Other applications mainly concern problems occurring in the context of optical image processing. Here we want to illustrate the method by a typical example concerning stereo vision [Po85, Ko86, He89].

Consider a two-dimensional stereographic image of a three-dimensional picture. The field depth d_i at a given position can be reconstructed from the relative lateral dislocation of corresponding points on the two photographs, which are taken at slightly different angles. Because of the presence of noise the relative dislocation of an individual point may not precisely correspond to its true value. The task now is to reconstruct the most likely correct structure by geometrical considerations, on the basis of the assumption that the depicted objects generally have smooth surfaces and edges. In order to simulate this problem on a neural network, it is useful to take analog-valued neural elements f_i (see Sect. 4.1) that describe the reconstructed depth at a given point i. To illustrate the principle, let us begin by considering a one-dimensional line of points. Spatial continuity of surface variations requires that the output values of neighboring neurons do not differ much. True representation of the noisy image demands that the output value f_i should not deviate strongly from the data d_i in the original image. Both requirements are simultaneously optimized by the minimum of the energy function

$$E[f] = \frac{A}{2} \sum_i (f_i - f_{i-1})^2 + \frac{B}{2} \sum_i (f_i - d_i)^2 \, . \tag{11.4}$$

The strict application of this principle would result in a relatively unstructured image of smoothly varying shades of gray everywhere. This is fine within a single continuous surface, but not at the border between the images of two objects. At these edges one must expect a discontinuous change in the depth. A simple way to allow for such behavior is to introduce binary neurons n_i, in addition to the analog-valued neural elements f_i. The value $n_i = 1$ signals that there is a discontinuity between the points $i - 1$ and i; $n_i = 0$ indicates that there is none. In the first case one has to remove the first term from the energy function (11.4), which forces the depth to change gradually; in the latter case it must remain in effect. However, an additional term in the energy function is needed, which inhibits the proliferation of edges. An appropriate choice for the energy function is

$$E[f, n] = \frac{A}{2} \sum_i (f_i - f_{i-1})^2 (1 - n_i) + \frac{B}{2} \sum_i (f_i - d_i)^2 + C \sum_i n_i . \quad (11.5)$$

This expression is easily generalized to more than one dimension by extending the sum to all neighboring pairs (ij) of points ($2N(N-1)$ pairs for a two-dimensional image composed of $(N \times N)$ points, $4N^2(N-1)$ for a three-dimensional image, etc.). For every pair a binary neuron $n_{(ij)}$ deciding on the existence of an edge is needed, yielding the energy function

$$E[f, n] = \frac{A}{2} \sum_{(ij)} (f_i - f_j)^2 (1 - n_{(ij)}) + \frac{B}{2} \sum_i (f_i - d_i)^2 + C \sum_{(ij)} n_{(ij)} . \quad (11.6)$$

More complicated terms depending solely on the edge variables $n_{(ij)}$ are often added in order to avoid point-like discontinuities or edges suddenly ending without corner [Ko86, Hu87]. Figure 11.6 shows the reconstruction of a black square without (a) and with (b) the line discontinuity process.

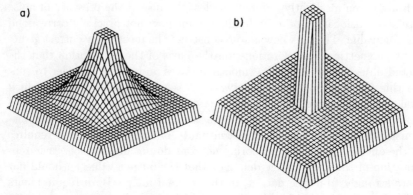

Fig. 11.6. Black square object reconstructed (a) without and (b) with the line discontinuity process (from [Hu87]).

We note that the specific task set by our example has led us beyond the framework of McCulloch–Pitts-type neural networks: the evolution law derived from (11.6) simultaneously couples more than two neurons, e.g. the elements f_i, f_j, and $n_{(ij)}$. Except for the purist, this poses no practical problems since these higher-order synaptic connections involve only locally neighboring elements. Such constructions may even occur in biological neural networks at synapses with presynaptic inhibition (see Fig. 1.6).

12. VLSI and Neural Networks

12.1 Hardware for Neural Networks

This book mainly deals with theoretical aspects and the general analysis of neural networks. However, to study this new concept in detail and to exploit it for practical and perhaps beneficial purposes, one must face the challenge of performing real-life neural computations. Here several approaches exist and have been widely explored. We enumerate the most important alternatives:

1. Simulation in software using sequential (von Neumann-type) computers.
2. Parallel computation using universal multiprocessor systems.
3. Realization of the network in hardware using special-purpose digital or analog electronic components.
4. Realization of the network using nonlinear optical construction elements.[1]

The first alternative at present is most easily accessible because of the widespread distribution and high standard of development of conventional computers. For instance, the demonstration programs accompanying this book are designed to operate on ordinary single-processor personal computers. However, this can be only a stopgap solution which manifestly contradicts the basic principle of the neural network, which lies in its massive parallelism of elementary operations. Serial computers cannot be expected to provide an efficient and cost-effective realization of artificial neural networks, and *neurocomputers* should follow a different design.

Ideally a neural network should be emulated by a very large number of simple processing elements with a high degree of interconnectivity. This is in some contrast with the mainstream trend in the development of parallel computers. Most of the general-purpose parallel computers which are operational at present consist of a relatively small number[2] of rather sophisticated processors. Popular examples are networks of transputers, which consist of fast

---

[1] We will not discuss the prospects of optical neural computers (see, e.g. [Mi89]). It should be mentioned, however, that such systems in principle can offer a very high degree of parallel connectivity, unmatched by electronic circuits which are constrained by the planar geometry of the chip.

[2] An extreme case is the *connection machine* which contains up to 65536 processing elements.

single-chip computers complete with memory and processing unit connected to a few neighboring units via high-speed data links. By using techniques of parallel programming such "transputer farms" can be used for fast simulations of neural networks.

However, such general-purpose parallel-processing architectures often do not represent the most convenient and cost-efficient way to implement a neural network. Often such processors are far too "intelligent" and thus too expensive to be used for the representation of a single neuron. They will have to multiplex in time a large number of individual processing elements (neurons). It would be much more in the original spirit of the neural network to have a large set of "artificial neurons" which individually are rather dumb, their faculties being confined merely to adding up the action potentials of the incoming synaptic connections and evaluating a nonlinear transfer function which generates the outgoing signal. To realize this concept considerable effort has gone into the development of dedicated "neuro-hardware". To date several dozen neuro-chips and neural computer boards have been described, many of which are commercially available. We cannot hope to do justice to these developments which would fill a specialized monograph (see [Gl94], [Ra91]). A review of neural hardware has been prepared in [Li95]; here we only can present a random selection.[3]

The first question a potential user of neural hardware has to decide is the choice between *analog* and *digital* circuitry. For general-purpose computing this battle was decided many years ago in favor of digital technology. As it turns out, neural networks by their very nature tend to favor the analog approach. In the next section we will describe in some detail how the neural processing elements and their connections can be realized as simple electronic circuits. There are certain practical drawbacks to the analog technology, however, and most neurochips presently in use are of digital design or operate in a *hybrid* mode where only part of the processing involves analog signals.

Having opted for a digital system one still can choose between different architectures. The traditional 'bit slice' architecture multiplexes part of the operations in time. Still higher speeds can be obtained using systems which operate fully in parallel. Here the most popular choice is the *SIMD* design (single instruction with multiple data) where an array of processors executes the same operation in parallel but on different data. As an alternative a *systolic array* [Me80] consists of a set of processors which perform a particular operation on their input data and then feed the result to the next processor. (The resulting data stream is thought to resemble the flow of blood in the circulatory system, hence the name).

---

[3] Much information on neural hardware has been collected at the CERN accelerator laboratory and is made available at the WWW address http://www1.cern.ch/NeuralNets/nnwInHepHard.html. The interest of high-energy physicists in this subject is understandable since they are in need of high-speed data acquisition and analysis systems which are at the cutting edge of technological progress.

An important criterion to distinguish among the various hardware implementations of neural networks is the way in which the connection weights are
determined. Except for hard-wired systems in which the weights are fixed during the manufacturing step, usually the networks can be trained to solve varying problems. Ideally this training (weight-adjustment) should be performed
on the same chip. Several neuro-chips are available which offer this feature;
they either implement a predefined learning rule (like back-propagation or
the stochastic Boltzmann algorithm which we will discuss later) or are freely
programmable, allowing for more flexibility. Those chips that are not able to
learn have to be fed with the final weight matrix by a host computer.

Table 12.1. A sample of presently available neuro-chips (adapted from [Li95] and
[Ho91]).

Type	Name	Architecture	Learn on-chip	Precision	Neurons	Synapses	Speed[a]
Analog	ETANN (Intel)	FeedFwd	-	6b×6b	64	10280	2000
Digital	MD-1220 (Micro Devices)	FeedFwd	-	1b×16b	8	2048	9
	NLX-420 (NeuraLogix)	FeedFwd	-	1–16b	16	off-chip	300
	Lneuro-1 (Philips)	FeedFwd	Hebb	1–16b	16 PE	64	26
	N64000 (Inova)	SIMD	program	1–16b	64 PE	128 k	870 (220)
	HNC 100 NAP (HNC)	SIMD	program	32b	100 PE	512k off-chip	250 (64)
	MA-16 (Siemens)	matrix operations	program	16 b	16 PE	16×16	400
	MT19003 (MCE)	FeedFwd	-	12b	8	off-chip	32
	WSI NAP (Hitachi)	SIMD pulse stream	Hebb	9b×8b	576	32 k	138
Hybrid	ANNA (AT&T)	FeedFwd	-	3b×6b	16–256	4096	3100
	CLNN-32 (Bellcore)	Hopfield	Boltzmann	6b×5b	32	992	100 (100)
	NeuroClassifier (Mesa Research)	FeedFwd	-	6b×5b	6	426	21000
	RN-200 (Ricoh)	FeedFwd	BackProp		16	256	3000

[a] Operating speed measured in 10^6 connections per second (MCPS).
  In brackets: Learning speed measured in 10^6 connection updates per second (MCUPS)

Table 12.1 collects information a subset of the available neuro-chips. They
are implemented using the techniques of VLSI (Very Large Scale Integration).
The last column of the table contains information on the speed of operation.

This is measured in terms of the number of *connections per second* (CPS) which will determine how fast the chip can perform mappings from input to output. The word "connections" stands for one term in the sum $\sum_j w_{ij} s_j$. Typically the speed of operation (measured in connection updates per second, CUPS) is somewhat lower in the learning phase. In digital implementations the calculation of the nonlinear sigmoidal function $f(h)$ is potentially time consuming. Often its value is taken from a discretized look-up table, but sometimes it is calculated outside the neuro-chip. To get a feeling for the speed values quoted in the table we note that an ordinary PC can operate at perhaps 0.2 MCPS while for the supercomputer Cray XMP a value of 50 MCPS has been reported. For specialized tasks neuro-chips clearly can surpass this speed at a fraction of the cost.

The user can acquire a neuro-chip and integrate it into an environment of his own design. Alternatively, a variety of complete systems is available commercially. These take the form of plug-in cards for personal computers or of complete *neurocomputers* which are interfaced to some host system which provides the input and output. The systems usually come with a programming environment and a full set of software tools. Let us mention two examples. The CNAPS neurocomputer manufactured by Adaptive Solutions Inc. consists of cards which host up to 8 neuro-chips of the SIMD type, each contining 64 processing elements. The cards also offer 16MBytes of data storage. The speed of a fully equipped system is reported [Mc91] as a remarkable 9.6 GCPS in the operating mode and 2.4 GCUPS in the learning mode (implementing a feed-forward network trained with back-propagation). The SYNAPSE-1 neurocomputer [Ra92] developed by Siemens AG combines MA-16 neuro-chips of the systolic architecture which can perform fast matrix operations. A system containing 8 such chips has a peak performance of 3.2 billion multiplications (16-bit×16-bit) and additions (48-bit) per second. With appropriate software the SYNAPSE computer can implement most of the common neural learning algorithms.

12.2 Networks Composed of Analog Electronic Circuits

It turns out that neural networks can be implemented in a very natural way by using *analog electronic circuits*. In principle all that is needed is a set of electronic amplifiers interconnected through a matrix of resistors. Let us discuss this in some detail.

As the basic building block of the network the "electronic neuron" consists of a nonlinear amplifier which transforms an input signal u_i into the output signal V_i. We neglect a possible intrinsic frequency dependence and assume that the response of the module is completely characterized by a voltage amplification function

$$V_i = g(u_i) . \tag{12.1}$$

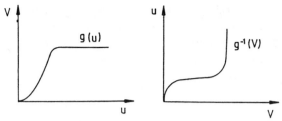

Fig. 12.1. The sigmoidal response function and its inverse of the electronic amplifier used to construct analog artificial neural networks.

This function is assumed to have an S-shaped (sigmoidal) form as depicted in Fig. 12.1. Large negative and positive input signals can steer the amplifier into saturation, thus providing for a degree of nonlinearity which is crucial for the operation of the network. Later it will become important that the function $g(u)$ increases monotonically so that it has a well-defined inverse $u_i = g^{-1}(V_i)$. The synaptic connections of the network are represented by resistors R_{ij} which connect the output terminal of the amplifier (neuron) j with the input port of the neuron i. The ensuing electronic circuit diagram depicted in Fig. 12.2 shows that much of the electronic neural network consists of a large matrix of resistors. These resistors must be able to take on different values, since in a distributed form they contain the stored information. In order that the network can function properly the resistances R_{ij} must also be able to take on negative values. This can be realized through a slight modification of Fig. 12.2. The amplifiers are supplied with an inverting output line which produces the signal $-V_j$. The number of rows in the resistor matrix is doubled, and whenever a negative value of R_{ij} is needed this is realized by using an ordinary resistor which is connected to the inverted output line.

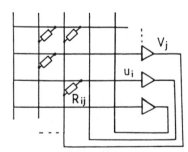

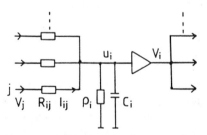

Fig. 12.2. The basic design of a Hopfield-type electronic neural network. The neurons and synapses are represented by nonlinear amplifiers and resistors.

Fig. 12.3. The wiring diagram of an artificial electronic neuron.

To understand the function of this network the earlier description has to be modified in two respects [Ho84]. The relation between the input and output signals of each neuron is determined by the *continuous function* $g(u)$ instead of the step function considered up to now. Furthermore, the time development of the signals has to be considered, which is determined by the time constant of the electronic circuit. To describe this quantitatively, in Fig. 12.3 we have depicted separately the circuit diagram of a single neuron. The input impedance of the amplifier unit is described by the combination of a resistor ρ_i and a capacitor C_i. The ability to perform calculations is bestowed upon this electronic circuit simply by Kirchhoff's law! The strength of the incoming and outgoing currents at the amplifier input port (which is assumed to draw no current) must balance, i.e.

$$C_i \frac{du_i}{dt} + \frac{u_i}{\rho_i} = \sum_j \frac{1}{R_{ij}}(V_j - u_i) \,. \tag{12.2}$$

Introducing the abbreviations

$$G_{ij} = \frac{1}{R_{ij}} \tag{12.3}$$

for the electric conductivity and

$$\frac{1}{R_i} = \frac{1}{\rho_i} + \sum_j G_{ij} \,, \tag{12.4}$$

we see that (12.2) reads

$$C_i \frac{du_i}{dt} + \frac{u_i}{R_i} = \sum_j G_{ij} V_j \,. \tag{12.5}$$

Multiplied by R_i this becomes

$$\tau_i \frac{du_i}{dt} + u_i = \sum_j w_{ij} g(u_i) \,. \tag{12.6}$$

Here $\tau_i = C_i R_i$ is a local time constant and $w_{ij} = R_i G_{ij}$ the synaptic strength.

Apart from the presence of time derivatives this system of equations is very similar to the discrete neural networks we have studied so far. It is not difficult to demonstrate that the system will relax to a stationary state in the limit $t \to \infty$ provided that the synaptic matrix is *symmetric*. This is true for any initial distribution of voltages $V_i(0)$. To prove this assertion one constructs an "energy" function[4] which has the property that its value decreases strictly with time. Such a function is given by

---

[4] Strictly speaking this is an abuse of language, since it is not related to the physical energy (which, e.g., is stored in the capacitors). Rather one should use the name *Lyapunov function*, following the usage of the stability theory of dynamical systems.

$$E = -\frac{1}{2}\sum_{i,j} G_{ij}V_iV_j + \sum_i \frac{1}{R_i}\int_0^{V_i} dV\, g_i^{-1}(V)\,. \tag{12.7}$$

Now we make use of the assumed symmetry of the synaptic matrix, i.e. $G_{ij} = G_{ji}$. Then the time derivative of E is given by

$$
\begin{aligned}
\frac{dE}{dt} &= -\sum_i \frac{dV_i}{dt}\left(\sum_j G_{ij}V_j - \frac{1}{R_i}g_i^{-1}(V_i)\right) \\
&= -\sum_i \frac{dV_i}{dt}C_i\frac{du_i}{dt} \\
&= -\sum_i C_i g^{-1\prime}(V_i)\left(\frac{dV_i}{dt}\right)^2\,,
\end{aligned}
\tag{12.8}
$$

where in the second step the equation of motion (12.5) was used. According to Fig. 12.1 the function g^{-1} increases monotonically so that its derivative is positive definite. This means that all the terms in the sum on the right-hand side. of (12.8) are positive. This leads to the conclusion that the "energy" will decrease monotonically:

$$\frac{dE}{dt} \leq 0\,. \tag{12.9}$$

On the other hand, the value of E is bounded from below since the voltages cannot exceed a maximum value determined by the saturation of the amplification function $g(u)$. This leads to the conclusion that the system has to approach an equilibrium value in the limit $t \to \infty$. Starting from an initial "pattern" of voltages u_i, if one waits long enough, the network settles into a stationary final state which is determined by the matrix of resistors R_{ij}. This is just the property which defines an associative memory.

To end this discussion let us draw attention to an analogy with the stochastic Hopfield network at finite temperature which we studied earlier. The stationary state of the electronic network satisfies the equation

$$u_i = \sum_j w_{ij}g(u_j)\,. \tag{12.10}$$

Writing $u_i = g^{-1}(V_j)$, we have

$$g^{-1}(V_j) = \sum_j w_{ij}g\big(g^{-1}(V_j)\big) = \sum_j w_{ij}V_j \tag{12.11}$$

which leads to the following nonlinear relation between the output voltages:

$$V_i = g\left(\sum_j w_{ij}V_j\right)\,. \tag{12.12}$$

This can be compared with the expression for the "magnetization" of the stochastic network in the mean-field approximation (4.7)

$$\langle s_i \rangle = \tanh\left(\beta \sum_j w_{ij} \langle s_i \rangle \right) . \tag{12.13}$$

Thus there exists an analogy between the continuous variables V_i and the expectation value $\langle \sigma_i \rangle$ of the discrete spin variables σ_i. Both networks can be expected to show similar behavior since the function $g(u)$ qualitatively resembles the tanh function. From this argument we conclude that the electronic network will function properly only if the amplification factor is sufficiently large. The opposite case, i.e. a slowly rising function $g(u)$, corresponds to a small value of the inverse temperature β in (12.13) in the stochastic model. We know, however, that at high temperatures there is a phase transition to a disordered phase where no patterns can be memorized.

The concept of a single-chip analog electronic neural network[5] along the lines discussed above has been implemented by several groups. We will mention a few of the pioneer implementations. Many details on the subject of VLSI electronic implementation of neural networks can be found in the monograph [Me89a].

- A group of workers at the AT&T Bell Laboratories [Gr87] has built a network consisting of 256 neurons using CMOS technology. A $6 \times 6\,\text{mm}^2$ silicon chip harbors a matrix of more than $130\,000$ (2×256^2) resistors consisting of amorphous silicon. The distribution of the resistance values R_{ij} is determined at production time; thus the network is programmed to store fixed patterns and cannot be trained. Furthermore, all resistors have the same value $R_{ij} = R$ (or ∞), there is only one value of the synaptic strength (or zero in those connections where no resistor is put into place). This is an extreme case of the clipping of synapses discussed in Sects. 10.2.2 and 20.2. It has been demonstrated that the network is able to operate despite this handicap, although the storage capacity is impaired. The network showed a fast convergence, typically of the order 300 ns.
- Carver Mead and collaborators at the California Institute of Technology [Si87] have constructed a neural VLSI circuit with only 22 neurons, which, however, has the ability to learn. The synaptic connections are formed by MOS transistors which in turn are driven by tristable flip-flop units. In consequence the synaptic strength can (as in the first implementation) take on three values $(+1, 0, -1)$. The state of the flip-flops is determined through the sequential presentation of patterns using addition units which are programmed according to Hebb's rule (with clipping). According to theoretical expectations, the chip was able to learn about three or four patterns. It showed remarkable tolerance against fluctuations in the properties of the electronic units and even against total malfunction of synapses.
- A group at Bell Communications Research [Al88] has developed a CMOS chip which is capable of learning classification tasks using both supervised

---

[5] The dynamics of analog neural circuits built from discrete elements were studied in [Ba87a, Ma89a].

and unsupervised stochastic learning paradigms as discussed in the next few chapters. The chip contains 6 neurons and 15 synapses. To achieve high flexibility each synapse contains 300 transistors. Despite the high integration, each of the synapses requires an area of $0.25\,\mathrm{mm}^2$ (which is about 3 orders of magnitude larger than the size of biological neurons!). The speed of learning of this network was found to be increased by a factor of a million compared to simulations on an ordinary sequential computer.

– The first commercially manufactured analog neuro-chip was the Electrically Trainable Analog Neural Network ETANN from Intel. It contains 64 neurons with a sigmoidal transfer function and can implement two-layer feed-forward networks with 64 inputs, 64 hidden neurons, and 64 output neurons using two 80×64 weight matrices. Weights are stored in non-volatile floating-gate synapses (Gilbert analog multipliers). On-chip learning is not available. It takes about 8 μsec for the signals to propagate through a two-layer network, which corresponds to about 2 billion connection updates per second (2 GCUPS).

– Mead and collaborators have developed an electronic representation of the *retina* found in the visual system of vertebrate animals. A set of 48×48 photosensors is arranged in a hexagonal array with fixed nearest-neighbor interconnections. Such a resistive grid is not a neural network in the conventional sense but it too performs parallel analog computation. The "silicon retina" is able to perform basic tasks of image processing, such as discontinuity and motion detection in a manner quite similar to the biological retina [Ko89a].

One potential drawback of analog implementations has been pointed out in [Ho93]: If on-chip training of the network is desired, say by using the backpropagation algorithm, one needs a certain numerical precision (at least about 12 significant bits) to guarantee convergence of the weights. This is hard to achieve with analog electronic circuits. To circumvent this problem one may, of course, perform the training on a separate (digital) computer and transfer the resulting weights to the analog chip.

13. Symmetrical Networks with Hidden Neurons

13.1 The Boltzmann Machine

The power of "hidden" neurons, i.e. neurons that do not directly communicate with the outside world, to perform complex tasks has become apparent in the case of layered feed-forward networks (perceptrons), discussed in Chapt. 6. On the other hand, the theoretical treatment of such neural networks with strongly asymmetric synapses is much more difficult than that of symmetrically connected networks, developed in Chapt. 3. It is therefore tempting to combine these two concepts, i.e. to study neural networks exhibiting hidden neurons *and* symmetric synapses.

The associative memory models studied in Chapts. 3 and 10 did not contain any hidden neurons. The patterns to be stored were imprinted on all N neurons of the network, the patterns to be recalled were presented to all these neurons, and the recovered patterns were exhibited by the complete network, and not only by a part of it. Even with the most powerful learning algorithms such a network can only store at most $2N$ patterns (see Part II). Hinton und Sejnowski [Hi83, Se86] have therefore proposed adding hidden neurons to the stochastic Hopfield network which operate according to the principles explained in Chapt. 4. Such networks form general computing machines based on stochastic computational rules, and have been termed *Boltzmann machines* as opposed to the computers based on von Neumann's principles of digital computing. The individual elements of a Boltzmann machine can take one of the two values $s_i = \pm 1$, assuming the positive value with probability $f(h_i)$, where as usual $h_i = \sum_k w_{ik} s_k$ and $f(h)$ is the Fermi function (4.3).

We learned in Sect. 3.3 that the dynamics of such a network can be described by the "energy" function

$$E[s] = -\frac{1}{2} \sum_{ik} w_{ik} s_i s_k \, . \tag{13.1}$$

The minima of this function correspond to stable configurations of the network. The concept of the Boltzmann machine stipulates that the neural network is first operated at a high temperature, which is gradually lowered until the network is trapped in an equilibrium configuration around a single minimum of the energy function. Averages, like $\langle s_i \rangle$ or $\langle s_i s_k \rangle$, are then easily computed by temporal sampling of configurations. By virtue of the symmetry of the synapses it is possible to derive many general results concerning the properties of Boltzmann machines by the methods of statistical physics. Here we refer the reader to Sect. 17.1, where a brief summary of statistical concepts is given. Since the equations (17.8, 17.10) for the probability $P[s]$ of finding the network in a given state $\{s_i\}$ and for the partition function Z of the network will play a special role in the following considerations, we repeat those here explicitly:

$$P[s] = Z^{-1}e^{-\beta E[s]} , \tag{13.2}$$

$$Z = \sum_{[s]} e^{-\beta E[s]} . \tag{13.3}$$

According to the concept of the Boltzmann machine we distinguish between *hidden* and *peripheral* neurons. The latter may be further subdivided into input neurons, which receive information from outside, and output neurons, which communicate the result of the "computation". E.g., in order to represent the exclusive-OR function, we would need three peripheral neurons, two (input) neurons to enter the two logical arguments and one (output) neuron to communicate the function value. Another characteristic task, which requires a division into input and output neurons, is the completion of patterns that are only partially known, e.g. if a stored picture of a human face is to be reconstructed from its left half. For such tasks one uses the term *heteroassociation* as opposed to the concept of autoassociation, where the complete, but maybe incorrect, pattern is presented to the network.

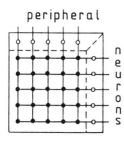

Fig. 13.1. Schematic view of a network with hidden (*black dots*) and peripheral neurons (*open circles*).

13.2 The "Boltzmann" Learning Rule

For symmetric neural networks without hidden neurons Hebb's rule provides a satisfactory solution of the question how the synaptic strengths w_{ik} should be determined so that the network is able to perfom a specific task. In the presence of hidden neurons Hebb's rule is no longer sufficient, since it does not tell us how to fix the synapses between peripheral and hidden neurons, and among the hidden neurons themselves. A completely new concept is required here. In the case of multilayered perceptrons we had found a successful algorithm on the basis of the gradient method with error back-propagation. This concept cannot be directly applied to the Boltzmann machine, because the synaptic connections are symmetric and not forward oriented.

Here we are saved by the fact that the principles of statistical physics can be applied to the Boltzmann machine. We first discuss a learning algorithm which is based on the concept of *entropy* [Ac85]. As a preparation we have to introduce an appropriate notation to distinguish between the hidden and the peripheral neurons. We denote the states of the output neurons by an index α, those of the input neurons by an index β, and those of the hidden neurons by an index γ. The state of the complete network is then denoted by the index combination $\alpha\beta\gamma$, etc.

For a certain training set the probability distribution of encountering a specific input configuration β may be described by Q_β. This distribution is given from outside, independent of the network dynamics. If the conditional probability of finding a certain state α of the output neurons for a given input configuration β is denoted as $P_{\alpha|\beta}$, the true distribution of output states is given by

$$P_\alpha = \sum_\beta P_{\alpha|\beta} Q_\beta \ . \tag{13.4}$$

Similar notations may apply if the states of the hidden neurons are also included. The index to the right of a vertical bar always indicates the fixed configuration for a conditional probability. All probability distributions are assumed to be normalized to unity, e.g.,

$$\sum_\alpha P_{\alpha|\beta} = \sum_{\alpha\gamma} P_{\alpha\gamma|\beta} = 1 \ . \tag{13.5}$$

The conditional probability $P_{\alpha|\beta}$ is determined by the synaptic strengths w_{ik} of the network; according to (13.2) we have

$$P_{\alpha|\beta} = \sum_\gamma P_{\alpha\gamma|\beta} = Z_{|\beta}^{-1} \sum_\gamma \exp(-E_{\alpha\gamma\beta}/T) \ , \tag{13.6}$$

where the energy $E_{\alpha\gamma\beta}$ is given by (13.1), and $Z_{|\beta}$ is the partition function for fixed input configuration β:

$$Z_{|\beta} = \sum_{\alpha\gamma} \exp(-E_{\alpha\gamma\beta}/T) \ . \tag{13.7}$$

The learning task is to choose the synapses in such a way that the conditional probability $P_{\alpha|\beta}$ takes on the desired value, which we denote by $Q_{\alpha|\beta}$. If a unique response $\alpha_0(\beta)$ is to be associated with every input β, this distribution is given by $Q_{\alpha|\beta} = \delta_{\alpha\alpha_0}$, but more general cases can be imagined where several different responses are possible with a graded preference rating.

In order to solve this problem with the help of the gradient method, we need to find an appropriate function D that describes the deviation between the desired response distribution $Q_{\alpha|\beta}$ and the actual distribution $P_{\alpha|\beta}$ for the given synaptic connections of the network. The simplest choice, namely the sum of squared deviations as in Sect. 5.2.2,

$$\frac{1}{2} \sum_{\alpha\beta} \left(Q_{\alpha|\beta} - P_{\alpha|\beta} \right)^2 Q_\beta , \tag{13.8}$$

is not conveniently treated by the methods of statistical physics. These are specifically designed to find that distribution which *minimizes the entropy* of a thermodynamic system. We therefore define the deviation function as follows [Ac85]:

$$D = \sum_{\alpha\beta} Q_{\alpha|\beta} Q_\beta \ln \frac{Q_{\alpha|\beta}}{P_{\alpha|\beta}} . \tag{13.9}$$

In the context of information theory D is called the *information gain* resulting from a measurement of the distribution $P_{\alpha|\beta}$. This gain vanishes if the measured distribution coincides with the expected distribution $Q_{\alpha|\beta}$. It is easy to show that the expression D is positive (semi-) definite and reaches its minimum value for $Q_{\alpha|\beta} = P_{\alpha|\beta}$. The proof is based on the relation $\ln(x) \geq (1 - x^{-1})$, and makes use of the normalization of the probability distribution:

$$D \geq \sum_{\alpha\beta} Q_{\alpha|\beta} Q_\beta \left(1 - \frac{P_{\alpha|\beta}}{Q_{\alpha|\beta}} \right) = \sum_{\alpha\beta} Q_\beta (Q_{\alpha|\beta} - P_{\alpha|\beta}) = 1 - 1 = 0. \tag{13.10}$$

We now apply the gradient method to the deviation function (13.8) and adjust the synaptic strengths according to the rule

$$\delta w_{ik} = -\epsilon \frac{\partial D}{\partial w_{ik}} = \epsilon \sum_{\alpha\beta} Q_\beta \frac{Q_{\alpha|\beta}}{P_{\alpha|\beta}} \frac{\partial P_{\alpha|\beta}}{\partial w_{ik}} . \tag{13.11}$$

Here we have taken into account that the desired conditional response probabilities $Q_{\alpha|\beta}$ do not depend on the values of the synaptic couplings w_{ik}. The essential point in the derivation is that the partial derivative of $P_{\alpha|\beta}$ with respect to the synaptic connection w_{ik} can be determined solely from the behavior of the two connected neurons i and k. For this purpose we express $P_{\alpha|\beta}$ explicitly in terms of the w_{ik} with the help of (13.6). The synaptic strengths enter the energy function

$$E_{\alpha\gamma\beta} = -\frac{1}{2} \sum_{ik} w_{ik} [s_i s_k]_{\alpha\gamma\beta} , \tag{13.12}$$

where $[s_i s_k]_{\alpha\gamma\beta}$ denotes the states of the two neurons under consideration, if the total network configuration is given by $(\alpha\gamma\beta)$. Taking into account (13.7), the partial derivative of $P_{\alpha|\beta}$ with respect to w_{ik} yields

$$\frac{\partial P_{\alpha|\beta}}{\partial w_{ik}} = \frac{1}{TZ_{|\beta}} \sum_{\gamma} \exp(-E_{\alpha\gamma\beta}/T)[s_i s_k]_{\alpha\gamma\beta} - \frac{\partial \ln Z_{|\beta}}{\partial w_{ik}} P_{\alpha|\beta}$$

$$= \frac{1}{T}\left(\sum_{\gamma} P_{\alpha\gamma|\beta}[s_i s_k]_{\alpha\gamma\beta} - \langle s_i s_k \rangle_{|\beta} P_{\alpha|\beta} \right), \tag{13.13}$$

where the index $|\beta$ indicates that the average is performed in the presence of a fixed input state β:

$$\langle s_i s_k \rangle_{|\beta} = \sum_{\alpha\gamma} P_{\alpha\gamma|\beta}[s_i s_k]_{\alpha\gamma\beta} = Z_{|\beta}^{-1} \sum_{\alpha\gamma} \exp(-E_{\alpha\gamma\beta}/T)[s_i s_k]_{\alpha\gamma\beta}. \tag{13.14}$$

The iterative adjustment of the synaptic couplings can now be derived from (13.10), considering the relation (13.4) for the conditional probability, giving the following result known as the *Boltzmann learning rule*:

$$\delta w_{ik} = \frac{\epsilon}{T} \sum_{\alpha\beta} Q_{\alpha|\beta} Q_{\beta} \left(\sum_{\gamma} \frac{P_{\alpha\gamma|\beta}}{P_{\alpha|\beta}} [s_i s_k]_{\alpha\gamma\beta} - \langle s_i s_k \rangle_{|\beta} \right)$$

$$= \frac{\epsilon}{T} \sum_{\beta} Q_{\beta} \left(\sum_{\alpha} Q_{\alpha|\beta} \langle s_i s_k \rangle_{|\alpha\beta} - \langle s_i s_k \rangle_{|\beta} \right). \tag{13.15}$$

The first term in parentheses describes the correlation between the states of the two neurons i and k, under the condition that the state of the input neurons is kept fixed during the operation of the network and an average is taken over the output states with the desired probabilities $Q_{\alpha|\beta}$. The second term expresses that the unconditional correlation between the states of the two neurons is subtracted from this result. As already mentioned above, it is essential that δw_{ik} can be determined from observations of the states s_i and s_k of the two connected neurons alone. The first term is obtained as the average value of the product $s_i s_k$ when the network evolves with fixed input and output states.[1] The synapses can therefore be adjusted on the basis of local observations, which greatly simplifies the network architecture, especially for large networks.

The associative memory networks studied in Chapts. 3 and 10 correspond to the special case that all neurons are both input and output neurons, and that there are no hidden neurons. In this case we have $Q_{\alpha|\beta} = \delta_{\alpha\beta}$ for all memorized patterns σ_i^μ, since the network is supposed to return the stored pattern when it is presented as an input. The first term in parentheses then yields precisely Hebb's rule (3.12), and the second term describes the intentional forgetting of patterns that have been invoked by random input. In this case the Boltzmann learning rule (13.15) reduces to the iterative learning algorithm (10.33). In other words, (13.15) can be considered a generalization of the concept of repeated forgetting and relearning to symmetric neural networks with hidden neurons.

---

[1] In electrical engineering this type of operation of an electrical circuit is called *clamping* of the input and output voltage.

The training process of a Boltzmann machine consists of a series of alternating steps. For each response to be learned, in the first step both input and output units are clamped, and the response of the network, after settling to thermal equilibrium, is measured. Subsequently the same procedure is repeated while letting the output units evolve freely. According to the rule of "contrastive Hebbian synapses" (13.15) the couplings w_{ik} are updated using the difference of both measurements. This has to be done for all input patterns β, and usually many such training cycles are required. This may be computationally expensive since correlations of stochastically fluctuating quantities $\langle s_i s_k \rangle$ have to be measured.[2]

A potentially faster learning *deterministic Boltzmann machine* might be constructed simply by running the network at zero temperature [Cu86]. A more flexible approach, which retains some of the thermodynamic properties of the system, is based on the *mean-field approximation* which averages over fluctuations [Pe87]. This replaces the stochastic neurons s_i by analog-valued deterministic units $\bar{s}_i$ with activation in the range $-1 \leq \bar{s}_i \leq +1$. They will settle to a stable fixed point of the system of equations

$$\bar{s}_i = \tanh\left(\frac{1}{T} \sum_j w_{ij} \bar{s}_j \right) \tag{13.16}$$

which minimizes the free energy $F = E - TS$. In this approach the correlations become simple products $\langle s_i \, s_k \rangle \rightarrow \bar{s}_i \, \bar{s}_k$.

13.3 Applications

Sejnowski et al. [Se86] have studied the performance of these Boltzmann machines, i.e. of symmetrically connected networks with hidden neurons, for a number of problems which also require the presence of hidden neurons if they are posed to networks of the perceptron type. Here we discuss two characteristic examples.

The first example is the already familiar exclusive-OR (XOR), which requires two input neurons and one output neuron. In addition, we need one hidden neuron, as in Sect. 6.1. In the present context, however, all four neurons are symmetrically connected by synapses, except that there is no direct synaptic connection between the two input neurons. After 30 learning sweeps over the training set with simultaneous gradual lowering of the temperature the rate of success reached 96%. After 255 further rounds and continued cooling the network performed perfectly, within the limitations set by the presence of the remaining small thermal fluctuations.

---

[2] However, a VLSI chip has been designed [Al88] as a hardware representation of a Boltzmann machine which employs naturally occurring electrical noise as a source of thermal fluctuations.

Since the Boltzmann network operates stochastically, it does not converge to the same solution at every attempt, if there is more than one valid solution of the problem. In total, the network found eight different strategies to represent the XOR function which made use of the hidden neuron in different ways. One example for the values of the synaptic strengths and the threshold potentials is shown in Fig. 13.2.

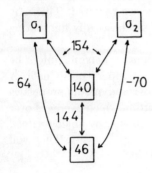

Fig. 13.2. Graphical representation of the synaptic connections for a specific solution of the exclusive-OR (XOR) function (after [Se86]).

Our second example is the task of *recognizing symmetries* present in a two-dimensional square pattern, which was represented on an array of $(N \times N)$ input neurons. The cases $N = 4$ and $N = 10$ were considered. In addition, the network contained 12 hidden neurons and three output neurons designed to signal the presence of one of the three considered symmetries (horizontal, vertical, and diagonal reflexion symmetry). It is well known that this task requires hidden neurons [Va84].

This example differs from the previous one in that the number of possible input patterns is very large. For $N = 4$ there are already $2^8 = 256$ possible patterns for each symmetry; for $N = 10$ the number grows to 2^{50} symmetrical patterns. It is clearly impossible to confront the network with all possible inputs during the learning phase, so the ability of the trained network to generalize can be tested.

It turned out that the success of learning progressed only slowly. Gradually lowering the temperature T enabled the network to attain a rate of success of 98.2% after 10^5 training sweeps in the case $N = 4$; for the larger network ($N = 10$) the success rate was only about 90% at this stage. Generally it was found that the performance of the network improved only very slowly after some initial rapid successes.

14. Coupled Neural Networks

14.1 Stationary States

Multilayered feed-forward networks (perceptrons) are special cases of the general McCulloch–Pitts neural network with arbitrarily interconnected neurons. On the other hand, any general "recurrent" neural network can be considered to be represente by a feed-forward perceptron, albeit one with possibly very many layers. The reason for this strange equivalence is that the temporal evolution (3.5) of an arbitrary network constructed from binary neurons

$$s_i(t+1) = \text{sgn}[h_i(t)] \equiv \text{sgn}\left(\sum_{j=1}^{N} w_{ij} s_j(t)\right) \qquad (14.1)$$

is necessarily periodic. This statement follows immediately from the observation that the N neurons can only assume 2^N configurations altogether, and hence some state of the network must reoccur after at most 2^N steps. Since only the present state of the network enters on the right-hand side of the evolution law (14.1), the subsequent evolution proceeds strictly periodically from that moment on. If one considers the neural network at a certain moment $t = n$ as the nth layer of a perceptron (with all layers identical!), the temporal-evolution law can be viewed as the law governing the flow of information from one layer to the next. It is then sufficient to take into account only a finite number of such layers, just as many as there are time steps leading up to the first repetition of a network configuration.

The length of the period can be anywhere between 1 and 2^N, which usually is a huge number. However, experience has shown that the period is short in most practical examples. In many cases the network even settles down into a stationary state, i.e. it reaches a configuration with the property $s_i(t+1) = s_i(t)$ for all neurons (period of length 1), and one has to work quite hard in order to guarantee a longer period [Cl85]. In the following we shall study a little more closely those networks which reach a stationary state in the course of the evolution. In many respects these networks are most closely related to the previously studied perceptrons. The stationary state here plays the same role as the output layer of neurons in a perceptron. In particular, we are interested in how we can teach the network to reach a specific stationary state.

We thus consider a fully connected neural network with threshold potentials I_i and an arbitrary activation function $f(x)$. Its temporal evolution is described by the law

$$s_i(t+1) = f[h_i(t)] \equiv f\left(\sum_k w_{ik}s_k(t) + I_i\right). \tag{14.2}$$

We consider the threshold potentials I_i to be given; in a sense they are assumed to represent the external stimulus received by the network.[1] If we now assume that the network reaches a stationary state s_i^0 after some time, this state represents a *fixed point* of (14.2):

$$s_i^0 = f[h_i^0] \equiv f\left(\sum_k w_{ik}s_k^0 + I_i\right). \tag{14.3}$$

The stationary state may be considered the response of the network to the external stimulus I_i, since it does not depend on the initial configuration $s_i(t=0)$ of the network according to (14.3).[2]

No general criterion, being both necessary and sufficient, is known which would guarantee the asymptotic stability of a network with arbitrary couplings. However, we have already encountered two special cases exhibiting stability: (1) purely feed-forward networks which are trivially stable since they have no loops which could give rise to oscillations; (2) networks with symmetric weights $w_{ij} = w_{ji}$. We noted in the discussion of the Hopfield network that the time evolution of a symmetric network is a relaxation process characterized by the decreasing value of a *Lyapunov function*. A third somewhat weaker condition for asymptotic stability has been found. A network of N neurons is stable if there exists a set of coefficients μ_i such that the weights satisfy the "principle of detailed balance" [Al89b, Sc89]:

$$\mu_i\, w_{ij} = \mu_j\, w_{ji} \qquad \text{for all} \quad i,j = 1,\ldots,N. \tag{14.4}$$

This condition guarantees the existence of a Lyapunov function. Cohen and Grossberg give a general discussion of the stability properties of neural networks [Co83].

---

[1] It is common to treat only some of the neurons as receptor neurons; then the I_i do not vanish only for those neurons.

[2] If the fixed-point equation (14.3) has more than one solution, it may well depend on the start configuration which fixpoint is reached. Here we do not consider this case of multi-stability further, and concentrate on a single task to be learned by the network.

14.2 Recurrent Back-Propagation

We once again ask the question how to adjust the synapses w_{ik} such that a given stationary state is associated with a certain choice of the threshold potentials I_i that represent the external input. For this purpose we again define the mean square deviation between the desired output values ζ_ℓ and the stationary state s_ℓ^0:

$$D = \frac{1}{2} \sum_{\ell \in \Omega} \left(\zeta_\ell - s_\ell^0 \right)^2 , \tag{14.5}$$

where the sum runs over a subset Ω of neurons regarded as output neurons, whose (stationary) state describes the reaction of the network to the external stimulus I_i. In the spirit of the gradient method with error back-propagation, discussed in Sect. 6.2, we obtain the following adjustment of the synaptic couplings:

$$\delta w_{ik} = -\epsilon \frac{\partial D}{\partial w_{ik}} = \epsilon \sum_{\ell \in \Omega} \left(\zeta_\ell - s_\ell^0 \right) \frac{\partial s_\ell^0}{\partial w_{ik}} . \tag{14.6}$$

The partial derivative on the right-hand side has a more complicated form than for the perceptron, because we have assumed here that all neurons of the network are interconnected.[3] Differentiating the fixed-point equation (14.3), we get two contributions:

$$\frac{\partial s_\ell^0}{\partial w_{ik}} = f'\left(h_\ell^0 \right) \left(\delta_{i\ell} s_k^0 + \sum_m w_{\ell m} \frac{\partial s_m^0}{\partial w_{ik}} \right) . \tag{14.7}$$

This is an implicit equation for the partial derivative. Transferring the second term on the right-hand side to the left-hand side of the equation, we obtain a system of linear equations for the wanted quantity:

$$\sum_m L_{\ell m} \frac{\partial s_m^0}{\partial w_{ik}} = f'\left(h_\ell^0 \right) \delta_{i\ell} s_k^0 \tag{14.8}$$

with the matrix

$$L_{\ell m} = \delta_{\ell m} - f'\left(h_\ell^0 \right) w_{\ell m} . \tag{14.9}$$

The system of equations can be formally resolved with the help of the inverse matrix L^{-1}:

$$\frac{\partial s_\ell^0}{\partial w_{ik}} = \left(L^{-1} \right)_{\ell i} f'\left(h_i^0 \right) s_k^0 . \tag{14.10}$$

We then can write the synaptic adjustments (14.6) again in the usual form

$$\delta w_{ik} = \epsilon \sum_{\ell \in \Omega} \left[\zeta_\ell - s_\ell^0 \right] \left(L^{-1} \right)_{\ell i} f'\left(h_i^0 \right) s_k^0 = \epsilon \Delta_i s_k^0 , \tag{14.11}$$

---

[3] The same complication occurs if lateral synaptic connections between the neurons contained in the same layer of a perceptron are allowed.

where

$$\Delta_i = f'\left(h_i^0\right) \sum_{\ell \in \Omega} \left(L^{-1}\right)_{\ell i} [\zeta_\ell - s_\ell^0] . \qquad (14.12)$$

This solves the task of teaching the network a certain combination of stimulus and reaction, but at the expense of inverting the matrix L. Surprisingly, this problem can be solved very elegantly by introducing a second similar network which does the job! [Pi87, Al87] Let us consider the sum in (14.12) without the prefactor $f'(h_i^0)$:

$$z_i \equiv \sum_{\ell \in \Omega} (L^{-1})_{\ell i} [\zeta_\ell - s_\ell^0] . \qquad (14.13)$$

Resolving the equation by multiplication from the right with the matrix L, and using the explicit definition (14.9) of this matrix, we get

$$[\zeta_k - s_k^0]\delta_{k \in \Omega} = \sum_i L_{ik} z_i = z_k - \sum_i f'(h_i^0) w_{ik} z_i \qquad (14.14)$$

where $\delta_{k \in \Omega} = 1$ for output neurons and $\delta_{k \in \Omega} = 0$ for hidden neurons. Finally, exchanging the names of the indices i and k, we can write (14.14) as implicit equation for z_i, which resembles in form the fixed-point equation for s_i^0 (14.3):

$$z_i = \sum_k f'(h_k^0) w_{ki} z_k + [\zeta_i - s_i^0]\delta_{i \in \Omega} . \qquad (14.15)$$

The quantities z_i can hence be represented as stationary states of a neural network whose synaptic coupling strengths $\overline{w}_{ik}$ and threshold potentials $\bar{I}_i$ are given by

$$\overline{w}_{ik} = f'(h_k^0) w_{ki} \quad , \quad \bar{I}_i = [\zeta_i - s_i^0]\delta_{i \in \Omega} , \qquad (14.16)$$

and whose evolution law is governed by the *linear* function $\bar{f}(x) = x$.

The strategy of "recurrent back-propagation" is thus as follows. One starts by running the original network until it reaches a stationary state s_i^0. Then one calculates the synaptic couplings and thresholds $\overline{w}_{ik}$ and $\bar{I}_i$ of the assistant network according to (14.16) and runs this, beginning with some initial configuration, until it, too, settles down into a stationary state z_i.[4] This stationary state is now used to determine the required modifications of the synapses of the original network according to the gradient method es expressed in Eqs. (14.11 – 14.13). This procedure is repeated until the deviation between the stationary state s_i^0 of the original network and the desired output ζ_i has disappeared or become sufficiently small.

---

[4] The existence of a stationary state of the assistant network is guaranteed if the original network has a fixed point, since its dynamics corresponds to that of the linearized original network in the vicinity of its fixed point, but run backwards in time. The matrix $\overline{w}_{ik}$ of the synaptic connections of the assistant network contains the transpose of the synaptic matrix w_{ki} of the original neural net, which means that the directions of all synapses have been reversed.

14.3 Back-Propagation Through Time

As we have learned in the last section the recurrent back-propagation algorithm allows us to train a network in such a way that a prescribed configuration of activations s_i^0 is a stable state (fixed point). This does not yet exploit the full potential of recurrent networks which, in contrast to unidirectional architectures, can exhibit a rich sprectrum of dynamical behavior. It is possible to exploit this fact and train a network so that its output signals *follow a prescribed trajectory in time* $s_i(t)$ [Pe89a, Pe90]. Apart from approaching a fixed point these trajectories may undergo periodic oscillations (limit cycles) or even non-periodic motion. The trajectory can depend on external inputs or on the initial condition $s_i(0)$.

In the following we will describe a network of fully coupled neurons (of course some of the couplings may be kept zero) with continuous-valued activations s_i. In contrast to, e.g., the Hopfield network introduced in Chapt. 3 the dynamics is now *continuous in time*. Instead of discrete updating steps the time development of the neuron activations is described by the following differential equation[5]:

$$T_i \frac{\mathrm{d}s_i}{\mathrm{d}t} = -s_i + f(h_i) \tag{14.17}$$

or

$$\frac{\mathrm{d}s_i}{\mathrm{d}t} = g_i(s, w) \quad \text{where} \quad g_i(s, w) = -\frac{1}{T_i}\left(s_i - f(h_i)\right) . \tag{14.18}$$

Here f is a nonlinear sigmoidal function and h_i is the usual "local field"

$$h_i(t) = \sum_k w_{ik} s_k(t) + I_i(t) . \tag{14.19}$$

The parameter T_i is a time constant which determines how fast neuron number i can change its state. If the values h_i were kept fixed the signal $s_i(t)$ would exponentially approach the limit $s_i(\infty) = f(h_i)$ with T_i as the exponential fall-off constant. Of course, because of the couplings contained in (14.19) the dynamics in reality will not be so trivial.

A "run" of the network consists of integrating the coupled set of differential equations (14.17) forward in time up to a maximum time τ. The initial condition is given by a chosen start vector

$$s_i(0) = S_i . \tag{14.20}$$

Now we want to teach the network to follow a prescribed trajectory in state space. As usual for all supervised learning methods this is achieved by minimizing an error which now is given as a functional

---

[5] In Chapt. 12 a similar differential equation was found to describe an electronic network of coupled nonlinear circuits.

$$E[s] = \frac{1}{\tau} \int_0^\tau dt\, D(s, t; w) \tag{14.21}$$

where D is a function of the neuron activations $s(t)$ and in general also depends on time explicitly. We also have indicated the parametric dependence on the fixed weights w. Typically the function D will be chosen as in (14.5) to measure the deviation of the state of the chosen output neurons $s_\ell, \ell \in \Omega$ from a target function $\zeta_\ell(t)$:

$$D(s, t; w) = \frac{1}{2} \sum_{\ell \in \Omega} (\zeta_\ell(t) - s_\ell)^2 \tag{14.22}$$

but also other choices for D are conceivable. In the spirit of the gradient-descent method for minimization, cf. (14.6), we now have to find the gradient of D with respect to the weights w (here we do not write down the indices of the weight matrix element). The error function has an implicit and an explicit weight dependence:

$$\frac{dE}{dw} = \frac{1}{\tau} \int_0^\tau dt\, \frac{dD}{dw} = \frac{1}{\tau} \int_0^\tau dt\, \left(\frac{\partial D}{\partial s_k} \frac{ds_k}{dw} + \frac{\partial D}{\partial w}\Big|_{\text{expl}} \right). \tag{14.23}$$

It turns out that the evaluation of (14.23) is facilitated if a set of time-dependent auxiliary variables $z_i(t)$ is introduced. This can be achieved by formulating a variational principle for minimization of the error function, imposing the equation of motion (14.18) as a *dynamical constraint* [Mi91, Bi94]. It is a standard procedure of variational theory to implement constraints in terms of Lagrange multipliers. The variational principle reads

$$\delta \int_0^\tau dt\, L(s, \dot{s}, t) = 0 \tag{14.24}$$

with the Lagrange function

$$L(s, \dot{s}, t) = \sum_i z_i(t) \big(\dot{s}_i - g_i(t) \big) - D(s, t; w) \tag{14.25}$$

where the $z_i(t)$ are the (time-dependent) Lagrange parameters. If the function D does not depend on the velocities $\dot{s}$, which we have assumed for simplicity, z corresponds to the "canonically conjugate variable"

$$p_i = \frac{\partial L}{\partial \dot{s}_i} = z_i . \tag{14.26}$$

Variational theory tells us that (14.24) lead to the Euler-Lagrange equation

$$\frac{d}{dt} \frac{\partial L}{\partial \dot{s}_i} - \frac{\partial L}{\partial s_i} = 0 . \tag{14.27}$$

For the Lagrangian (14.25) this leads to

$$\frac{\mathrm{d}z_i}{\mathrm{d}t} = -\sum_j z_j \frac{\partial g_j}{\partial s_i} - \frac{\partial D}{\partial s_i} \ . \tag{14.28}$$

For this linear differential equation we will impose the boundary condition at the upper end of the time interval

$$z_i(\tau) = 0 \tag{14.29}$$

which will turn out useful later. The function $z_i(t)$ can be obtained by integrating (14.28) backward in time, using the activations $s_i(t)$ obtained from the forward integration of the differential equation (14.17).

We will now show how the gradient of the error function (14.23) can be expressed in terms of the conjugate variables z. For this purpose we solve (14.28) for $\partial D/\partial s_k$ and insert this into the first term of (14.23):

$$\sum_k \int_0^\tau \mathrm{d}t \, \frac{\partial D}{\partial s_k} \frac{\mathrm{d}s_k}{\mathrm{d}w} = -\sum_k \int_0^\tau \mathrm{d}t \, \dot{z}_k \frac{\mathrm{d}s_k}{\mathrm{d}w} - \sum_{j,k} \int_0^\tau \mathrm{d}t \, z_j \frac{\partial g_j}{\partial s_k} \frac{\mathrm{d}s_k}{\mathrm{d}w} \ . \tag{14.30}$$

The first integral on the right-hand side can be integrated by parts, discarding the surface terms because of (14.29) and $\mathrm{d}s_k/\mathrm{d}w(0) = 0$. Then the equation of motion (14.18) is used for $\dot{s}_k$:

$$-\sum_k \int_0^\tau \mathrm{d}t \, \dot{z}_k \frac{\mathrm{d}s_k}{\mathrm{d}w} = \sum_k \int_0^\tau \mathrm{d}t \, z_k \frac{\mathrm{d}\dot{s}_k}{\mathrm{d}w} = \sum_k \int_0^\tau \mathrm{d}t \, z_k \frac{\mathrm{d}g_k}{\mathrm{d}w}$$

$$= \sum_k \int_0^\tau \mathrm{d}t \, z_k \left(\sum_j \frac{\partial g_k}{\partial s_j} \frac{\mathrm{d}s_j}{\mathrm{d}w} + \frac{\partial g_k}{\partial w} \right) \ . \tag{14.31}$$

The double sum drops out in (14.30) and the implicit part of the gradient of the error function becomes

$$\left. \frac{\mathrm{d}E}{\mathrm{d}w} \right|_{\mathrm{impl}} = \frac{1}{\tau} \sum_k \int_0^\tau \mathrm{d}t \, z_k \frac{\partial g_k}{\partial w} \ . \tag{14.32}$$

This leads to the gradient learning rule

$$\delta w = -\epsilon \frac{\mathrm{d}E}{\mathrm{d}w} = \frac{1}{\tau} \int_0^\tau \mathrm{d}t \left(\sum_k z_k \frac{\partial g_k}{\partial w} + \left. \frac{\partial D}{\partial w} \right|_{\mathrm{expl}} \right) \ . \tag{14.33}$$

While the treatment so far has been general we now use the special error function (14.22) and also insert (14.18). The equation of motion (14.28) becomes

$$\dot{z}_i = \frac{1}{T_i} z_i - \sum_k \frac{1}{T_k} w_{ki} f'(h_k) z_k + \left(\zeta_i(t) - s_i \right) \delta_{i \in \Omega} \ . \tag{14.34}$$

Since D does not explicitly depend on the weights the second terms in (14.33) drops out. The gradient learning rule takes the simple form

$$\delta w_{ik} = -\epsilon \frac{1}{T_i} \frac{1}{\tau} \int\limits_0^\tau dt\, z_i f'(h_i) s_k \;. \tag{14.35}$$

Let us summarize the *Back-Propagation Through Time* (BTT) algorithm [Pe89a, Wi89] which consists of three steps.

1. The differential equations (14.17) are integrated from time $t = 0$ up to $t = \tau$.
2. The differential equations (14.34) are integrated backward in time from $t = \tau$ down to $t = 0$, starting with the initial condition (14.29). The conjugate variables $z_i(t)$ are driven by the inhomogeneous term in (14.34), i.e. by the deviation $s_i - \zeta_i(t)$ for the output signals.
3. The weights are updated according to the rule (14.35). This procedure is repeated until convergence is achieved.

This algorithm has been demonstrated to work very well for various practial applications. The convergence, however, is not guaranteed in general since the state-space trajectory traced by the network may depend on the weights in a discontinuous way [Do92]. The presence of bifurcations separating, e.g., fixed points and limit cycles as a function of the system parameters is a common feature of nonlinear dynamics.

The BTT algorithm is strongly reminiscent of Pineda's recurrent back-propagation algorithm discussed in the last section. Indeed the latter can be obtained as a limiting case of BTT. Assume that the network settles to a *fixed point* $s_i(t) \to s_i^0$ for $t \to \infty$ (of course BTT is more generally applicable) which according to (14.17) is given by

$$s_i^0 = f(h_i^0) \tag{14.36}$$

in agreement with (14.3). Then also the conjugate variables $z_i(t)$ will settle down to a constant values. If we rescale the variable according to $z_i = T_i \tilde{z}_i$ the fixed-point equation reads

$$\tilde{z}_i = \sum_k w_{ki} f'(h_k^0) \tilde{z}_k + (\zeta_i - s_i^0)\delta_{i \in \Omega} \tag{14.37}$$

which coincides with (14.15). If the upper integration limit τ is chosen large enough the transient parts of the signals can be ignored and the gradient learning rule reduces to the old result (14.11):

$$\delta w_{ik} = -\epsilon\, \tilde{z}_i f'(h_i^0) s_k \;. \tag{14.38}$$

To conclude the discussion we mention a variant of the BTT algorithm which is known as *teacher forcing*. The standard algorithm may run into problems if, at the beginning of the training phase, the trajectory $s_i(t), i \in \Omega$ differs very markedly from the target function $\zeta_i(t)$ so that the error signal may be unsuitably large. The teacher-forcing strategy tries to avoid this problem by clamping the output neurons, forcing them to prescribe the prescribed trajectory $s_i(t) = \zeta_i(t)$. The other variables $s_i(t)$ are allowed to evolve freely. Then, of course, the error function (14.22) is unusable and gets replaced by

$$D(s, t; w) = \frac{1}{2} \sum_{\ell \in \Omega} \left(\dot{\zeta}_\ell(t) - g_\ell(s, w) \right)^2 \tag{14.39}$$

which according to (14.18) measures the error of the output "velocities" $\dot{s}_\ell$. Error signals z_i now only have to be calculated for the hidden units $i \notin \Omega$. They are describe by the differential equation

$$\dot{z}_i = \frac{1}{T_i} z_i - \sum_{k \notin \Omega} \frac{1}{T_k} w_{ki} f'(h_k) z_k$$
$$- \sum_{k \in \Omega} \frac{1}{T_k} w_{ki} f'(h_k) \left[\frac{1}{T_k} \left(f(h_k) - s_k \right) - \dot{\zeta}_k(t) \right] \tag{14.40}$$

instead of (14.34). The gradient learning rule (14.35) remains unchanged for the hidden neurons $i \notin \Omega$. For the output units $i \in \Omega$ the implicit derivative of the error function $\partial D / \partial w_{ik}$ vanishes since the corresponding weights w_{ik} are not used during the learning phase. Instead there now is an explicit weight dependence contained in $g_\ell(s, w)$. This leads to the learning rule

$$\delta w_{ik} = -\epsilon \frac{1}{T_i} \int_0^\tau dt \left[\frac{1}{T_i} \left(f(h_i) - s_i \right) - \dot{\zeta}_i(t) \right] f'(h_k) s_k \quad \text{for } i \in \Omega . \tag{14.41}$$

Use the demonstration program BTT to train a fully connected network to trace out trajectories in a two-dimensional plane using several variants of the BTT algorithm (see Chapt. 28).

14.4 Network Hierarchies

The concept of using a second neural network to control the learning process of the original network is very attractive and can be implemented in various different ways. The method discussed above, due to Pineda and Almeda, applies the gradient formula to adjust the synaptic connections. It has the advantage that the assistant network is not more complicated than the original one, both having the same number of neurons. This is quite different in the method devised by Lapedes and Farber to link the operation of two neural networks, which they call "*master*" and "*slave*" networks [La86, La87c]. Here the slave network is the one that learns the desired task, while the master network controls the synaptic coupling strengths of the slave network. This is achieved by using as master net a network containing N^2 neurons whose stationary state determines the N^2 synapses of the slave net, which consists of N neurons. This is a clear disadvantage compared to the method of Pineda and Almeda, especially for large values of N. On the other hand, the master network has symmetric synapses and it yields the correct synaptic couplings of the slave network after a single run. Clearly, the gain in computational speed is bought at the expense of greater complexity.

In order to understand how the hierarchical "team" of networks operates, we start from the fixed-point equation (14.3) for the desired stationary state $s_i^0 = \zeta_i$. We rewrite this equation by applying the inverse excitation function $f^{-1}(x)$ to both sides:

$$f^{-1}(\zeta_i) = \sum_k w_{ik}\zeta_k + I_i \; . \tag{14.42}$$

The inverse function always exists for analog-valued neurons, because the neural excitation function $f(x)$ grows smoothly with x. In order to determine the synapses w_{ik} of the slave network such that (14.42) is satisfied, we define a deviation function D, which assumes its minimum when the relation (14.42) is fulfilled. An appropriate choice is

$$D = \frac{1}{2}\sum_i \left(f^{-1}(\zeta_i) - \sum_k w_{ik}\zeta_k - I_i \right)^2 \; . \tag{14.43}$$

Performing the square and ordering the resulting terms according to powers of the synaptic couplings w_{ik}, we find

$$D = \frac{1}{2}\sum_{ik\ell} s_k s_\ell w_{ik} w_{i\ell} - \sum_{ik}[f^{-1}(\zeta_i) - I_i]s_k w_{ik} + \frac{1}{2}\sum_i [f^{-1}(\zeta_i) - I_i]^2 \; . \tag{14.44}$$

The last term is simply a constant that does not depend on the w_{ik} and hence can be neglected. The task is now to choose the w_{ik} in such a way that the quantity

$$\begin{aligned}
D' &= \frac{1}{2}\sum_{ik,j\ell}[s_k s_\ell \delta_{ij}]w_{ik}w_{j\ell} - \sum_{ik}[f^{-1}(\zeta_i) - I_i]s_k w_{ik} \\
&\equiv -\frac{1}{2}\sum_{ik,j\ell} T_{ik,j\ell} w_{ik} w_{j\ell} - \sum_{ik} I_{ik} w_{ik}
\end{aligned} \tag{14.45}$$

becomes minimal. At first glance this appears quite difficult, but a comparison with (17.12) shows that D' is just the "energy" function for a symmetrically connected network with synapses

$$T_{ik,j\ell} = -s_k s_\ell \delta_{ij} \tag{14.46}$$

and neural activation thresholds

$$I_{ik} = [f^{-1}(\zeta_i) - I_i]s_k \; . \tag{14.47}$$

The N^2 neurons of this *master* network exactly correspond to the N^2 synapses of the original (the *slave*) network. Equation (14.45) expresses the fact that in the minimum of the function D the neurons of the master network just take on the values needed to satisfy the fixed-point equation (14.42).

We conclude that the wanted synaptic strengths are obtained by simply running the neural net defined by relations (14.46, 14.47) until it reaches a stationary state. The resulting synapses directly yield a network with the desired stationary state ζ_i. No iteration of this procedure is required, and one can say that the relation between the two networks is one-sided or hierarchical, whence the names "master" and "slave".[6] Aside from the potentially very large size of the master network, it is not clear how the method can be extended to find synapses simultaneously yielding *several* stationary states.

---

[6] It is not difficult to extend the procedure to "slave" networks with hidden neurons. Lapedes and Farber have shown [La86] that the procedure is then closely related to the gradient method with error back-propagation for perceptrons.

Unsupervised Learning

15.1 "Selfish" Neurons

Although the gradient method with error back-propagation has proved to be successful at teaching multilayered neural networks to perform many tasks, it has a number of rather unrealistic aspects, especially concerning the comparison with biological nerve nets. It is particularly troublesome in this respect: that complete knowledge of the deviation of the output from the desired reaction is required to determine the adjustment even of neurons in hidden layers far separated from the output layer. It is hard to believe that such extended back-coupling mechanisms can operate in complex biological neural networks.

One may therefore ask whether the learning goal cannot also be achieved by methods involving less-detailed back-propagation of information about the success of the network. To take a comparison with experience from school: the gradient method with error back-propagation corresponds to a situation where the teacher tells the student exactly what she does wrong and how she can do better. This procedure may be practical in private lessons, but certainly not for teaching a large class of students. In that case the teacher will be forced to grade the students in a much more cursory fashion according to their overall performance. If the grades are good, the student will continue as before; if they are bad, she may take this as an indication to improve her performance.

Barto et al. [Ba85a, Ba85b] have proposed a strategy to apply this concept of *reward* and *penalty* to teach multilayered perceptrons. Since the (hidden) neurons are designed to react to the grading signal in such a way as to maximize their reward, but not to optimize the overall performance of the network, they are also called *selfish* neurons.[1] This method of reward and penalty learning has also been described as learning through *criticism* [Wi73].

In oder to understand how the method works, let us consider a three-level perceptron. The N_0 output neurons are assumed to be analog-valued and deterministic with states n_i given by

---

[1] The concept of *selfish* or *"hedonistic"* neurons was apparently first introduced by A.H. Klopf [Kl82].

$$n_i = f(h_i) = [1 + e^{-h_i}]^{-1}, \qquad h_i = \sum_j w_{ij}\overline{n}_j \, . \tag{15.1}$$

The neurons of the hidden layer are assumed to operate stochastically with the binary states $\overline{n}_j = 0, 1$, which are taken with probabilities

$$P(\overline{n}_j = 1) = f(\overline{h}_j) \, , \quad P(\overline{n}_j = 0) = f(-\overline{h}_j) \, , \quad \overline{h}_j = \sum_k \overline{w}_{jk}\nu_k \, , \tag{15.2}$$

where ν_k are the states of the input neurons. The desired output states are again denoted by ζ_i, taking analog values between 0 and 1, and the deviation of the actual output states from the desired ones is denoted by

$$D = \frac{1}{N_0} \sum_i (\zeta_i - n_i)^2 \, . \tag{15.3}$$

Here the normalization is chosen in such a way that the value of D also lies between 0 and 1. A small value of D indicates that the desired output is well reproduced.

The synaptic strengths at the output neurons are adjusted according to the gradient method, which is both convenient and reasonable since the required information, i.e. the remaining deviation, is locally available. The modifications are then given by our familiar expression (5.10):

$$\delta w_{ij} = \epsilon(\zeta_i - n_i)f'(h_i)n_j \, . \tag{15.4}$$

For the modifications of the synapses at the hidden neurons, which we have denoted by $\overline{w}_{jk}$, we now deviate from our previous strategy. Here we do not want to use the complete information from the output layer, but only a grading signal r which indicates the overall degree of success of the network. Following Barto et al., we find two alternative choices at our disposal. (1) We can use a binary signal, which assumes the value $r = 1$, signaling success, with probability $(1 - D)$, and the value $r = 0$, indicating failure, with probability D. (The normalization of D to the range $0 \le r \le 1$ allows us to interpret D as a probability!) The smaller D becomes, the better the agreement with the desired output is, and the more likely $r = 1$ becomes. (2) Alternatively, we can define an analog-valued grading signal by means of the assignment $r = (1 - D)$. This has been found to give better practical results and is also known as the method of *linear reward*.

The synapses at the hidden neurons are then modified as follows:

$$\delta\overline{w}_{jk} = r\epsilon[\overline{n}_j - f(\overline{h}_j)]\nu_k + (1 - r)\lambda\epsilon[1 - \overline{n}_j - f(\overline{h}_j)]\nu_k \, . \tag{15.5}$$

In order to comprehend the result of this modification, we first consider the case of success, $r = 1$. Since $f(\overline{h}_j)$ has a value between 0 and 1, $\delta\overline{w}_{jk}$ is positive if $\overline{n}_j = 1$ and negative if $\overline{n}_j = 0$. Hence the synaptic connection is reinforced if the hidden neuron was active, otherwise it will be weakened. But this is precisely the content of Hebb's rule! In the case of failure, $r = 0$, everything works in the opposite way: if the hidden neuron was inactive, the synaptic coupling is strengthened; if it was active, the connection is weakened. Again this is an implementation of Hebb's rule. The parameter λ is an expression of the importance of failure for the learning process. It has been found to be useful to select a value $\lambda \ll 1$, i.e. to weight success much more strongly than failure.[2]

A mathematical analysis of this strategy of learning by reward and penalty with the binary grading signal and $\lambda = 0$ has shown that the synaptic correction (15.5) leads to an increase in the expectation value of the grading signal. In other words: the network tries to improve its own performance as measured by the grading signal r. As just mentioned, it is even better to choose a small positive value of λ. Moreover, one can sample the output of the network several times before adjusting the synapses in order to reduce the stochastic fluctuations and to obtain a more reliable measure of success. General convergence theorems for the various variations of this learning technique have been proved [Ba85b, La81], but in practice the question is: How fast does the learning algorithm converge?

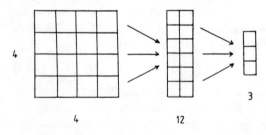

4

4 12 3

Fig. 15.1. Architecture of the network designed to detect symmetries in a (4×4) pattern.

The reward–penalty method has been compared with the method of error back-propagation for the task of finding symmetries in a 4×4 pattern [Ba87c]. As discussed in Sect. 13.3, the network was composed of 16 input, 12 hidden, and 3 output neurons (see Fig. 15.1). To reach a 95% success rate required about half a million learning steps! The method based on the linear grading signal required about 30% fewer steps than that using a binary grading signal. By contrast, the same accuracy of response was achieved in less than 5000 learning steps by the gradient method with error back-propagation. The very slow progress during the training phase is generally regarded as a major shortcoming of reward–penalty learning. On the other side of the balance we have the fact that transmission of a single, scalar grading signal to all neurons of a complex network is a much simpler technical task than the transmission of an entire error gradient. However, since this consideration is of little relevance for software realizations of neural networks on conven-

---

[2] This is a well-known educational principle: in most cases reward of successful behavior makes a much better teacher than punishment of failures. Nevertheless, punishment can serve its educational purpose, if applied sparingly and wisely.

tional computers, reward–penalty learning has found relatively few practical applications so far.

15.2 Learning by Competition

15.2.1 "The Winner Takes All"

In the learning protocols discussed so far we always assumed the existence of a "supervisor" who judges the training progress and controls the synaptic plasticity. In biological neural nets such a supervisor is not readily discernible. Moreover, it is not always clear in real life what the learning goal precisely is. Certainly, the ultimate goal is well defined: to improve the chances of survival of the individual or the species, but how exactly this goal is best served by a particular neural reaction is often hard to define.

Such questions have been addressed in connection with studies of nerve centres that process optical patterns. It was first recognized in experiments with cats that the visual cortex contains specific neurons which recognize particular patterns, such as vertical or horizontal lines, etc. [Hu62]. These neurons are activated when several receptor cells lying on a straight line in the retina of the eye are stimulated. Since it is unlikely that this assignment of neurons to specific stimuli is determined genetically in all details, there must exist a biological mechanism establishing this relation during the formative period of the brain.

A similar problem is posed in connection with the evolution of biological species. As Darwin realized, the animal and plant species did not develop toward a predetermined goal – e.g. towards the human species, but under the pressure of selection the species fittest for survival developed from a diversity created by successive genetic mutations. It is thus a promising thought to apply the principle of evolution by selection, i.e. by competition of neurons among each other, to learning processes in neural networks. To be precise, the competition is not to be applied towards the neurons themselves but to their synaptic connection s: *successful* synapses, i.e. those which have activated the post-synaptic neuron, are strengthened; the others are weakened. Because many synapses are located on a single neuron, this principle potentially leads to cooperative behavior of entire groups of neurons. Such a development can be encouraged by special modifications of the selection process.

The simplest selection rule is that only a single neuron from a certain group of neurons, e.g. from a given layer of neurons in a perceptron, is allowed to become active [Gr76, Ca87c]. This can be enforced by strong inhibitory connections among the neurons within this group. The neuron with the strongest synaptic excitation potential h_m will fire and thus prevent all other neurons becoming active as well. For illustrative purposes we consider a two-layered perceptron, with post-synaptic potentials

$$h_i = \sum_k w_{ik}\sigma_k, \qquad h_m = \max_i(h_i) .\tag{15.6}$$

The combined strength of all synapses attached to a single neuron is assumed to be normalized to a fixed value according to $\sum_i w_{ik}^2 = 1$. All synapses connecting to the winning neuron m, i.e. to the neuron with the maximal value of h_i, are first enhanced according to Hebb's rule and then renormalized to combined unit strength:

$$\delta w_{ik} = \epsilon(\sigma_k^\mu - w_{ik})\delta_{im} .\tag{15.7}$$

The success of this concept is best visualized by conceiving the aggregate of all N_i synapses w_{ik} connecting to a given neuron i as an N_i-dimensional unit vector whose components are enumerated by the second index k: $1 \leq k \leq N_i$. The states of the input neurons, σ_k, also represent such a vector. The post-synaptic potential h_i can now be considered to be the scalar product of these two vectors for the neuron i. In other words, the learning rule (15.7) implies that the synaptic unit vector of the successful neuron m is turned towards the direction of the input vector σ_k, thus further increasing the scalar product of the two vectors.

If we are dealing with a number of different input vectors σ_k^μ, those neurons whose synaptic vectors are accidentally neighbors in the direction of one of the input vectors begin to align themselves with it more and more. This realignment is significantly accelerated if the input vectors fall into groups of similar patterns, which unite to pull a neuron into their common range of attraction. In this way individual neurons emerge that specialize in the recognition of a certain type of pattern, or of a certain property of the patterns. Often a few neurons remain which are winners for none of the patterns presented to the network, because in each single case another neuron is better positioned. Here one has to choose between two alternatives: either one can let them "die" by the gradual weakening of their synapses or, better, one may force their participation in the adjustment process by slowly lowering their excitation thresholds.

15.2.2 Structure Formation by Local Inhibition

A slightly modified version of this strategy was utilized by von der Malsburg [Ma73] to construct a model of the learning processes leading to the structuring of the cat's visual cortex. For this purpose he replaced the extreme selection principle of "winner takes all" by a clever combination of local inhibition and enhancement. In the output layer of his two-layered network all neighboring neurons were connected by excitatory synapses, where the synaptic connections to more distant neurons were of the inhibitory type, their strength diminishing with increasing distance. In the two-dimensional layer this yields a pattern of connectivity, where an inner positive area is surrounded by a broad negative ring, similar to the shape of a mexican hat.

If one of the output neurons becomes active, this results in an enhanced sensitivity of the neighboring neurons, while the activation threshold of all other neurons is raised. It is intuitively obvious that this principle supports the formation of local regions of neurons responding to similar stimuli.

The specific model studied by von der Malsburg consisted of a triangular input layer composed of 19 neurons. A directional line pattern was represented by seven neurons lying on a straight line. The output layer contained 169 neurons, which also formed a triangle. During the course of the learning phase local groups of neurons were formed in this layer, which reacted to one specific directional pattern.

15.2.3 The Kohonen Map

A related problem was posed by studies of the structure of the motor centre in the human cortex. The investigations showed that the neural regions controlling the motion of the various parts of the human body are arranged side by side as on a map of the body surface. However, this map is strangely distorted (see Fig. 15.2): The size of a certain region does not correspond to the size of that part of the body which it controls, but rather to the importance and the complexity of its activities. For example, the neurons controlling the movements of the lips occupy a much larger area than those responsible for the entire torso! Nonetheless, adjacent parts of the body are controlled by neighboring regions of the motor cortex. How could this remarkable organization have been formed?

Kohonen has proposed a mechanism by means of which a neural network can organize *itself* in such a way that neurons are allocated to a region in the parameter space of external stimuli according to its amount of activity. What is meant by this is best explained by a simple example. Let us consider a two-layered neural network composed of two analog-valued input neurons and N^2 output neurons arranged in a square. The values of the two input neurons, which are assumed to lie in the range from 0 to 1, are denoted by x and y. The aim is to teach each output neuron to respond to a particular pair (x, y) of input values, or rather to all input values in the vicinity of a particular pair. A trivial solution would be obtained by "mapping" the unit square of possible inputs (x, y) with $0 \leq x, y \leq 1$ on the square output layer of neurons. However, we are here particularly interested in cases where the input values (x, y) that really occur are not evenly distributed over the unit square, but only over a certain geometric subset, e.g. over the lower-left and the upper-right quarters of the square, or over a ring-shaped region around the center of the square.

We can identify the neurons in the square output layer by a two-digit index (ik), $1 \leq i, k \leq N$, which indicates their position in the square. The pairs of numbers $\left(\frac{i}{N+1}, \frac{k}{N+1} \right)$ then define a regular grid on the unit square. Each output neuron receives signals from two synapses originating at the

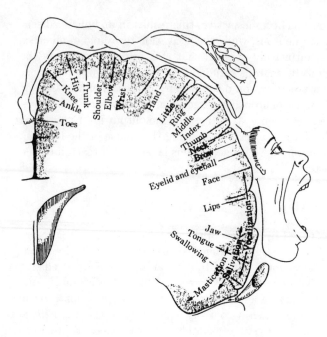

Fig. 15.2. Representation of various parts of the human body in the motor center of the cortex. This "homunculus" was deduced by electrically stimulating the cortex of conscious patients undergoing brain surgery (from [Wo63]).

input neurons. We denote their synaptic strengths by $X_{(ik)}$ and $Y_{(ik)}$, and we assume also that each strength has a value between 0 and 1. Through the pair $(X_{(ik)}, Y_{(ik)})$ the neuron (ik) is associated with the input value (x, y) given by

$$x = X_{(ik)}, \qquad y = Y_{(ik)} . \tag{15.8}$$

We now demand that the post-synaptic potential $h_{(ik)}$ takes its maximal value for this naturally associated input pair defined by (15.8). This can be realized, e.g., by setting

$$h_{(ik)} = \frac{xX_{(ik)} + yY_{(ik)}}{\sqrt{x^2 + y^2}\sqrt{X_{(ik)}^2 + Y_{(ik)}^2}} . \tag{15.9}$$

The goal is to adjust the synapses to the distribution of input values (x, y), so that no neuron is idle because it is associated with an input that never occurs. In order to achive this, the neural "grid" is to be smoothly deformed, starting from its original square shape, so that it just covers the region of values that really occur in the unit square. For every pair of values (x, y) that occur there must exist a neuron (ik) with synaptic weights $(X_{(ik)}, Y_{(ik)})$ in the vicinity of the input, and no neuron should be left waiting idly to respond to an input which never occurs. The adjustment process may be envisaged as the task of pulling at the strings of an elastic net in such a way that it takes on a certain geometrical shape without tearing.

The task can be solved with the help of the procedure discussed in Sect. 15.2.1. One identifies the neuron (ik) with synaptic strengths $(X_{(ik)}, Y_{(ik)})$

closest to the present input pair (x, y), and turns the synaptic vector (X, Y) of this neuron and those in its vicinity by a small amount into the direction of the vector (x, y). This process is performed for all input vectors. Then the procedure is repeated several times, under gradual reduction of the size of the neighborhood of the successful neuron, in which the synapses are adjusted, until the desired result is obtained. The mapping

$$(ik) \rightarrow (X_{(ik)}, Y_{(ik)}) \tag{15.10}$$

then represents a topologically contiguous mapping of the unit square onto the region covered by the input vectors (x, y), which is called the *Kohonen map* [Ko82, Ko84].

It is a remarkable property of the Kohonen map that the dimension of the sensory input space and of the array of neurons need not agree. If the dimensions differ (usually the input space will have the higher dimension) the learning algorithm leads to an "automatic selection of feature dimensions": While preserving the neighborhood relations between neurons the mapping is folded in such a way that the input space is uniformly covered by associated neurons. Details of the model using concepts from the physics of stochastic processes are discussed in [Ri86, Ri88b] and in the monograph [Ri92].

In addition to its appeal in describing aspects of the organization of the neural cortex in biological brains the Kohonen map has also found technical applications. E. g., it may be used for pattern recognition like the identification of spoken phonemes [Ko84] or for the control of motor tasks. In a typical application of the latter type the input might arise from optical sensors and the neurons might drive servo motors controlling a robot arm in response to the sensory input.[3]

 Use the demonstration program KOHOMAP to generate Kohonen maps of (1) the upper-right and lower-left quarters of the unit square or (2) an annular region inside the unit square (see Chapt. 27).

15.3 Implementations of Competitive Learning

15.3.1 A Feature-Sensitive Mapping Network

The concepts of competitive and Hebbian learning can be combined into a mapping neural network that exploits the presence of characteristic structural

---

[3] A variant of the Kohonen algorithm has also been used to solve the traveling salesman problem [Fo88c]: an arrangement of neurons originally representing a closed circle is continuously deformed under the pull of the cities to be visited.

aspects in the input data [He88b]. The network is designed to represent a mapping $(F : \mathbb{R}^n \to \mathbb{R}^m)$ of an n-dimensional input vector $\mathbf{x}$ into an m-dimensional output vector $\mathbf{y}$: $\mathbf{y} = F(\mathbf{x})$. In its simplest form the network contains three layers of neurons, which are linked by forward-feeding synaptic connections. The input layer contains $(n + m)$ neurons to feed the input–output pairs x_k, y_k of the training set into the network. The m neurons y_i' of the output layer represent the result of the mapping in the normal operating phase, when the values $\mathbf{x}$ and $\mathbf{y} = 0$ are entered into the input neurons. The hidden layer consists of N binary, competitive neurons z_j, which obtain signals through synapses $u_{jk}(j = 1,\dots,N; k = 1,\dots,n)$ from each one of the first group of input neurons x_k (*not* from y_k) and feed into all output neurons. The synaptic connections w_{ij} between the hidden neurons and the output neurons are influenced by the state of the second group of input neurons, y_k, during the training phase. The architecture of the network is schematically illustrated in Fig. 15.3.

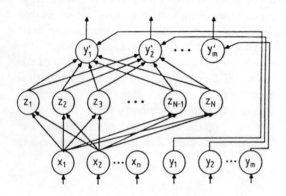

Fig. 15.3. Architecture of the Hecht-Nielsen feature--sensitive mapping network (from [He87a]).

During the learning phase the hidden neurons are trained to respond to one of several characteristic features present in the sample input vectors x_k. The response of the hidden neurons is defined by the "winner takes all" principle, $z_j = \delta_{jj_m}$, where j_m is the number of the neuron receiving the largest synaptic excitation:

$$I_j = \sum_{k=1}^{n} u_{jk}x_k \ . \tag{15.11}$$

The synaptic connections at the hidden layer are updated according to the learning rule (15.7), which rotates the synapse vector u_{kj_m} of the winning neuron closer to the input vector x_k:

$$\delta u_{jk} = \epsilon(x_k - u_{jk})z_j \ . \tag{15.12}$$

The synapses at the output layer learn to represent the desired result of the mapping when a particular feature has been detected by a neuron of the hidden layer:

$$\delta w_{ij} = \alpha(y_i - \lambda w_{ij})z_j \,, \tag{15.13}$$

where a slow-forgetting mechanism has been added through the parameter $\lambda \ll 1$. In normal-operation mode the output neurons respond linearly according to:

$$y_i' = \sum_{j=1}^{N} w_{ij}z_j \,. \tag{15.14}$$

In this form, the network works somewhat like a look-up table, emitting a definite prediction $\mathbf{y}'$ for each one of the features detected in the input data $\mathbf{x}$. The predicted output value is the average of the vectors y_k encountered during the learning phase, when the particular feature was detected in the input data. In order to endow the neural network with the ability to interpolate, one can modify the strict "winner takes all" rule for the hidden neurons in the operational mode. One possible way is to allow several of the hidden neurons z_j, e.g. those with the two or three highest inputs I_j to become active, emitting signals z_j in proportion to the size of their excitation potentials I_j, and normalized to total strength one. Other modifications are also possible. In particular, the network may be duplicated for invertible mappings into two interconnected three-layered networks sharing the hidden, competitive neuron layer. This structure has been described as a *counterpropagation* network [He87a].

The advantage of this type of network over multilayered-perceptron learning with the error back-propagation algorithm is that it learns to represent the training data much faster. Although the representation of the data contained in the training set is less accurate than for a fully converged back-propagation network, the predictions of the feature-sensitive mapping network are explicitly based on characteristic structures contained in the input data. This seems to ensure a certain consistency of extrapolation or generalization, for which there is no guarantee in the case of a back-propagation network. The network structure discussed above has been applied to the classification of patterns, statistical analysis, and data quantization, and even to the prediction of commodity trading prices [He87a].

15.3.2 The Neocognitron

An early attempt to apply concepts of competitive learning to pattern recognition was made by Fukushima [Fu80, Fu88a], who called his neural-network structure the NEOCOGNITRON, indicating that it was an improved version of its precursor, the so-called COGNITRON [Fu75]. The basic principle is to combine feature-extracting neurons ("S cells"), which are trained competitively, with neurons ("C cells") that correct spatial shifts in the position of the detected features. The full network consists of several layers containing interconnected neurons of both types. The idea is that a C-cell neuron looks at several S-cell neurons which detect the same feature at slightly different positions in space, and is activated independently of the precise position of the feature. This reduces the sensitivity on position step by step. The detailed structure of a single layer is shown in Fig. 15.4.

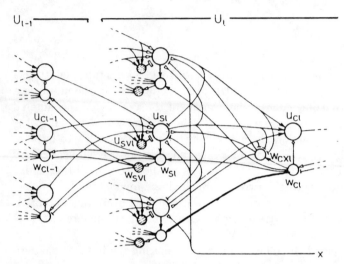

Fig. 15.4. Architecture of the NEOCOGNITRON (detail). Excitatory synapses are indicated by *arrows*, inhibitory ones by *flat ends* (from [Fu80]).

Training of the network is based on the principle of reinforcement of synapses receiving the largest signal, i.e. a synaptic connection is strengthened if (1) the post-synaptic neuron shows the strongest response among neurons in the vicinity, and (2) the presynaptic neuron is active. In order to achieve the desired sensitivity to characteristic features, an elaborate structure of excitatory and inhibitory neurons is arranged around each S-cell neuron, with synaptic connections upstream ("W_S cells") or downstream ("U_S cells"), as indicated in Fig. 15.4. Connections leading to the S cells exhibit plasticity and are fixed during the training phase, while connections from S cells to C cells are predetermined and fixed.

In the more recent applications [Fu88a] the network has been trained by a teacher (not by competition among neurons!) to recognize handwritten numerals from "0" to "9". In this configuration the network contained 19×19 input neurons to represent the digitized numerals, 10 output neurons indicating which numeral was recognized, and a total of 34970 hidden neurons. After completion of training the network was able to recognize even rather distorted or noisy numerals (see Fig. 15.5). Of course, it must be stressed that this success was achieved at the expense of considerable human effort in the structural design and training of the neural network, which may be regarded as a substitute for the "preprocessing" needed in other realizations of invariant pattern recognition (see Sect. 9.3).

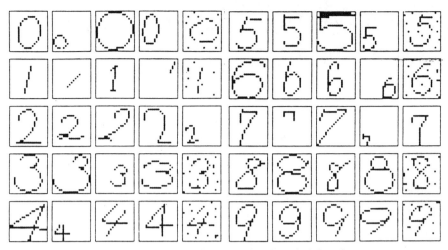

Fig. 15.5. Some distorted numerals correctly identified by the NEOCOGNITRON (from [Fu88a]).

16. Evolutionary Algorithms for Learning

Biological neural networks have developed over time through genetic evolution. It therefore seems plausible that evolutionary concepts could also be effective when applied to artificial neural networks. The central concepts of natural evolution are selection and mutation; however, when modeling biological systems these concepts must be augmented by the mechanisms of genetic reproduction, such as genetic recombination or *crossover*. Of course, the use of genetic operators need not be restricted to biological modeling. In fact, the application of these concepts to machine learning, which was pioneered by John Holland in the 1970s [Ho75], has found widespread interest recently [Ho88a, Go88, Mo89, An94, Ma94a, Mc95a, Mi94, Qi94, Pr94].

Technically, evolutionary algorithms are a form of multi-agent stochastic search. Single-agent stochastic searches (SASS), such as simulated annealing, use only one solution vector. In the case of neural networks, this solution vector could simply be the network weights and activation thresholds or some other specification of the network configuration and/or dynamics. In contrast to SASS, multi-agent stochastic searches (MASS) maintain a *population* of solution vectors. This population is then used to map out the energy surface of the problem at hand.

In an evolutionary algorithm the population of solution vectors is processed by means of competitive selection and individual mutation. A genetic algorithm further specifies that the solution vector is encoded in a *"chromosome"* on which genetic operators such as crossover act. Although the genetic operators used can vary from model to model, there are a few standard or canonical operators: crossover and recombination of genetic material contained in different parent chromosomes (inspired by meiosis), random mutation of data in individual chromosomes, and domain-specific operations, such as migration of genes.

In order to apply selection, the central concept of evolution, one also needs a way of rating different chromosomes according to their performance on a given problem or problem set. This performance is sometimes called the individual's *"fitness"*. For feed-forward networks this may simply be the standard mean squared deviation between the actual and the desired output, which also forms the basis for the error back-propagation algorithm discussed in Sect. 6.2.

The genetic algorithm then operates as follows. The population is initialized and all chromosomes are rated by means of the chosen evaluation function. The procedure now calls for (possibly repeated) reproduction of the population of chromosomes. The parents are reproduced by rating each according to its fitness and placing them in the next generation with a probability proportional to their fitness. Once a parent has passed the selection criterion, its genetic information is copied into the next generation either individually by mutation or in pairs by crossover of genes. The performance of the children is rated and they are inserted into the population. In some versions, the entire population is replaced in each cycle of reproduction; in others, only a certain fraction is replaced. By the application of this prescription, and with a judicious choice of genetic operators, the algorithm produces a population of increasingly better individuals. Experience has shown that genetic algorithms are very effective in converging to the *global* optimum, because they explore a large region of the available space of network models. Other algorithms, such as those based on gradient descent, may converge faster, but they often get stuck in local minima of the evaluation function. It is the action of the genetic operators like crossover and mutation which maintains population diversity, allowing the population to avoid being captured by local minima, similar to the action of thermal fluctuations in the method of simulated annealing.

Early research by Montana and Davis [Mo89] compared the performance of gradient learning by error back-propagation with that of a specifically designed genetic algorithm on training a feed-forward neural network to recognize structures in sonar data from underwater acoustic receivers, sometimes called lofargrams. The network contained four input neurons, one output neuron, and two hidden layers of seven and ten neurons, respectively, yielding a total of 126 synaptic weights. The training set contained 236 examples.

After optimizing the parameters of the genetic algorithm and including only mutation and crossover, Montana and Davis found that it outperformed the back-propagation algorithm by about a factor of two over 10 000 iterations. This result is important, since the genetic algorithm has the advantage that it can be directly applied to networks with discrete, binary neurons and recurrent networks. The authors also tried to incorporate certain elements of gradient learning in their algorithm based on the gradient of the evaluation function averaged over the entire population, but they found that this operation carried a large risk of being caught in local minima. Therefore its application was only useful when the global minimum was certain to be near. We will discuss further the delicate balance between local and global optimization.

16.1 Why Do Evolutionary Algorithms Work?

Traditional theoretical analysis of genetic algorithms relies both on a binary string representation of the solutions and the notion of schemata for comparing the similarities of chromosomes (see e.g., [Ho75]). However, in keeping with the statistical physics approach applied thus far we intend to perform a statistical analysis of evolutionary algorithms. The key assumption in this analysis is that the population is very large, allowing the construction of a continuous fitness distribution function. To some this may seem a crippling assumption since in practice only finite populations are used; however, as will be shown below, an infinite-population limit can aid in understanding the roles of various genetic operators. The treatment presented below is a modified version of one presented by J. Shapiro and A. Prügel-Bennett [Pr94]. For simplicity we choose to analyze the performance of a genetic algorithm in finding the ground state of a one-dimensional Heisenberg spin chain of length N with an energy functional given by

$$E[s] = \frac{1}{2}\text{Abs}\left[\sum_{ij}^{N} w_{ij}s_i s_j\right] \tag{16.1}$$

where $-1 \leq s_i \leq 1$, $w_{ij} = \delta_{j(i+1)} + \delta_{j(i-1)}$, and periodic boundary conditions are chosen. The absolute value has been inserted to move the minimum from $-\sum_{ij}^{N} w_{ij}$ to zero in order to simplify the analysis. As our genetic encoding we choose to set each bit i along the chromosome equal to the spin at site i — s_i. The population is initialized by choosing P random spin chains/strings.

Within this framework we will study the effects of selection, crossover, and mutation on the genetic population. To study the evolution of the population it is sufficient to study the evolution of the moments of the population's energy distribution $P(E,t)$, which is a function of the energy E and the "time" or generation t. This distribution function, when multiplied by the infinitesimal dE gives the probability of finding a member of the population with energy between E and $E + dE$. The mth moment of the probability distribution $P(E)$ is defined as

$$P_m(t) \equiv \int dE\, P(E,t)E^m , \tag{16.2}$$

with $P_0 = 1$, $P_1 = <E> = \mu$, $P_2 = <E^2> = \sigma^2 + \mu^2$, ...

1. Selection: At the end of each generation we rank the population according to their fitness (the string with the lowest energy is the fittest string in this case) and place them in the next generation with a probability that decreases with decreasing fitness according to a selection function $S(E)$. In this treatment we choose the selection function to be

$$S(E_i) = \frac{e^{-\beta E_i^2}}{Z} \qquad Z = \sum_{i=1}^{N} e^{-\beta E_i^2} , \tag{16.3}$$

where i indexes the members of the population. This is not the standard selection function used in the literature, but this function is somewhat arbitrary and choosing the above form simplifies the final expressions considerably. To obtain the probability that a member with energy E will be present in the new population we evaluate the product $AS(E)P(E)dE$, where A is a normalization constant. Under n applications of this selection the probability distribution function will be given by

$$P(E, t = n) = A^n S^n(E) P(E, t = 0) . \qquad (16.4)$$

This equation specifies the evolution of all the moments of P. For instance, the population's mean and standard deviation at time t are given by

$$\mu(t) = A' \int_{-\infty}^{\infty} dE\ E e^{-\beta n E^2} P(E, t = 0) ,$$

$$\sigma^2(t) = A' \int_{-\infty}^{\infty} dE\ E^2 e^{-\beta n E^2} P(E, t = 0) - \mu^2(t) . \qquad (16.5)$$

Substituting a Gaussian form for the initial distribution gives

$$\mu(n) = \frac{\mu_0}{1 + 2\beta n \sigma_0^2} ,$$

$$\sigma^2(n) = \frac{\sigma_0^2}{1 + 2\beta n \sigma_0^2} . \qquad (16.6)$$

From this it can be seen that selection always reduces the mean energy of the population. Unfortunately, the standard deviation or diversity of the population also always decreases under selection. When diversity is small the population may only find a local extremum not a global one. Clearly, other genetic operators are needed to maintain population diversity.

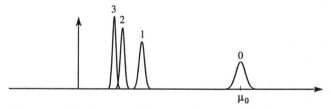

Fig. 16.1. Evolution of the population probability distribution under selection alone.

2. Crossover: In practice crossover involves selecting two random parents (possibly weighted according to their fitness), exchanging complementary coding sections, and placing one or both of them into the next generation. For our example we choose to cut each parent gene at a random site L, exchange the appropriate pieces between parents, and put both new

genes into the next generation. This procedure can be represented by a convolution

$$P(E, t+1) = A \int_{-\infty}^{\infty} dE' \, C(E, E') P(E', t) , \tag{16.7}$$

where $C(E, E')$ is the probability that a member with energy E' will have energy E after recombination. The action of selection followed by crossover can be represented as

$$P(E, t+1) = A \int_{-\infty}^{\infty} dE' \, C(E, E') S(E') P(E', t) . \tag{16.8}$$

Although we will have to calculate the specific form of $C(E, E')$, we can make some qualitative statements by assuming

$$C(E, E') \propto e^{-(E-E')^2/(2\sigma_c^2)} . \tag{16.9}$$

Performing the integration (16.8) and evaluating the moments gives

$$\begin{aligned}
\mu(n+1) &= \frac{\mu(n)}{1 + 2\beta\sigma^2(n)} , \\
\sigma^2(n+1) &= \sigma_c^2 + \frac{\sigma^2(n)}{1 + 2\beta\sigma^2(n)} .
\end{aligned} \tag{16.10}$$

From this we can see that if $\sigma_c^2 \equiv \bar{\sigma}_c^2 = 2\beta\sigma_0^4/(1 + 2\beta\sigma_0^2)$ the population diversity will remain constant throughout the evolution. Comparing Fig. 16.2 with Fig. 16.1 we see that crossover not only leads to more diversity, it also increases the convergence of the mean.

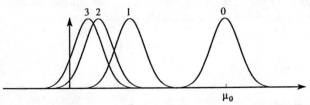

Fig. 16.2. Evolution of the population probability distribution under selection and crossover with $\sigma_c^2 = \bar{\sigma}_c^2$.

3. Mutation: Mutation can be treated in the same way as crossover but with a different convolution function $M(E, E')$. If a Gaussian convolution function is appropriate for both crossover and mutation then the diversity enhancements from crossover and mutation simply add as $\sigma_c^2 + \sigma_m^2$. If $\sigma_c^2 \gg \sigma_m^2$, then the effects of mutation are negligible. Of course, these widths depend on the particular problem chosen and so it is unwise to make general statements about the relative importance of the two operations.

Now we would like to evaluate $C(E', E)$ and $M(E', E)$ for the one-dimensional Heisenberg spin chain with nearest neighbor interactions. For crossover, we consider two strings a and b that are both cut at a random site L and end pieces are interchanged, e.g.,

$$
\begin{aligned}
E^a &= \sum_{i=1}^{N} J_i s_i^a s_{i+1}^a \\
&= \sum_{i=1}^{L-2} J_i s_i^a s_{i+1}^a + J_L s_{L-1}^a s_L^a + \sum_{i=L}^{N} J_i s_i^a s_{i+1}^a \\
&= E_{L-1}^a + \Delta^{aa} + E_{N-L}^a ,
\end{aligned} \tag{16.11}
$$

which under crossover with string b becomes

$$
E'^a = E_{L-1}^a + \Delta^{ab} + E_{N-L}^b . \tag{16.12}
$$

The probability that $E^a \to E'^a$ is given by the product of the probability to find a subchain with energy E_{L-1}^a times the probability of having interface energy Δ^{ab} times the probability that the final subchain has energy $E - E_{L-1}^a - \Delta^{ab}$ integrated over all possible subchain lengths, subchain energies, and interface energies. This can be written as

$$
\begin{aligned}
C(E', E) = \int_{-1}^{1} d\Delta^{ab} \int_{0}^{1} dl \int_{-\infty}^{\infty} dE_L\, \mathcal{P}_E(E_L, l) \\
\mathcal{P}(\Delta^{ab}) \mathcal{P}_E(E' - E_L - \Delta^{ab}, 1 - l)
\end{aligned} \tag{16.13}
$$

where $\mathcal{P}_E(E_L, l)$ is the probability that a subchain of length $L = lN$ will have energy E_L and $\mathcal{P}(\Delta^{ab})$ is the probability of having interface energy Δ^{ab}. The probability $\mathcal{P}_E(E_L, l)$ can be calculated in the large P limit giving

$$
\mathcal{P}_E(E_L, l) = \frac{e^{-(E_L - lE)^2 / 2l(1-l)\sigma^2}}{\sqrt{2\pi l(1-l)\sigma^2}} . \tag{16.14}
$$

After performing the E_L integration equation (16.13) becomes

$$
\begin{aligned}
C(E', E) = \int_{-1}^{1} d\Delta^{ab} \mathcal{P}(\Delta^{ab}) \int_{0}^{1} dl \sqrt{\frac{E^2}{4\pi l(1-l)\sigma^2}} \\
\times \quad e^{-(E' - (E - \Delta^{ab}))^2 / (4l(1-l)\sigma^2)} .
\end{aligned} \tag{16.15}
$$

Making an approximation for the dl integration and assuming a Gaussian form for the Δ^{ab} distribution gives

$$
C(E', E) \propto e^{-(E - E')^2 / (2\sigma_c^2)} , \tag{16.16}
$$

where $\sigma_c^2 \sim \frac{1}{4}\sigma^2$. With this we can specify the evolution of the mean and standard deviation of a large genetic population under selection and crossover. The mutation width can be calculated in a similar manner giving $\sigma_m \sim \frac{1}{20}\sigma^2$. Therefore, for the model considered crossover is much more important in

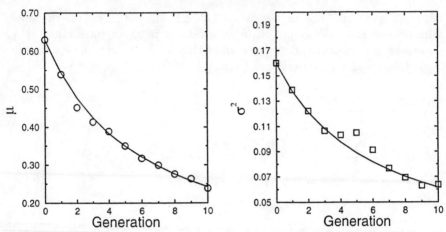

Fig. 16.3. Comparison between the prediction of equation (16.17) and numerical simulation. $P = 1000, N = 1000, \beta = 0.5, P_m = 0.01$, and $P_c = 0.33$.

maintaining population diversity. Using the results for these two widths, we can rewrite equation (16.10) as

$$\mu(n+1) = \frac{\mu(n)}{1 + 2\beta\sigma^2(n)}$$

$$\sigma^2(n+1) = \left(\frac{1}{20}P_m + \frac{1}{4}P_c\right)\sigma^2(n) + \frac{\sigma^2(n)}{1 + 2\beta\sigma^2(n)} , \qquad (16.17)$$

where P_m and P_c are the probilities of mutation and crossover, respectively. Figure 16.3 shows a comparison between the predicition of equation (16.17) and numerical simulation of a population of 1 000.

The discrepancies present in Fig. 16.3 are due to finite-size effects. In fact, the theory prediction will be completely wrong for small population sizes as shown in Fig. 16.4.

The techniques outlined above can be used to determine the relative importance of various genetic operators in the large P limit, but for small populations one must usually resort to numerical simulations. Some additional theoretical analyses of the convergence of evolutionary algorithms include: [Pr94, Qi94, Da91, Ei90, Fo92], and references contained therein.

16.2 Evolving Neural Networks

In recent years there has been considerable research into the application of genetic algorithms to neural networks. The ability of GAs to find global minima within high-dimensional spaces, coupled with their possible application in training binary and recurrent networks makes them especially appealing. Historically, the weights and/or topology of the network are encoded, and the

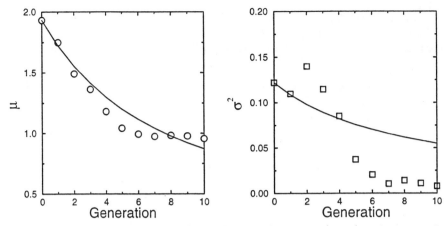

Fig. 16.4. Comparison between the prediction of equation (16.17) and numerical simulation. $P = 50, N = 1000, \beta = 0.5, P_m = 0.01$, and $P_c = 0.33$.

genetic operations included are selection, mutation, and crossover; however, it has been shown that the inclusion of other network parameters and genetic operators can enhance GA performance.

16.2.1 Feed-Forward Networks

In a recent paper by V. Maniezzo [Ma94a] both the weights and network topology are encoded in the chromosomes. In addition, the bit length of the weight encoding, or the coding *granularity*, is also allowed to evolve and a local optimization operator called the GA simplex [Be92] is applied. Increasing the bit length allows for increased precision in the specification of weights, e.g. an n bit weight encoding means that each weight may have 2^n distinct values. It has been shown that the proper choice of the granularity is at least as important as the network topology. If the granularity is too coarse a suitable solution may not exist, and if the granularity is too fine the search space becomes large, slowing down the learning process considerably. By allowing the coding granularity to evolve, the system can exploit coarse grained networks early on for rapid identification of promising weight configurations. In addition, once a coarse grained solution has been found, the ability to increase the coding length allows for more accurate solutions.

This idea was implemented in both a serial and local parallel implementation, indicated by SGA and PGA, respectively. In the parallel version each member of the population was assigned to a cell on a toroidal grid. The genetic operators were then applied on a local three-by-three or five-by-five grid, e.g. selection probabilities are calculated by comparing the fitness a member with its nearest or next nearest neigbors.

The performance of both the serial and parallel implementations was measured by comparing the performance of the GAs with standard error back-progation. Six problems were chosen from Rumelhart's test suite [Ru86b]:

1. XOR: 2 inputs, 1 output. Output should be the exclusive or of the inputs.
2. Parity: 4 inputs, 1 output. Output should be 1 if there is an odd number of 1s in the input pattern, and 0 otherwise.
3. Encoding: 8 inputs, 8 outputs, 3 hidden layers. Output should be equal to the input for any of the 8 combinations of seven 0s and one 1. Network topology is fixed.
4. Symmetry: 4 inputs, 1 output. Output should be 1 if the input is symmetric, 0 otherwise.
5. Addition: 4 inputs, 3 outputs. Output should be the sum of two 2-bit input numbers.
6. Negation: 4 inputs, 3 outputs. Output should be equal to the three rightmost input bits if the leftmost bit is 0, otherwise the output should be equal to their negation.

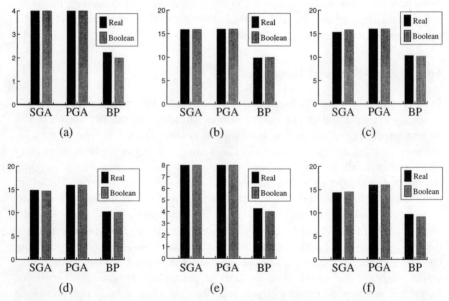

Fig. 16.5. Comparison of the fitness of the best member of the population for serial GA, parallel GA, and back-propagation on Rumelhart's test suite [Ru86b]: (a) the XOR problem, (b) the symmetry problem, (c) the parity problem, (d) the addition problem, (e) the encoding problem, and (f) the negation problem. The Boolean fitness is obtained by interpreting the output values as 0 or 1. Data taken from [Ma94a].

In Fig. 16.5 the fitness of the best member of the population for the SGA, PGA, and a standard back-propagation algorithm (BP) are compared. Each

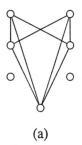 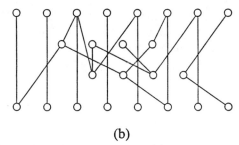

(a)   (b)

Fig. 16.6. Topology of one of the evolved nets for (a) the *modified* XOR problem and (b) the *modified* encoding problem. In (a) two neurons have been identified as useless and pruned from the network. In (b) evolution has discovered that the most efficient coding involves direct connections from inputs to output.

problem was tested ten times, both on a VAX (serial) and on a CM2 (parallel). Each run was allowed to evolve through 3000 generations. For comparison, the back-propagation algorithm was run on the fully connected version of each network and trained for an equal number of floating point operations. This is unfair to back-propagation since if coupled with an efficient pruning algorithm back-propagation might outperform the GAs. Clearly further benchmarking is needed. On the other hand, it is clear from this figure that the serial and parallel GAs with local optimization outperform standard back-propagation. In fact, in every test of all of the problems the parallel GA found the optimal solution within about 200 generations. The ability of the GAs to optimize the topology of the networks was also studied. Figure 16.6 shows one of the network configurations evolved for the *modified* XOR and encoding problems. Some of the connections illustrated were not allowed in the benchmark tests against back-propagation. In the *modified* problems layers are connected to *all* layers below them, not just the next layer down. It has been shown that improved performance can be had on most problems if connections directly from the input to the output are allowed. This is trivially true for the encoding problem, which evolution has discovered as well (see Fig 16.6b). For the XOR problem (Fig. 16.6a) we see that an unnecessary hidden layer was eliminated by evolution.

16.2.2 Recurrent Networks

Since there are currently no general algorithms for training recurrent neural networks, evolutionary algorithms appear a possible choice. There have been several frameworks proposed for the construction of recurrent networks [Be89, Pa88, An94] using evolutionary paradigms. A particularly nice example has been given by J.R. McDonnell and D. Waagen [Mc94]. They use a combination of single-agent stochastic search and evolutionary programming to train simple infinite impulse response (IIR) filters to predict time series. Figure 16.7a shows the architecture of a simple nonlinear IIR filter and Fig.

16.7b shows the architecture of a linear-nonlinear IIR filter. In their model, both the order and the tapped-delay line weights are optimized by a hybrid single-agent/multi-agent stochastic search.

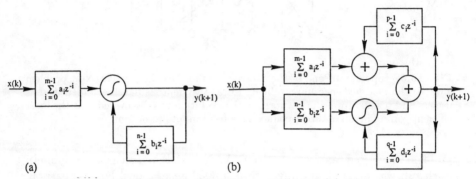

(a)   (b)

Fig. 16.7. (a) The nonlinear IIR filter architecture. (b) The parallel linear-nonlinear IIR filter architecture.

As a benchmark, McDonnell and Waagen compared their program with previous work by A. Weigend et al. [We90], Tong and Lim [To80], and C. Svarer et al. [Sv92] in the prediction of the sunspot series for the years 1700–1988. The data set was partitioned into a training set over years 1700–1920 and a test set for years 1921–1955 and 1956–1979. The second test set was created since the series during this time period is unlike the previous test set and training set [Sv92]. Weigend et al. used a 12-8-1 fully-connected feed-forward network which was reduced to a 12-3-1 network by weight elimination. The number of inputs was chosen to allow comparison with the autoregressive model of Tong and Lim. Svarer et al. used the Optimal Brain Damage Method to generate a partially connected 5-3-1 network.

Table 16.1 lists the normalized error of each model. The results from McDonnell are for an evolved transversal and recurrent network, respectively. A transversal network only includes a tapped-delay input signal allowing the network to be sensitive to temporal patterns, but without having fully recurrent units. Also shown are some more recent results of McDonnell and Waagen which are discussed below. Although the performance of the recurrent network is not better than traditional methods, these results indicate that with further research recurrent networks trained in this manner could be effective tools for time series prediction. Other applications of evolutionary algorithms in training recurrent neural networks are given in [Be89, Pa88, An94, Mc95a]. A review of traditional time-series methods can be found in M.B. Priestly's book [Pr88]. In addition, the Sante Fe Institute has published the results of a recent time series prediction contest [We94a].

McDonnell and Waagen have written a more recent paper [Mc95b] in which they studied an evolutionary strategy for evolving a cascade-correlation

Table 16.1. A comparison of normalized error on the prediction of the sunspot time series.

Model	Train	Test I	Test II	# of parameters
Tong and Lim	0.097	0.097	0.28	16
Weigend et al.	0.082	0.086	0.35	43
Svarer et al.	0.090	0.082	0.35	12-16
McDonnell 1	0.0987	0.0971	0.3724	14
McDonnell 2	0.1006	0.0972	0.4361	22
CCLA-EP (not pruned)	0.084	0.082	0.36	58
CCLA-EP (pruned)	0.094	0.083	0.25	25-27

networks for time series prediction. With the cascade-correlation algorithm [Fa90] networks are constructed by continually adding more hidden units until the desired accuracy is achieved. The new hidden units are selected from a population of candidates which have their weights and connectivities optimized by the application of genetic operators. The fitness of a new hidden unit is determined by measuring the correlation between its output and the residual output error of the network. The unit with the maximum correlation is considered the fittest. Table 16.1 lists the performance of McDonnell and Waagen evolutionary cascade-correlation model (CCLA-EP) in the prediction of the sun spot time series and Fig. 16.8 shows a picture of the time series and the model prediction.

 Use the program NEUROGEN to study the effects are various parameter choices on prediction of the logistic and sunspot time series.

16.3 Evolving Back-Propagation

Another interesting application of evolutionary algorithms comes from a paper by David Chalmers [Ch90] in which he explores the possibility of evolving the learning algorithm rather than the network itself. In this treatment a prescription for the changes in the weights of the network is encoded in the genome. As a constraint, only local information to that connection is used and the same function is applied for every connection. For a given connection, from input unit i to output unit j, the local information used is:

1. a_j : the activation of the input unit j,
2. o_i : the activation of the output unit i,
3. t_i : the training signal on output unit i,
4. w_{ij} : the current value of the connection.

The change in weights was chosen to be a *linear* function of the four dependent variables and their six pairwise products so that

$$\Delta w_{ij} = k_o(k_1 w_{ij} + k_2 a_j + k_3 o_i + k_4 t_i + k_5 w_{ij} a_j + k_5 w_{ij} a_j$$
$$+ k_6 w_{ij} o_i + k_7 w_{ij} t_i + k_8 a_j o_i + k_9 a_j t_i + k_{10} o_i t_i) . \qquad (16.18)$$

For example, the delta rule of standard back-propagation is given by

$$\Delta w_{ij} = 4(-2a_j o_i + 2a_j t_i)$$
$$= 8a_j(t_i - o_i) . \qquad (16.19)$$

Table 16.2. The best learning algorithms evolved in 10 runs.

k_0	k_1	k_2	k_3	k_4	k_5	k_6	k_7	k_8	k_9	k_{10}	Fitness
0.25	0	0	0	0	0	0	0	-4	4	0	89.6%
-2.00	0	0	0	0	0	0	0	2	-2	0	98.0%
0.25	0	-1	-2	4	0	0	0	-2	4	-2	94.3%
0.25	0	-1	-2	4	0	0	0	-2	4	-2	92.9%
-0.25	0	0	1	-1	0	1	-1	4	-4	0	89.8%
-1.00	0	0	-1	1	0	0	0	4	-4	0	97.6%
4.00	0	0	1	-1	0	0	0	-2	2	0	98.3%
-0.06	0	0	0	-2	-1	2	2	4	-4	2	79.2%
-0.25	0	0	2	-1	0	-1	-1	2	-4	0	89.8%
0.25	0	-1	-2	4	0	0	0	-2	4	-2	93.2%

Table 16.2 lists the best learning algorithms produced in 10 evolutionary runs. The fitness of a learning rule is determined by evaluating its performance with 10 training cycles on a set of given tasks. A fitness of 50% represents chance performance, or no learning, and a fitness of 100% represents perfect learning, i.e. after 10 cycles each task is learned perfectly. The second run generated the delta or Widrow-Hoff rule, but none of the runs generated anything superior to this. It seems unlikely that these evolutionary methods will provide an improvement to the learning rules already available for feed-forward multi-layer networks with supervised learning; however, for reinforcement learning, unsupervised learning, and recurrent networks these methods could provide important results.

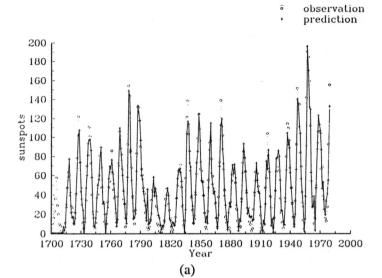

(a)

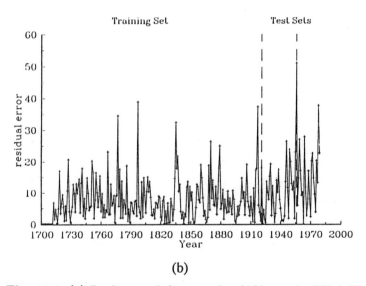

(b)

Fig. 16.8. (a) Prediction of the pruned, 4 hidden node CCLA-EP, (b) residual error.

Part II

Statistical Physics of Neural Networks

17. Statistical Physics and Spin Glasses

17.1 Elements of Statistical Mechanics

We have learned in the first part of this book that there is a complete analogy between neural networks having symmetric synaptic efficacies w_{ij} and a certain type of magnetic system characterized by a lattice of discrete spin variables $s_i = \pm 1$. Such systems are known as *Ising systems*. If the spin–spin interaction, i.e. the coupling coefficients w_{ij}, extend over large distances and take on irregular values one speaks of a *spin glass*. The peculiar properties of such spin glasses have caught the attention of physicists and have been studied closely during the last decade. In the following chapters we will make ample use of the results and the methods developed in the course of these investigations. Despite the close magnetic analogy, however, we will always keep in mind that we intend to describe neural networks.

In order to develop the formal theory of stochastic neural networks, we have to make heavy use of the language of statistical mechanics. A familiarity with this subject will therefore be of great advantage to the reader.[1] For those not actively familiar with it, but having some prior education in the subject, we shall briefly develop the basic tools in this section.[2]

As mentioned already in Sect. 3.3.2, in principle one has to distinguish between sequential dynamics (with the spins being updated one after the other according to the Hopfield prescription) and synchronous dynamics (where all spins are updated at the same time as postulated by Little). In the following we will study sequential dynamics, which is easier to analyze theoretically. A treatment of synchronous dynamics can be found in the literature, see [Am85b].

In a stochastic network the probability that a given neuron s_i changes its state is determined by

$$W(s_i \rightarrow -s_i) = \frac{1}{1 + e^{2\beta s_i h_i}} \ . \tag{17.1}$$

---

[1] Modern introductions to the methods of statistical physics are, e.g., those by Feynman [Fe72] and Reichl [Re80].

[2] We partly follow here the presentation of Hertz, Krogh, and Palmer [He91].

Here the full excitation potential h_i acting on the neuron number i is given as a linear combination of all the synaptic influences by the full set of other neurons in the net, i.e.

$$h_i = \sum_j w_{ij} s_j .\tag{17.2}$$

We will assume that the synaptic connections are symmetric, i.e. $w_{ij} = w_{ji}$, since this is a necessary condition for the existence of equilibrium.[3]

Now we will investigate situations where such a network is in equilibrium. By this we mean, of course, a dynamical equilibrium quite distinct from a completely static situation where the state of the network, i.e. the configuration of all the neurons $\{s_i\}$ does not change in time. This is immediately ruled out by the stochastic nature of the time evolution law (17.1) which, at finite temperature $T = 1/\beta$, will always induce fluctuations. Rather, the equilibrium is characterized by the property that the state of a neuron will not change on the average. Thus the probability that a neuron changes its state of activity (the "spin is flipped" in the language of the magnetic analogy) has to be independent of the state of the neuron. To state it once again: the probability of exciting a neuron that presently is inactive ($s_i = -1$) must be equal to the probability that the active neuron ($s_i = +1$) becomes deexcited. Under this condition the flow of activity between the active and inactive state is balanced and the value of s_i is constant on the average (*detailed balance*).

Denoting the probability distribution for a certain activity pattern of the net (in all there are 2^N such configurations) by $P(s_1...s_N)$ this condition for equilibrium takes the form

$$W(s_i \to -s_i)P(...s_i...) = W(-s_i \to s_i)P(...-s_i...) .\tag{17.3}$$

According to the updating rule (17.1) the ratio of transition probabilities is given by

$$\frac{W(-s_i \to s_i)}{W(s_i \to -s_i)} = \frac{1 + e^{2\beta s_i h_i}}{1 + e^{-2\beta s_i h_i}} = e^{2\beta s_i h_i} ,\tag{17.4}$$

which means that also the probability distribution for the network configuration in the equilibrium state obeys

$$\frac{P(s_1...s_i...s_N)}{P(s_1...-s_i...s_N)} = e^{2\beta s_i h_i} .\tag{17.5}$$

One immediately verifies that this condition is satisfied by the following expression for the distribution function

$$P(s_1...s_N) = Z^{-1} e^{\frac{1}{2}\beta \sum_i s_i h_i} = Z^{-1} e^{\frac{1}{2}\beta \sum_{ij} w_{ij} s_i s_j} .\tag{17.6}$$

The factor $\frac{1}{2}$ in the exponent is required to avoid double counting since each spin s_i enters twice in (17.6). Here it is essential that the interaction

---

[3] For a statistical treatment of asymmetric neural networks, see [Cl88].

coefficients w_{ij} are symmetric. Z^{-1} is a normalization factor; it is determined by the condition that the summed probability must be equal to unity

$$\sum_{\{s_i\}} P(s_1...s_N) = \sum_{\{s_i\}} Z^{-1} e^{\frac{1}{2}\beta \sum_{ij} w_{ij} s_i s_j} = 1 . \tag{17.7}$$

The summation extends over all the 2^N configurations of the whole network. Since this type of summation over all configurations in the future will arise quite often it is useful to introduce an abbreviation. Thus in the following we will denote the dependence on the complete configuration of all neurons $\{s_i\}$ by square brackets $[s]$, i.e. we will write $P[s]$ instead of $P(s_1...s_N)$ or $P(\{s_i\})$.

The function (17.6) is seen to be just a *Boltzmann distribution*

$$P[s] = Z^{-1} e^{-\beta E[s]} \tag{17.8}$$

if we introduce an "energy function" $E[s]$ of the network according to

$$E[s] = -\frac{1}{2} \sum_{ij} w_{ij} s_i s_j \tag{17.9}$$

and identify the parameter $T = \beta^{-1}$ as a "temperature". In the case of the magnetic spin system, $E[s]$ can indeed be identified with the magnetic interaction energy. This direct physical interpretation, of course, does not hold for neural networks, where $E[s]$ is an abstract quantity related to the information content of the system.

The fact that the function $P[s]$ is just given by a Boltzmann distribution opens the way to using all the techniques which are offered by statistical physics for the determination of equilibrium properties of thermodynamic systems. The normalization factor

$$Z \equiv e^{-\beta F} = \sum_{[s]} e^{-\beta E[s]} , \tag{17.10}$$

plays a central role in statistical physics. It is known as the canonical *partition function* and allows one to determine the average values of various physical quantities. The quantity F is known as the *Helmholtz free energy* of the system, $F = -T \ln Z$. For processes occurring at a fixed temperature $T = \beta^{-1}$ the free energy plays the role of a thermodynamic potential. If the Helmholtz free energy depends not just on temperature, but also on other variables, the equilibrium state corresponds to a minimum of the free energy. A quantity that is closely related to the free energy is the *entropy S*, which is defined as

$$S = -\partial F/\partial T . \tag{17.11}$$

Since Z usually grows with temperature (the sum on the r.h.s. of (17.10) increases with T) and is much larger than unity, the free energy F is negative and *decreases* with temperature. Hence the entropy is a positive definite function which vanishes only for $T = 0$.

The calculation of statistical averages is much simplified if one modifies the energy function by an auxiliary "source term" according to

$$E[s] = -\frac{1}{2} \sum_{ij} w_{ij} s_i s_j + \sum_i \vartheta_i s_i . \tag{17.12}$$

The new parameter ϑ_i can be interpreted as a threshold potential for the excitation of neuron number i. This means that the probability for the change of the state of the neuron is no longer given by (17.1) but rather by

$$W(s_i \to -s_i) = \frac{1}{1 + e^{2\beta s_i (h_i - \vartheta_i)}} . \tag{17.13}$$

The additional ϑ_i-dependent terms will be kept in all intermediate calculations and made to vanish only at the end. As an example of the use of this trick let us calculate the average value of the activity of neuron number i. We can make use of the fact that $\partial E / \partial \vartheta_i = s_i$ and determine the average value $\langle s_i \rangle$ by

$$\langle s_i \rangle \equiv \sum_{[s]} s_i P[s] = -\lim_{\vartheta_i \to 0} \frac{T}{Z} \sum_{[s]} \frac{\partial}{\partial \vartheta_i} e^{-\beta E[s]} = -\lim_{\vartheta_i \to 0} T \frac{\partial \ln Z}{\partial \vartheta_i} , \tag{17.14}$$

using the chain rule of differentiation and identifying $T = \beta^{-1}$. In just the same fashion the average energy can be calculated. Here the source term ϑ_i is not required and can be left out of the calculation.

$$\langle E \rangle = \sum_{[s]} E[s] P[s] = -\frac{\partial \ln Z}{\partial \beta} . \tag{17.15}$$

The expectation values of various other quantities of interest (e.g. spin correlations $\langle s_i s_j \rangle$) can also be calculated provided that the functional dependence of the partition function Z on these quantities is known.

17.2 Spin Glasses

17.2.1 Averages and Ergodicity

Spin glasses are magnetic systems in which the interactions between the magnetic moments are "in conflict" with each other, because of some frozen-in structural disorder [Bi86b, Ko89b]. As a result of these conflicting interactions no long-range order in the conventional sense can be established among the spins; nonetheless, a transition to a new kind of ordered phase is observed at low temperatures. Interest in spin glasses focused on the nature of this new type of phase transition, in particular on the question which order parameter describes the ordered state.

The structural disorder or randomness of spin glasses lies at the heart of a formal problem which greatly complicates the theoretical treatment of such materials. In order to define a specific sample of material, the interactions of the random magnetic moments should be completely specified, in principle. However, it goes against common sense (and against observational evidence) that the macroscopic properties of the material depend on these detailed assignments at the atomic level. In fact, these properties depend only on the *average* values of the distribution of magnetic-moment strengths. The question then arises: which quantities should be averaged?

The answer depends on the physical properties of the material. If the random interaction variables fluctuate quickly compared with the observation time, it is correct to take the average of the partition function Z and calculate the free energy as

$$\langle\!\langle F \rangle\!\rangle = -T \ln\langle\!\langle Z \rangle\!\rangle , \qquad (17.16)$$

where the double brackets $\langle\!\langle \cdots \rangle\!\rangle$ indicate an average over the random interaction parameters. This procedure is called an *annealed average*. However, this is not what occurs in spin glasses, because at low temperature the magnetic interactions of the atoms situated on a given lattice site vary on extremely long time scales. In this case one can only average extensive variables, such as the free energy, and one speaks of a *quenched average*:

$$\langle\!\langle F \rangle\!\rangle = -T \langle\!\langle \ln Z \rangle\!\rangle . \qquad (17.17)$$

This is always allowed, since a macroscopic sample can be divided into many smaller subsystems, which all have the same free energy per atom, although their detailed atomic structure will be quite different. One also says that the system is *"self-averaging"*. The calculation of the average expectation value of a physical quantity A then requires two separate averaging procedures: first over the Gibbs ensemble of all possible orientations of the atomic spins for a given set of magnetic interaction weights w_{ij}, which is usually indicated by brackets $\langle \cdots \rangle$, followed by the average over the interaction parameters w_{ij}: $\langle\!\langle [\langle A \rangle] \rangle\!\rangle$.

A second difficult problem in the theory of random magnetic materials (spin glasses) concerns the average over all possible configurations of the system that is implicit in the definition of the partition function (17.10). In the experiment one averages over a certain observation time and not over all possible microscopic configurations (here we mean the orientations of the spins, not the detailed atomic-interaction structure discussed above). If the two averages coincide, the system is called *ergodic*. In random materials one usually finds a very broad distribution of relaxation times, some of which are much longer than any reasonable observation time.[4] Such materials are – for all practical purposes – nonergodic, at least in certain aspects.

In spin glasses the breaking of ergodicity occurs because of the existence of a very large number of almost degenerate ground states, which are not related

---

[4] The best-known example of such materials is *window glass*, which is a "liquid" with virtually infinite viscosity.

to each other by any simple symmetry. In principle, one should only average over the fluctuation around one of these ground states. Since the ground states are degenerate in energy, one does not incur an error in averaging over all of them. Detailed studies have shown that this property extends to all observable macroscopic quantities [En84], but problems can occur if one calculates certain ("non-self-averaging") quantities, which are not directly measurable.

Another concept that has been found to be useful in attempts to understand the properties of spin glasses is *frustration*. Here one looks at a minimal ensemble of nearest-neighbor spins, a so-called elementary *plaquette*, and asks whether the spins can be oriented in such a way that all interactions among nearest neighbors contribute the smallest possible value to the total energy. If this is not the case, i.e. if at least one nearest-neighbor interaction ("bond") must remain repulsive, one says that the plaquette is *frustrated*. Since every bond is shared by two adjacent plaquettes, frustrated plaquettes always appear pairwise.

Fig. 17.1. A frustrated plaquette. *Solid lines* indicate attractive interactions between equally oriented spins; *dashed lines* indicate repulsive interactions. The bond on the right-hand side is frustrated.

If a magnetic system has many frustrated bonds, as spin glasses generally do, the configuration of frustrated plaquettes forms a complicated network. In general, there are many different ways of arranging the frustrated plaquettes, which do not differ in their total energy. Different solutions correspond to different ground states of the spin glass. This observation provides a natural explanation for the multitude of nonequivalent ground states in a spin glass [To79]. It also explains why spin-glass models capture many features of combinatorial optimization problems and have led to new ideas for obtaining solutions of such problems ("simulated annealing").

17.2.2 The Edwards–Anderson Model

Even with the most powerful methods of statistical mechanics it is difficult to study realistic models of random magnetic materials, except by numerical simulation. These usually provide numbers that can be compared with experimental data, but relatively little insight. Edwards and Anderson [Ed75] have therefore proposed a simplified model of a spin glass which is accessible to treatment by analytic methods. In this model the magnetic moments i are located on a regular lattice and their interaction strengths w_{ij} are random variables determined by a suitable distribution $P(w_{ij})$. The standard choice is a Gaussian distribution:

$$P(w_{ij}) = [2\pi(\Delta w_{ij})^2]^{-1/2} \exp\left[-\frac{(w_{ij} - \bar{w}_{ij})^2}{2(\Delta w_{ij})^2}\right].$$ (17.18)

The average in (17.17) is then defined as

$$\langle\!\langle \ln Z \rangle\!\rangle = \prod_{i,j} \int dw_{ij} \, P(w_{ij}) \ln Z[w].$$ (17.19)

A further simplification arises if one assumes that the distribution is the same for all pairs (ij) $(\bar{w}_{ij} = \bar{w})$, independent of the distance between atoms i and j (the infinite-range model of Sherrington and Kirkpatrick [Sh75]). For such a model, mean-field theory ought to be exact, but even this extreme simplification (compared with the realistic case of a distance-dependent interaction) has been a major challenge of theoretical physics.

The interesting order parameter is the mean square magnetization at each lattice site: $q = \langle\!\langle [\langle s_i \rangle^2] \rangle\!\rangle$. When q becomes nonzero, this signals a phase transition. However, the existence of many different ground-state configurations ("valleys") separated by large energy barriers causes a problem here, because an experiment or a computer simulation running only over a short time will not sample the whole Gibbs ensemble[5] implied by the configurational average $\langle s_i \rangle$. Counting the partial ensembles corresponding to configurations in a particular valley by an index ℓ, the order parameter q takes the form

$$q = \left\langle\!\!\left\langle \left(\sum_\ell P_\ell \langle s_i \rangle_\ell\right)^2 \right\rangle\!\!\right\rangle,$$ (17.20)

where P_ℓ denotes the statistical weighting of the different ground states. Clearly, the square in (17.20) implies that terms involving two *different* valleys occur in the definition of q that are not measured in a short-time experiment. Edwards and Anderson have therefore proposed considering an order parameter q_{EA}, which is defined as the average over the diagonal contributions in (17.20):

$$q_{EA} = \left\langle\!\!\left\langle \sum_\ell P_\ell (\langle s_i \rangle_\ell)^2 \right\rangle\!\!\right\rangle.$$ (17.21)

The difference between the two definitions becomes most apparent when q and q_{EA} are defined in terms of the dynamical evolution of the system through a time average of the quantity $q(t) = \langle\!\langle s_i(0) s_i(t) \rangle\!\rangle$. The energy barriers between valleys diverge in the thermodynamic limit (number of spins $N \to \infty$), so that no communication between different valleys is possible in this limit (the bar indicates the time average):

$$q = \lim_{N\to\infty} \lim_{t\to\infty} \overline{q(t)}, \qquad q_{EA} = \lim_{t\to\infty} \lim_{N\to\infty} \overline{q(t)}.$$ (17.22)

---

[5] Experiments on the magnetic properties of spin glass materials have shown relaxation times of 10^5 seconds and more to reach the equilibrium state[Ko89b].

For the Sherrington–Kirkpatrick (SK) model one can show that all contributions to q_{EA} from different valleys are the same, so that it is sufficient to calculate $(\langle s_i \rangle_\ell)^2$ for a single valley. In the formal treatment of the SK model given later (in Chapt. 19) the multitude of ground states separated by infinite-energy barriers has the consequence that different copies or "replicas"[6] of the same system do not need to have the same thermodynamic properties. Denoting the various replicas by a latin index a or b, the order parameter q becomes a matrix q_{ab} with vanishing diagonal elements. When $q_{ab} = q$ for all $a \neq b$, one has the so-called *replica-symmetric* solution corresponding to $q = q_{EA}$. Otherwise one speaks of *replica-symmetry breaking*, and q corresponds to the average over all off-diagonal elements, whereas the Edwards–Anderson definition is obtained in the limit $q_{EA} = \lim_{a \to b} q_{ab}$.

Numerical solutions of the SK model of spin glasses, as well as the analytic treatment in the framework of mean-field theory with replica-symmetry breaking, have shown that it exhibits four different phases depending on the values of the parameters $T/\Delta w$ and $\bar{w}/\Delta w$ (see Fig. 17.2). Two of these phases are relatively standard states of magnetic systems: the paramagnetic phase (P) corresponds to complete disalignment among spins, whereas the ferromagnetic phase (F) has all spins aligned along the preferred direction defined by the sign of the interactions of a spin with its nearest neighbors.[7] These two phases are replica symmetric, i.e. $q = q_{EA}$. The third phase (SG) with replica symmetry breaking, which occurs for small values of the parameters $T/\Delta w$ and $\bar{w}/\Delta w$, is peculiar to spin glasses and has no analog in magnetic materials with regular interactions. Finally, a small region of the phase diagram is occupied by a phase (F+SG), where the ground state is ferromagnetic although replica symmetry is broken.

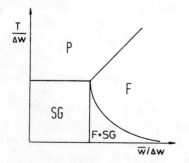

Fig. 17.2. Phase diagram of the Sherrington–Kirkpatrick model of spin glasses. (P) and (F) denote the paramagnetic and ferromagnetic phases, respectively, while (SG) denotes the region of genuine spin–glass behavior characterized by an infinity of ground states. (F+SG) denotes that region of the ferromagnetic phase, where replica symmetry is broken.

---

[6] The replicas are introduced to facilitate the calculation of the quenched average of $\ln Z$.

[7] The ferromagnetic phase is the one most interesting for analogous neural network models of associative memory, since its ground state corresponds to a memorized pattern.

According to Parisi [Pa80, Me87] the true spin-glass phase[8] is characterized by an infinity of ground states ℓ with nonvanishing off-diagonal elements of the *overlap* matrix

$$q_{\ell\ell'} = \lim_{N\to\infty} \frac{1}{N} \sum_{i=1}^{N} \langle s_i \rangle_\ell \langle s_i \rangle_{\ell'} \ . \tag{17.23}$$

Because there are infinitely many such matrix elements, it is useful to define their distribution, which is usually denoted by $P(q)$ with $0 < q < 1$. In the Parisi ansatz the free energy of the spin glass becomes a functional of $P(q)$, and the function $P(q)$ can be calculated by looking for the maximum of the free energy.

The property that the overlap of different valleys does not vanish implies that there is a certain similarity among them, the degree of likeness being measured by the magnitude of $q_{\ell\ell'}$. In order to discuss the relation between the various states it is better to define a *distance* that measures how much they differ from each other:

$$d(\ell, \ell') = \lim_{N\to\infty} \frac{1}{N} \sum_{i=1}^{N} \Big(\langle s_i \rangle_\ell - \langle s_i \rangle_{\ell'} \Big)^2 \ . \tag{17.24}$$

Of course, $q_{\ell\ell'}$ and $d(\ell, \ell')$ are closely related; in fact, $d(\ell, \ell') = 2(q_{\text{EA}} - q_{\ell\ell'})$. One finds, however, that this distance measure has very peculiar properties. E.g., for three states the distance between two of them never exceeds the larger of the distances to the third state:

$$d(\ell_1, \ell_2) \le \max\big[d(\ell_1, \ell_3), d(\ell_2, \ell_3) \big] \qquad \text{for all } \ell_3 \ . \tag{17.25}$$

This relation is known as the *ultrametricity* inequality, in analogy to the triangle inequality satisfied by all distance measures. As is well known to mathematicians, it implies that the set of ground states can be divided into clusters, each of which can be further subdivided into subclusters, and so on. Measured by the distance $d(\ell, \ell')$ the space of ground states of a spin glass assumes the same structure as a family tree. This immediately provides an explanation for the wide range of relaxation times found in spin glasses: although all the barriers between the different ground-state valleys are very high (they are of finite height for a finite sample), some are higher than others, and it takes longer to cross over from one valley into the next.

Whether these and other properties are peculiar aspects of the SK model or whether they are generic features of spin glasses is still a matter of debate. Other, essentially different models have been proposed which do not exhibit some of the features discussed above (see [Bi86b]). It is also a valid question how relevant these aspects are to neural network models, in particular to

---

[8] Actually, this region of the phase diagram is characterized by the *coexistence of an infinite multitude of phases* corresponding to the various ground states, and one faces the unfamiliar task of doing a statistical mechanics of phases!

such models that try to capture realistic features of the brain. The ultrametric structure of the spin-glass ground states has been vaguely associated with our brain's ability to classify objects and events according to their resemblance, but the insight into this question is still incomplete.

Note: A vivid first introduction, also readable by nonexperts, into the physics of spin glasses can be found in a series of "Reference Frame" columns by Philip W. Anderson, which appeared in Physics Today (issues 1,3,5,9/1988, 7,9/1989).

18. The Hopfield Network for $p/N \to 0$

In this chapter we will analyze the statistical properties of a Hopfield network which has been trained with p patterns σ_i^μ using Hebb's learning rule. The analysis will be valid in the *thermodynamic limit* of a system of infinite size, $N \to \infty$. However, the number of patterns p at first will be kept fixed so that the number of stored patterns per neuron, $\alpha = p/N$, becomes vanishingly small. The more complicated case of finite pattern loading, $\alpha =$ constant, will be treated in Chapt. 19.

18.1 Evaluation of the Partition Function

In Sect. 3.3 we have shown that synaptic efficacies defined according to Hebb's rule

$$w_{ij} = \frac{1}{N} \sum_{\mu=1}^{p} \sigma_i^\mu \sigma_j^\mu \qquad (i \neq j) \tag{18.1}$$

lead to the energy function

$$E[s] = -\frac{1}{2N} \sum_{\mu=1}^{p} \left(\sum_{i=1}^{N} \sigma_i^\mu s_i \right)^2 + \frac{p}{2}, \tag{18.2}$$

where the constant term serves to cancel the diagonal term $(i = j)$ in the sum. This means that we have to evaluate the partition function

$$Z = e^{-\frac{1}{2}\beta p} \sum_{[s]} \exp \left[\frac{\beta}{2N} \sum_\mu \left(\sum_i \sigma_i^\mu s_i \right)^2 \right]. \tag{18.3}$$

Now it would be an easy task to evaluate the sum over all configurations

$$\sum_{[s]} \equiv \sum_{s_1=\pm 1} \cdots \sum_{s_N=\pm 1} \tag{18.4}$$

if the terms to be summed could be split into factors which each depend on only one of the variables s_i. This is not true in our case, however, since the spins s_i enter the exponent of (18.3) quadratically in such a way that there are mixing terms. In this situation it turns out that a formal artifice is helpful which is often used in statistical physics. We introduce additional *auxiliary variables* m^μ for each of the stored patterns and rewrite the term in (18.3) as an integral over m^μ in such a way that the variables s_i enter the exponent in a linear fashion. To achieve this we make use of the *Gaussian integral* formula

$$\int_{-\infty}^{\infty} dz \, \exp(-az^2 + bz) = \left(\frac{\pi}{a}\right)^{1/2} \exp\left(\frac{b^2}{4a}\right) , \tag{18.5}$$

which will be used in (18.3) to replace the exponential function on the right-hand side by the Gaussian integral.

Setting $a = \frac{1}{2}\beta N$ and $b = \beta \sum_i \sigma_i^\mu s_i$ the following transformation applies for each of the stored patterns μ

$$\exp\left[\frac{\beta}{2N}\left(\sum_i \sigma_i^\mu s_i\right)^2\right]$$

$$= \left(\frac{\beta N}{2\pi}\right)^{1/2} \int dm^\mu \exp\left[-\frac{1}{2}\beta N(m^\mu)^2 + \beta m^\mu \sum_i \sigma_i^\mu s_i\right] . \tag{18.6}$$

Each term in the sum of (18.3) contains N such integral factors. To avoid undue complexity of the formulas, at this stage it is useful to introduce a vector notation for the index μ. That is, we introduce the p-component vectors $\mathbf{m} = \{m^\mu\}$ und $\boldsymbol{\sigma}_i = \{\sigma_i^\mu\}$ to designate the set of patterns. Then the partition function Z can be written in the following manner:

$$Z = \left(\frac{\beta N}{2\pi}\right)^{p/2} e^{-\frac{1}{2}\beta p} \int d\mathbf{m} \, e^{-\frac{1}{2}\beta N \mathbf{m}^2} \prod_i \left(\sum_{s_i = \pm 1} e^{\beta \mathbf{m} \cdot \boldsymbol{\sigma}_i s_i}\right) . \tag{18.7}$$

We have interchanged the sum over the states of the network and the integrations over the auxiliary variables. Since the Gaussian integrals are absolutely convergent this is justified.

Now the summation over each of the s_i is really trivial since it contains just two terms:

$$\sum_{s_i = \pm 1} e^{\beta \mathbf{m} \cdot \boldsymbol{\sigma}_i s_i} = e^{\beta \mathbf{m} \cdot \boldsymbol{\sigma}_i} + e^{-\beta \mathbf{m} \cdot \boldsymbol{\sigma}_i} = 2 \cosh(\beta \mathbf{m} \cdot \boldsymbol{\sigma}_i) . \tag{18.8}$$

The partition function thus takes the form

$$Z = e^{-\beta F} = \left(\frac{\beta N}{2\pi}\right)^{p/2} \int d\mathbf{m} \, e^{-\beta N f(\mathbf{m})} , \tag{18.9}$$

with the function

$$f(\mathbf{m}) = \frac{\mathbf{m}^2}{2} - \frac{1}{\beta N} \sum_i \ln\left[2\cosh(\beta \mathbf{m} \cdot \boldsymbol{\sigma}_i)\right] + \frac{p}{2N} , \tag{18.10}$$

which loosely speaking can be interpreted as the free energy per neuron.

A comparison of (18.3) and (18.9) reveals that the Gaussian-integration trick has led to the replacement of an N-fold summation by a p-fold integration. This might seem to be no great achievement (although indeed the number of operations is greatly reduced in the limiting case $p \ll N$ under discussion here). The exponent in the integrand, i.e. the function $f(\mathbf{m})$, is quite a complex expression. However, (18.9) differs in a small but important detail from the original formulation: the number of neurons N here enters the exponent in the numerator, in contrast to (18.3) where it stood in the denominator. Since the limit of large N is to be studied, this has the effect that the integrand is a function rapidly falling off with increasing $\mathbf{m}$. Appreciable contributions to the integral arise only from a small region in the vicinity of the minimum of the function $f(\mathbf{m})$. Thus a good approximation to the integral will be obtained by expanding the function $f(\mathbf{m})$ into a Taylor series close to its minimum and keeping only the lowest-order terms in the expansion.

This approximation method is known under the name *method of steepest descent* or *saddle-point approximation*. To gain experience with the method we first consider a one-dimensional integral of the form

$$I = \sqrt{N} \int_{-\infty}^{\infty} dz\, e^{-Ng(z)} \,. \tag{18.11}$$

Let us assume that the function $g(z)$ has its minimum at the point z_0. The Taylor-series expansion around this point reads

$$g(z) = g(z_0) + \frac{1}{2}g''(z_0)(z - z_0)^2 + \cdots \,, \tag{18.12}$$

with $g''(z_0) > 0$. If we truncate this expansion after the quadratic term the integral I is of Gaussian type and can be immediately solved using (18.5):

$$I \simeq \left(\frac{\pi}{g''(z_0)}\right)^{1/2} e^{-Ng(z_0)} \,. \tag{18.13}$$

The value of the integral is thus essentially determined by the minimum value of the function $g(z)$, which enters the exponent multiplied by the large coefficient N. The detailed dependence of the function, which here is characterized by the curvature $g''(z_0)$, only enters the preexponential factor, which is of order one. The same would hold true if we had included higher terms of the Taylor expansion.

To be able to apply this method to the integral over $\mathbf{m}$ in (18.9) we have to search for the minimum with respect to all of the p variables. Demanding that the gradient of $f(\mathbf{m})$ vanishes leads to a set of p coupled nonlinear equations.

$$\frac{\partial f}{\partial m^\mu} = m^\mu - \frac{1}{N} \sum_i \sigma_i^\mu \tanh(\beta \mathbf{m} \cdot \boldsymbol{\sigma}_i) = 0 \,. \tag{18.14}$$

In vector notation this equation for the saddle point reads

$$\mathbf{m} = \frac{1}{N} \sum_i \boldsymbol{\sigma}_i \tanh(\beta \mathbf{m} \cdot \boldsymbol{\sigma}_i) \,. \tag{18.15}$$

We can get an idea of the meaning of this saddle point by calculating the average activity of neuron number i. How this can be done was discussed at the end of Sect. 17.1. A fictitious threshold potential ϑ_i is introduced which we define to be the scalar product of the spin vector $\boldsymbol{\sigma}_i$ with a p-component constant vector $\mathbf{h}$. The energy function thus takes the form

$$E'[s] = E[s] - \sum_i (\mathbf{h} \cdot \boldsymbol{\sigma}_i) s_i \,. \tag{18.16}$$

Since the additional term is linear with respect to the variables s_i it poses no problem for the evaluation of the partition function. Repeating the steps that have led to (18.9), (18.10) using the modified energy function leads to the following result:

$$f(\mathbf{m}, \mathbf{h}) = \frac{\mathbf{m}^2}{2} - \frac{1}{\beta N} \sum_i \ln\{2 \cosh[\beta(\mathbf{m} + \mathbf{h}) \cdot \boldsymbol{\sigma}_i]\} \,. \tag{18.17}$$

The equation which determines the saddle point, i.e. (18.15), is not modified since at the end of the calculation $\mathbf{h} = 0$ will be taken.

The average activity of neuron number i is obtained by differentiating the partition function, cf. (17.14),

$$\sum_i \boldsymbol{\sigma}_i \langle s_i \rangle = \lim_{\mathbf{h} \to 0} T \frac{\partial}{\partial \mathbf{h}} \ln Z \,. \tag{18.18}$$

Here the angular brackets denote the average over the thermal distribution $P[s]$. The derivative of the function $\ln Z$ at the saddle point follows from (18.9)

$$\lim_{\mathbf{h} \to 0} T \frac{\partial}{\partial \mathbf{h}} \ln Z = -\lim_{\mathbf{h} \to 0} N \frac{\partial}{\partial \mathbf{h}} f(\mathbf{m}, \mathbf{h}) = \sum_i \boldsymbol{\sigma}_i \tanh(\beta \mathbf{m} \cdot \boldsymbol{\sigma}_i) = N\mathbf{m}. \tag{18.19}$$

Thus we have found that the saddle point is given just by the normalized overlap of a pattern and the corresponding average equilibrium state of the network

$$\mathbf{m} = \frac{1}{N} \sum_i \boldsymbol{\sigma}_i \langle s_i \rangle \,. \tag{18.20}$$

Comparison of this result with (18.15) shows that the thermal average of the activity of neuron number i is related to the saddle point by the nonlinear equation

$$\langle s_i \rangle = \tanh(\beta \mathbf{m} \cdot \boldsymbol{\sigma}_i) \,. \tag{18.21}$$

For a given finite value N of the size of the system the solution of (18.15), which determines the saddle point, depends on the choice of the stored patterns $\{\sigma_i^\mu\}$. Going to the limit $N \to \infty$, keeping the number of patterns p

fixed, however, it can be shown [He82, Pr83] that the average over all neurons i can be replaced by an average over the distribution of the variables σ_i^μ, keeping μ fixed, i.e. for a fixed pattern. This property leads to a great simplification since the solution is independent of the special properties of the stored patterns. This is made plausible by the following argument. In the limit $N \to \infty$ only the frequency of occurrence of the values $\sigma_i^\mu = \pm 1$ matters and this determined by the probability distribution $P(\sigma)$. The particular distribution of these values on the neuron net becomes irrelevant. The average over the statistical distribution of the values ± 1 of the pattern variable $\boldsymbol{\sigma}_i$ is analogous to the averaging over synaptic weights w_{ij}, and will be denoted by double angular brackets $\langle\!\langle \cdots \rangle\!\rangle$ in contrast to the thermodynamic average $\langle \cdots \rangle$. This is the concept of *self-averaging* introduced in Sect. 17.2.

In this way we arrive at the following expression for the saddle point which is valid in the limit $p/N \to 0$:

$$f(\mathbf{m}) = \frac{\mathbf{m}^2}{2} - \frac{1}{\beta} \langle\!\langle \ln\left[2\cosh(\beta\mathbf{m}\cdot\boldsymbol{\sigma})\right] \rangle\!\rangle , \tag{18.22}$$

$$\mathbf{m} = \langle\!\langle \boldsymbol{\sigma}\tanh(\beta\mathbf{m}\cdot\boldsymbol{\sigma}) \rangle\!\rangle . \tag{18.23}$$

Applying the same procedure to the average over the neurons (18.20) can be written in the form

$$\mathbf{m} = \langle\!\langle \boldsymbol{\sigma}_i\langle s_i\rangle \rangle\!\rangle . \tag{18.24}$$

18.2 Equilibrium States of the Network

If we are interested in the detailed properties of the network we have to decide on a specific choice of the values $\sigma_i^\mu = \pm 1$ of the stored patterns. In this section we will study the simplest case of *unbiased patterns* assuming that active and inactive neurons in all stored patterns occur with equal probabilities, i.e. for all values of μ the probability is given by

$$Pr(\sigma^\mu) = \frac{1}{2}\delta_{\sigma^\mu,1} + \frac{1}{2}\delta_{\sigma^\mu,-1} . \tag{18.25}$$

As both signs of σ are equally probable we can conclude immediately that the average value of any odd power of a component of the vector $\boldsymbol{\sigma}$ will vanish, while even powers lead to the result 1:

$$\langle\!\langle (\sigma_i^\mu)^n \rangle\!\rangle = \frac{1 + (-1)^n}{2} . \tag{18.26}$$

Furthermore, the probability distributions associated with different patterns μ and ν will be assumed to be statistically independent, i.e. the correlation function should satisfy

$$\langle\!\langle \sigma^\mu\sigma^\nu \rangle\!\rangle = \delta_{\mu\nu} . \tag{18.27}$$

To evaluate (18.22) we expand the transcendental function with respect to powers of **m**, or equivalently powers of $\boldsymbol{\sigma}$. To the second order we obtain

$$f(\mathbf{m}) = \frac{\mathbf{m}^2}{2} - T\Big\langle\!\Big\langle \ln 2 + \frac{1}{2}(\beta\mathbf{m}\cdot\boldsymbol{\sigma})^2 + \cdots]\Big\rangle\!\Big\rangle$$

$$= -T\ln 2 + \frac{1}{2}(1-\beta)\mathbf{m}^2 + \cdots . \qquad (18.28)$$

From this equation it is obvious that the minimum of the function $f(\mathbf{m})$ is reached at the value $\mathbf{m} = 0$, provided that $\beta < 1$. This means that at high temperatures $(T > 1)$ the network has only the trivial equilibrium state $\langle s_i \rangle = 0$. No recallable information can be stored under this condition.

At lower temperatures, $T < 1$, (18.23) possesses nontrivial solutions. The simplest class of such solutions can be characterized by vectors **m** which contain n equal nonvanishing components. As the index μ just counts the patterns which can be assigned an arbitrary ordering, we may assume without loss of generality that the first n components of **m** are are those with finite value. Then the symmetrical solutions have the form

$$\mathbf{m} = \frac{m_n}{\sqrt{n}}(\underbrace{1,\ldots,1}_{n},\underbrace{0,\ldots,0}_{p-n}) . \qquad (18.29)$$

In the simplest case, $n = 1$, the $\mu = 1$ component of (18.23) reads

$$m = \big\langle\!\big\langle \sigma^1 \tanh(\beta m \sigma^1) \big\rangle\!\big\rangle = \tanh(\beta m) , \qquad (18.30)$$

which we have already encountered in Sect. 4.2. There are two ways of arriving at this result. On the one hand, it follows immediately if we use (18.26), since the Taylor expansion of the function $x\tanh x$ consists only of even powers of x. Alternatively, one can explicitly perform the averaging over the probability distribution (18.25):

$$\big\langle\!\big\langle \sigma^1 \tanh(\beta m \sigma^1) \big\rangle\!\big\rangle$$
$$= \frac{1}{2}\big[(+1)\tanh(+\beta m) + (-1)\tanh(-\beta m)\big] = \tanh(\beta m) . \qquad (18.31)$$

According to the discussion in Sect. 4.2 (18.30) has two nonvanishing solutions provided that $\beta > 1$, i.e. $T < 1$. These solutions are identical, except for the sign, and rapidly approach the value ± 1 as the temperature approaches zero, $T \to 0$. According to (18.21) the resulting equilibrium state of the network is characterized by the average values of neural activation

$$\langle s_i \rangle = \sigma_i^1 \tanh(\beta\mathbf{m}) = m\sigma_i^1 . \qquad (18.32)$$

Here we have made use of the fact that $\sigma_i^1 = \pm 1$. In this way we have constructed the simplest equilibrium states of the network, which are just proportional to one of the stored patterns. They have been given the name *Mattis states*. Since there are p independent patterns and since m can take on two signs, in all there are $2p$ different states of this type.

In the more general case $n \geq 1$ the first n components of (18.23) have to be summed, leading to the condition

$$m_n = \langle\!\langle \, \zeta_n \tanh(\beta m_n \zeta_n) \, \rangle\!\rangle \,, \tag{18.33}$$

where

$$\zeta_n = \frac{1}{\sqrt{n}} \sum_{\mu=1}^{n} \sigma^\mu \,. \tag{18.34}$$

According to (18.22) the corresponding value of the free energy per neuron is given by

$$f_n = \frac{1}{2} m_n^2 - T \langle\!\langle \, \ln[2\cosh(\beta m_n \zeta_n)] \, \rangle\!\rangle \,. \tag{18.35}$$

It can be shown [Am85b] that all values of f_n are higher than f_1. Equations (18.33–18.35) are easily evaluated in the zero-temperature limit, leading to $m_n = \langle\!\langle \, |\zeta_n| \, \rangle\!\rangle$ and $f_n = -\frac{1}{2} m_n^2$. An explicit evaluation yields $f_1 = -\frac{1}{2}$, $f_2 = -\frac{1}{4}$, $f_3 = -\frac{3}{8}$, etc. It is no coincidence that the free energy of the state $n = 2$ lies above that of the state $n = 3$. It turns out that values f_n obey the following ordering :

$$f_1 < f_3 < f_5 < \cdots < f_6 < f_4 < f_2 \,. \tag{18.36}$$

Now we have to address the question whether the states for $n > 1$ are stable with respect to small fluctuations. It can be shown in general that the states for even values of n correspond to an unstable equilibrium [Am85b]. Thus the case of greatest interest, which we will study in some detail, corresponds to $n = 3$. A state is thermodynamically stable if it satisfies the following criterion: the saddle point determined by (18.14) has to be a *minimum* of the function $f(\mathbf{m})$. In mathematical terms this means that the stability matrix of second-order partial derivatives

$$A_{\mu\nu} = \frac{\partial^2 f}{\partial m^\mu \partial m^\nu} = \delta_{\mu\nu} - \beta \left\langle\!\!\left\langle \frac{\sigma^\mu \sigma^\nu}{\cosh^2(\beta \mathbf{m} \cdot \boldsymbol{\sigma})} \right\rangle\!\!\right\rangle = \delta_{\mu\nu} - \beta(\delta_{\mu\nu} - Q_{\mu\nu}) \tag{18.37}$$

with

$$Q_{\mu\nu} = \langle\!\langle \, \sigma^\mu \sigma^\nu \tanh^2(\beta \mathbf{m} \cdot \boldsymbol{\sigma}) \, \rangle\!\rangle \tag{18.38}$$

has only positive eigenvalues. In the special case of symmetric states of type (18.29) the $(n \times n)$ matrix A has a very simple form. All its diagonal elements are equal to $1 - \beta(1 - q)$ while all off-diagonal elements are given by βQ. We have introduced the abbreviations

$$\begin{aligned} q &= \langle\!\langle \, \tanh^2(\beta m_n \zeta_n) \, \rangle\!\rangle \,, \\ Q &= \langle\!\langle \, \sigma^1 \sigma^2 \tanh^2(\beta m_n \zeta_n) \, \rangle\!\rangle \,. \end{aligned} \tag{18.39}$$

The critical eigenvalue of this matrix A is $(n-1)$-fold degenerate; it has the value

$$a_c = 1 - \beta(1 - q) - \beta Q . \tag{18.40}$$

If the order n is odd we have in the zero-temperature limit $q \to 1$, $Q \to 0$, and thus $a_c(T = 0) = 1$. The states with odd n are stable. With increasing temperature the stability is reduced until finally only the Mattis states ($n = 1$) remain stable.

To give an example let us calculate the critical temperature T_3 related to the states with $n = 3$ defined as the point where the eigenvalue a_c changes its sign. Requiring $a_c = 0$ and multiplying (18.40) with T we get

$$T_n = 1 - (q - Q) = 1 - \langle\!\langle (1 - \sigma^1 \sigma^2) \tanh^2(\beta m_n \zeta_n) \rangle\!\rangle . \tag{18.41}$$

Since $n = 3$ there are eight possible spin combinations $\sigma^1, \sigma^2, \sigma^3 = \pm 1$. It is an easy task to perform the averaging over these combinations explicitly, which leads to

$$T_3 = 1 - \tanh^2\left(\frac{1}{\sqrt{3}}\beta_3 m_3\right) . \tag{18.42}$$

Since the value of m_3 is not yet known we need additional information. For this we can make use of the equation determining the saddle point, (18.33). Averaging over the eight possible spin combinations this reads

$$m_3 = \frac{\sqrt{3}}{4}\left[\tanh\left(\sqrt{3}\beta_3 m_3\right) + \tanh\left(\frac{1}{\sqrt{3}}\beta_3 m_3\right)\right] . \tag{18.43}$$

To write these equations in a more compact form we divide (18.43) by $\beta_3 m_3$ and introduce the abbreviation $x = \frac{1}{\sqrt{3}}\beta_3 m_3$. Thus we end up with the pair of equations

$$\begin{aligned}
T_3 &= 1 - \tanh^2 x, \\
T_3 &= \frac{1}{4x}(\tanh 3x + \tanh x) .
\end{aligned} \tag{18.44}$$

The transcendental equation obtained by equating the two right-hand sides has to be solved numerically, yielding the solution $x = 0.9395$. The result for the critical temperature is $T_3 = 0.4598$. The values for the next higher symmetric states are $T_5 = 0.39$, $T_7 = 0.35$, etc. Above the temperature T_3 all states with $n \geq 3$ are unstable.

In conclusion, we have found that the network possesses many symmetric spurious attractors of the type defined in (16.29). From combinatorial considerations the number of these states is found to be about $\frac{1}{2}3^p$. However, at sufficiently high operating temperature $T_3 < T < 1$ these spurious states are destabilized and only the retrieval states having finite overlap with one of the stored patterns survive. This remains true even if one goes beyond the symmetric ansatz (16.29) and allows for more complicated mixed spurious states [Am85b].

19. The Hopfield Network for Finite p/N

19.1 The Replica Trick

Up to now we have studied the case $N \to \infty$, keeping the number of stored patterns fixed. This is a rather academic problem since for practical purposes one is interested in networks with a finite pattern loading; indeed one would like to have a memory with a capacity as large as possible. Thus we endeavor to attack the more relevant problem of a network in the thermodynamic limit $N \to \infty$, keeping the fraction of stored patterns per neuron at a finite value $\alpha = p/N$. This is a hard problem for which the method introduced in the last section cannot be applied, because the number of patterns and thus the number of auxiliary integrations increases without bound. In particular the averaging procedure over the distribution of stored patterns σ_i^μ requires much greater care [Am85a, Am87a].[1]

Our investigation of the storage properties will be based on the fact that the state of the network in equilibrium can be characterized by the free energy F. In the thermodynamic limit it will be useful to study the free energy per element (i.e. per neuron, spin, etc.) $f = F/N$. We will not be interested in properties which depend on the special choice of the stored pattern. Instead statements will be derived which are valid on the average if a set of patterns is selected arbitrarily out of a huge *ensemble of patterns*. The process of averaging over this ensemble again will be denoted by the use of double angular brackets $\langle\!\langle \cdots \rangle\!\rangle$. The mean free energy per element is given by

$$f = -N^{-1}T \langle\!\langle \ln Z \rangle\!\rangle = -\frac{1}{\beta N} \left\langle\!\!\!\left\langle \ln \sum_{[s]} e^{-\beta E[s]} \right\rangle\!\!\!\right\rangle . \tag{19.1}$$

We remind the reader that the energy function $E[s]$ (17.9) depends on the stored patterns since the coupling parameters (synaptic efficacies) w_{ij} are determined from these patterns via Hebb's rule.

An equilibrium state s_i of the network can be characterized by the mean overlap with the set of stored patterns σ_i^μ according to

---

[1] From a technical point of view the reason for these problems originates from the fact that the number of Gaussian integrations over the auxiliary variables m^μ is of the order of N. Then the fluctuations at the saddle point are large and cannot be neglected even in the limit $N \to \infty$.

$$m^\mu = \frac{1}{N} \sum_i \langle s_i \rangle \sigma_i^\mu .$$ (19.2)

It can be shown that in the limit $N \to \infty$ only a finite number k of components of the overlap vector m^μ is nonvanishing. k is known as the number of *condensed patterns*, i.e. those patterns which in the given state of the network are represented with finite strength. We will concentrate on the case $k = 1$, which is of special interest since in such a state the network uniquely 'recalls' one pattern without admixtures. Besides the overlap vector m^μ it is useful to introduce two further order parameters. One of them is the sum of the ensemble-averaged squares of those overlaps which correspond to patterns which are not condensed:

$$r = \frac{N}{p} \sum_{\mu > k}^{p} \langle\!\langle (m^\mu)^2 \rangle\!\rangle .$$ (19.3)

Furthermore, the Edwards–Anderson parameter [Ed75]

$$q = q_{\text{EA}} = \left\langle\!\!\left\langle \frac{1}{N} \sum_i (\langle s_i \rangle)^2 \right\rangle\!\!\right\rangle$$ (19.4)

is introduced, which in terms of the magnetic analogy has the meaning of the average magnitude of local magnetization.

To perform the averaging over the variables σ_i^μ in the expression for the specific free energy (19.1) is a very demanding task, complicated by the fact that the partition function Z stands in the argument of a logarithm. Nevertheless a clever trick has been devised [Ed75, Sh75] to overcome this difficulty. The trick is based on the representation of the logarithm in terms of a power function through the identity

$$\ln Z = \lim_{n \to 0} \frac{Z^n - 1}{n} .$$ (19.5)

Now averaging of a power of Z is not much more involved than the task of averaging Z itself since the latter is a sum of exponential functions. Thus Z^n is just the partition function of a system which consists of n identical copies (called *replicas*) of the original network. This is immediately seen if one writes down the definition of Z^n:

$$Z^n = \left(\sum_{[s]} e^{-\beta E[s]} \right)^n = \sum_{[s^1]} \cdots \sum_{[s^n]} e^{-\beta(E[s^1] + \cdots E[s^n])} .$$ (19.6)

Here and in the following development we denote the number of the copy by an additional *upper index* to the variable s_i^a.

This expression for Z^n of course makes sense only for positive *integer* values of n. In order to employ (19.5) we thus have to evaluate (19.6) for these values and then perform an analytic continuation to arbitrary *real* values. Finally the limit $n \to 0$ has to be performed. It is not surprising that mathematical problems are encountered when following this procedure. We will come back briefly to this point at the end of the discussion. However, let us first begin fearlessly with the analytic evaluation of the average $\langle\!\langle\, Z^n \,\rangle\!\rangle$.

19.1.1 Averaging over Patterns

When attempting to evaluate the partition function raised to nth power Z^n we face the same problem of nonfactorizability encountered in Sect. 18.1, only slightly aggravated by the presence of n independent replicas. The cure again is to introduce Gaussian integrals over auxiliary variables, cf. (18.6). In all we now need a set of $n\,p$ auxiliary variables m_a^μ, where a is the replica index and counts the number of the copy. The result (18.7) for the partition function Z is easily generalized to our problem. It now reads

$$\langle\!\langle\, Z^n \,\rangle\!\rangle = \mathrm{e}^{-\frac{1}{2}\beta np} \left\langle\!\!\!\left\langle \sum_{[s^1]\cdots[s^n]} \left(\frac{\beta N}{2\pi}\right)^{np/2} \int \left(\prod_{a=1}^{n}\prod_{\mu=1}^{p} dm_a^\mu\right), \prod_\mu X_\mu \right\rangle\!\!\!\right\rangle, \quad (19.7)$$

where X_μ stands for the following abbreviation

$$X_\mu = \exp\left(-\beta N \sum_a \left[\frac{1}{2}(m_a^\mu)^2 - \frac{1}{N}m_a^\mu \sum_i \sigma_i^\mu s_i^a\right]\right). \quad (19.8)$$

By introducing the double angular brackets we have already indicated that we are interested in the average over an ensemble of patterns μ. Since summation and integration are linear operations they commute with the averaging procedure, which thus acts directly on the integrand

$$\langle\!\langle\, X_\mu \,\rangle\!\rangle = \exp\left[-\frac{\beta N}{2}\sum_a (m_a^\mu)^2\right] \left\langle\!\!\!\left\langle \exp\left[\beta\sum_a m_a^\mu \sum_i \sigma_i^\mu s_i^a\right] \right\rangle\!\!\!\right\rangle. \quad (19.9)$$

We now average over a large ensemble of statistically independent 'unbiased' patterns so that the variables σ_i^μ take on the values ± 1 with equal probability. The averaging can be carried out explicitly, leading to

$$\left\langle\!\!\!\left\langle \exp\left[\beta\sum_a m_a^\mu \sum_i \sigma_i^\mu s_i^a\right] \right\rangle\!\!\!\right\rangle = \frac{1}{2}\prod_i \sum_{\sigma_i^\mu=\pm 1} \exp\left[\beta\sum_a (m_a^\mu s_i^a)\sigma_i^\mu\right]$$

$$= \prod_i \cosh\left(\beta\sum_a m_a^\mu s_i^a\right). \quad (19.10)$$

If the cosh function is raised into the exponent, the averaged integrand of (19.7) can be written as

$$\langle\!\langle\, X_\mu \,\rangle\!\rangle = \exp\left[-\frac{\beta N}{2}\sum_a (m_a^\mu)^2 + \sum_i \ln\cosh\left(\beta\sum_a m_a^\mu s_i^a\right)\right]. \quad (19.11)$$

The task now is to evaluate the various sums and integrations over this expression. The averaging over patterns has been carried out already (note that this deviates from the course pursued in the last section). What remains to be done is the integrations over the np auxiliary variables m_a^μ and the summation over the spin configurations $[s_i^a]$ which represents the core of the partition-function calculation. Now it becomes essential to distinguish between the condensed and the noncondensed patterns. For simplicity we will assume that only one of the patterns, named $\mu = 1$, is condensed (i.e. 'recalled' by the network). The saddle point for the non-condensed patterns is located at $m_a^\mu = 0$. This means that the exponential function in the saddle-point integral (18.13) reduces to unity, leaving only the preexponential factor, which is independent of N. The number of such factors is $n(p-1)$ and their contribution does not vanish in the limit $N \to \infty$, provided that the number of patterns grows proportionally to N. As already mentioned in the beginning of this section this is a major obstacle which has to be overcome in the case of finite pattern loading of the network.

In the case of the noncondensed patterns, $\mu > 1$, only the neighborhood of zero contributes to the m_a^μ integration. Therefore we expand the function $\ln \cosh x \approx 1 + x^2/2$ in the exponent around the origin up to second order in x. The corresponding part of our integrand becomes

$$X \equiv \prod_{\mu > 1} \langle\!\langle X_\mu \rangle\!\rangle \approx \exp\left[-\frac{\beta N}{2} \sum_{\mu > 1} \sum_{a,b} B_{ab} m_a^\mu m_b^\mu\right] \tag{19.12}$$

with the matrix

$$B_{ab} = \delta_{ab} - \frac{\beta}{N} \sum_i s_i^a s_i^b . \tag{19.13}$$

So we again have obtained Gaussian integrals over the variables $m_a^\mu, \mu > 1$! Naively we might proceed by analytically solving these integrals. It is a standard problem to generalize the one-dimensional integral formula (18.5) to the case of several variables. The general formula for a *Gaussian integral* in M dimensions reads

$$\int dz_1 \cdots dz_M \, \exp\left(-\frac{1}{2} \sum_{i,j}^M z_i A_{ij} z_j\right) = (2\pi)^{M/2} (\det A)^{-1/2} , \tag{19.14}$$

where A is a real symmetric positive-definite matrix. The proof of (19.14) is based on the fact that an orthogonal transformation of the variables can be found that diagonalizes the matrix A. In this way the the multiple integral is factorized into M independent Gaussian integrals, each containing one of the eigenvalues in the exponent. According to (18.5) this leads to the product of square roots of all the eigenvalues in the denominator, which is just the square root of the determinant of the matrix. For later use we also note the extension of the integral formula (19.14) to the case of a general quadratic form in the exponent

$$\int dz_1 \cdots dz_M \, \exp\left(-\frac{1}{2} \sum_{i,j}^M z_i A_{ij} z_j + \sum_i^M b_i z_i\right)$$

$$= (2\pi)^{M/2} (\det A)^{-1/2} \exp\left(\frac{1}{2} \sum_{i,j}^{M} b_i (A^{-1})_{ij} b_j\right). \tag{19.15}$$

Applied to our case of $n(p-1)$ variables the integral reads

$$\int \left(\prod_a \prod_{\mu>1} dm_a^{\mu}\right) X = \left(\frac{2\pi}{\beta N}\right)^{\frac{1}{2} n(p-1)} (\det B)^{-\frac{1}{2}(p-1)}. \tag{19.16}$$

Unfortunately this is a highly complex function of the spin variables since it contains the determinant over the matrix B which depends on all the s_i^a. Insertion of the result (19.16) into the sum over all configurations (19.7) leads us nowhere; we would really like to have a product of independent exponential factors.

Perhaps it is not surprising any more to the reader that this can be achieved by introducing a collection of integrations over auxiliary variables! The configuration of the network enters the function X only via the quadratic expression $\sum_i s_i^a s_i^b$ contained in B_{ab}. This expression is symmetric under the exchange of a and b and it is trivial for $a = b$ since $(s_i^a)^2 = 1$ for all values of i and a. Thus there are $n(n-1)/2$ independent nontrivial combinations of replica spin variables. For each of them we will now introduce an auxiliary integration. To prepare for this let us write out (19.12) making use of the symmetry property of the matrix B_{ab}:

$$X = \left(\prod_{a,b}^{a<b} \exp\left[\beta^2 \sum_i s_i^a s_i^b \sum_{\mu>1} m_a^{\mu} m_b^{\mu}\right]\right)$$
$$\times \left(\prod_a \exp\left[\frac{1}{2}\beta(\beta-1)N \sum_{\mu>1} (m_a^{\mu})^2\right]\right). \tag{19.17}$$

Each of the factors with $a \neq b$ now will be multiplied by a rather intricate representation of unity, namely

$$1 = \int_{-\infty}^{\infty} dq_{ab}\, \delta\left(q_{ab} - \frac{1}{N} \sum_i s_i^a s_i^b\right) =$$
$$= \int_{-\infty}^{\infty} dq_{ab} \int_{-\infty}^{\infty} \frac{dr'_{ab}}{2\pi} \exp\left[ir'_{ab}\left(q_{ab} - \frac{1}{N} \sum_i s_i^a s_i^b\right)\right]. \tag{19.18}$$

Then the factors are given by

$$\int dq_{ab} \int \frac{dr'_{ab}}{2\pi} \exp\left[ir'_{ab}q_{ab} - i\left(r'_{ab} + i\beta^2 N \sum_{\mu>1} m_a^{\mu} m_b^{\mu}\right)\frac{1}{N} \sum_i s_i^a s_i^b\right]. \tag{19.19}$$

The decisive step now is to substitute the integration variable r'_{ab} through the linear transformation

$$\bar{r}_{ab} = r'_{ab} + i\beta^2 N \sum_{\mu>1} m_a^{\mu} m_b^{\mu}, \tag{19.20}$$

We will shift the integration contour so that it extends again over the real $\bar{r}_{ab}$ axis from $-\infty$ to $+\infty$. With this substitution (19.19) reads

$$\int dq_{ab} \int \frac{d\bar{r}_{ab}}{2\pi} \exp\left[i\bar{r}_{ab}\left(q_{ab} - \frac{1}{N}\sum_i s_i^a s_i^b \right) + q_{ab}\beta^2 N \sum_{\mu>1} m_a^\mu m_b^\mu \right]. \quad (19.21)$$

What did we gain by introducing the auxiliary integrations? Well, the new expression (19.21) has the key advantage that it factorizes in a term which depends on the spin variables s_i^a and one that contains the variables m_a^μ separately. Thus the Gaussian integration now can be performed using (19.14) without producing an intractable dependence of the spin variables! The integration leads to

$$\int \left(\prod_a \prod_{\mu>1} dm_a^\mu \right) \exp\left[-\frac{1}{2}\beta N \sum_{ab} \Lambda_{ab} \sum_{\mu>1} m_a^\mu m_b^\mu \right]$$
$$= \left(\frac{2\pi}{\beta N} \right)^{\frac{1}{2}n(p-1)} (\det \Lambda)^{\frac{1}{2}(p-1)}, \quad (19.22)$$

with a matrix Λ_{ab} which does not depend on s_i^a, namely

$$\Lambda_{ab} = (1-\beta)\delta_{ab} - \beta q_{ab} . \quad (19.23)$$

Note that in the exponent of (19.22) all combinations of indices a and b have been admitted, which introduced an additional factor of $1/2$ in the nondiagonal terms. If we raise the determinant into the exponent using the well-known matrix identity

$$\det \Lambda = \exp[\text{Tr} \ln \Lambda] , \quad (19.24)$$

the result for the average value of Z^n has the following form:

$$\langle\!\langle Z^n \rangle\!\rangle = e^{-\frac{1}{2}\beta np} \left(\frac{\beta N}{2\pi} \right)^{n/2} \int \left(\prod_{a=1}^n dm_a^1 \right) \int \left(\prod_{ab}^{a<b} dq_{ab}d\bar{r}_{ab} \right)$$

$$\times \exp\left[-\frac{1}{2}\beta N \sum_a (m_a^1)^2 - \frac{p-1}{2}\text{Tr} \ln \Lambda - i\sum_{ab}^{a<b} \bar{r}_{ab}q_{ab} \right]$$

$$\times \left\langle\!\!\!\left\langle \sum_{[s^1]\cdots[s^n]} \exp\left[\beta \sum_a m_a^1 \sum_i \sigma_i^1 s_i^a + \frac{i}{N}\sum_{ab}^{a<b} \bar{r}_{ab} \sum_i s_i^a s_i^b \right] \right\rangle\!\!\!\right\rangle. $$
$$(19.25)$$

The term in angular brackets involves averaging over the condensed pattern. The expression in the exponent of the term to be averaged is local in the spin variables, therefore the exponential function splits into N independent factors, each of which depends only on the variables s_i^a at one given position i. Hence the sum over configurations is easy to perform, leading to

$$\left\langle\!\!\left\langle \sum_{[s^1]\cdots[s^n]} \prod_{i=1}^{n} \exp\left[\beta \sum_a m_a^1 \sigma_i^1 s_i^a + \frac{i}{N} \sum_{ab}^{a<b} \bar{r}_{ab} s_i^a s_i^b\right] \right\rangle\!\!\right\rangle$$

$$= \left\langle\!\!\left\langle \sum_{s^1\cdots s^n} \exp\left[\beta \sum_a m_a^1 \sigma^1 s^a + \frac{i}{N} \sum_{ab}^{a<b} \bar{r}_{ab} s^a s^b\right] \right\rangle\!\!\right\rangle^N . \tag{19.26}$$

Note that now the variables s^a no longer carry a lower index i, the summation extends only over the 2^n possible configuration of a single neuron and its replica copies.

19.1.2 The Saddle-Point Approximation

Now we have reached the stage at which the averaged 'free energy per neuron' f can be evaluated employing the saddle-point approximation introduced in Sect. 18.1. For a start, let us write down the full expression for f derived in the last section. For simplicity of notation we will drop the index $\mu = 1$ on m_a^μ and σ since there is only one condensed pattern.

$$\begin{aligned}
f &= -\frac{T}{N}\langle\!\langle \ln Z \rangle\!\rangle = \lim_{n\to0} \frac{1 - \langle\!\langle Z^n \rangle\!\rangle}{n\beta N} \\
&= \lim_{n\to0} \frac{1}{n\beta N}\left[1 - \left(\frac{\beta N}{2\pi}\right)^{n/2} \int \left(\prod_{a=1}^{n} dm_a\right) \int \left(\prod_{ab}^{a<b} dq_{ab}d\bar{r}_{ab}\right) e^{-\beta N\Phi}\right];
\end{aligned} \tag{19.27}$$

here the exponent under the integral is given by the rather lengthy function

$$\begin{aligned}
\Phi &= \frac{np}{2N} + \frac{1}{2}\sum_a (m_a)^2 + \frac{(p-1)}{2\beta N}\text{Tr}\ln\Lambda + \frac{i}{\beta N}\sum_{ab}^{a<b}\bar{r}_{ab}q_{ab} \\
&\quad -\frac{1}{\beta}\ln\left\langle\!\!\left\langle \sum_{s^1\cdots s^n} \exp\left[\beta\sum_a m_a\sigma s^a + \frac{i}{N}\sum_{ab}^{a<b}\bar{r}_{ab}s^a s^b\right]\right\rangle\!\!\right\rangle . \tag{19.28}
\end{aligned}$$

As before, the saddle point is given by the condition that the first derivatives of the exponent with respect to all the integration variables vanish:

$$\frac{\partial\Phi}{\partial m_a} = 0, \qquad \frac{\partial\Phi}{\partial\bar{r}_{ab}} = 0, \qquad \frac{\partial\Phi}{\partial q_{ab}} = 0 . \tag{19.29}$$

Up to corrections of the order of N^{-1}, which vanish in the thermodynamic limit $N \to \infty$, the value of f is given by

$$f = \lim_{n\to0} \frac{1}{n}\Phi , \tag{19.30}$$

which has to be evaluated at the saddle point. The values of the parameters m_a, q_{ab} and $\bar{r}_{ab}$ thus determine the equilibrium properties of the network. According to the usage of statistical physics they are called the *order parameters* of the network.

The task to differentiate the last term in (19.28) with respect to a variable contained in the exponent requires some further deliberation. To begin with, differentiation of the logarithm leads to the reciprocal of its argument. Furthermore, following the chain rule, differentiation of the exponential function simply leads to a supplemental factor under the summation and averaging operations. Taking a close look at the result, we note that it is just the (thermal) average over this supplemental factor. In highly abbreviated language this is expressed in the form

$$\frac{\partial}{\partial m_a} \ln\!\left\langle\!\!\left\langle \sum \exp\cdots \right\rangle\!\!\right\rangle = \frac{\left\langle\!\!\left\langle \sum \beta\sigma s^a \exp\cdots \right\rangle\!\!\right\rangle}{\left\langle\!\!\left\langle \sum \exp\cdots \right\rangle\!\!\right\rangle} = \beta\!\left\langle\!\!\left\langle \sigma\langle s^a\rangle \right\rangle\!\!\right\rangle . \tag{19.31}$$

With this result the conditions (19.29) for a saddle point for the parameters m_a, $\bar r_{ab}$ and q_{ab} lead to a set of three equations:

$$m_a = \left\langle\!\!\left\langle \sigma\langle s^a\rangle \right\rangle\!\!\right\rangle , \tag{19.32}$$

$$q_{ab} = \left\langle\!\!\left\langle \langle s^a\rangle\langle s^b\rangle \right\rangle\!\!\right\rangle , \tag{19.33}$$

$$\bar r_{ab} = \mathrm{i}\frac{p-1}{2}\!\left\langle\!\!\left\langle \frac{\partial}{\partial q_{ab}}\mathrm{Tr}\ln\Lambda(q) \right\rangle\!\!\right\rangle . \tag{19.34}$$

In the second relation the thermal average $\langle s^a s^b\rangle$ has been replaced by the uncorrelated product $\left\langle\!\!\left\langle \langle s^a\rangle\langle s^b\rangle \right\rangle\!\!\right\rangle$ since $a \neq b$ and the replicas are statistically independent. To evaluate the r.h.s. of (19.34) we have to return to the representation of the determinant of the matrix Λ in terms of Gaussian integrals, cf. (19.22–19.24). Differentiating the integrand of (19.22) yields

$$\beta^2 N \sum_{\mu>1} m_a^\mu m_b^\mu , \tag{19.35}$$

where a factor of 2 has been included, since each pair of indices a, b occurs twice. In addition the inverse power of the determinant in (19.22) enters as a factor $-2/(p-1)$. Thus (19.34) is transformed into

$$\bar r_{ab} = -\mathrm{i}\beta^2 N\!\left\langle\!\!\left\langle \sum_{\mu>1} m_a^\mu m_b^\mu \right\rangle\!\!\right\rangle . \tag{19.36}$$

Absorbing a normalization factor, this relation reduces to the standard expression for the order parameter r for the overlaps of the noncondensed patterns that was introduced at the beginning of this section, cf. (19.3):

$$\bar r_{ab} = -\mathrm{i}p\beta^2 r_{ab} . \tag{19.37}$$

The equations (19.2) and (19.4) for the two additional order parameters follow from the saddle-point conditions (19.32, 19.33) through averaging over the neurons.

Taken together, (19.32,19.33,19.36) form a set of n^2 equations for determining the order parameters m_a, q_{ab}, r_{ab}. These equations do not single out any one of the copies of the network denoted by the replica index a. Therefore it is tempting to assume that the solution will not depend on the value of the replica index. This conjectured property is called *replica symmetry*. If it is satisfied by the equilibrium state then all copies of the network are indistinguishable, i.e. they have identical properties[2]. Taking the replica symmetry for granted, we write

$$m_a = m, \qquad q_{ab} = q, \qquad r_{ab} = r , \qquad (19.38)$$

and begin to evaluate the function Φ at the saddle point.

One of the ingredients of (19.28) is the trace of the matrix $\ln \Lambda$. This can be expressed in terms of the n eigenvalues λ_a of the matrix Λ_{ab} through

$$\operatorname{Tr} \ln \Lambda = \sum_{a=1}^{n} \ln \lambda_a . \qquad (19.39)$$

This relation is easily proved by diagonalization of Λ, since $\ln \Lambda$ is invariant under orthogonal transformations. According to its definition, (19.23), the matrix Λ has a simple structure if we assume replica symmetry. All the diagonal elements are equal to $(1 - \beta)$, while the nondiagonal elements are given by $-\beta q$:

$$\Lambda = \begin{pmatrix} 1 - \beta & -\beta q & \cdots & -\beta q \\ -\beta q & 1 - \beta & \cdots & -\beta q \\ \vdots & \vdots & \ddots & \vdots \\ -\beta q & -\beta q & \cdots & 1 - \beta \end{pmatrix} . \qquad (19.40)$$

The eigensystem of this matrix can be easily read off. One eigenvector is $(1, 1, \ldots, 1)$ having the eigenvalue

$$\lambda_1 = (1 - \beta) - (n - 1)\beta q . \qquad (19.41)$$

In addition there is a system of $n - 1$ degenerate eigensolutions which can be spanned, e.g., by the set of linearly independent vectors $(1, -1, 0, \ldots, 0)$, $(1, 0, -1, 0, \ldots, 0)$, up to $(1, 0, \ldots, 0, -1)$. They all correspond to the same eigenvalue

$$\lambda_2 = \cdots = \lambda_n = (1 - \beta) + \beta q . \qquad (19.42)$$

Thus the trace of $\ln \Lambda$ is given by

$$\begin{aligned} \operatorname{Tr} \ln \Lambda &= \ln \det \Lambda = \ln\big[1 - \beta - (n - 1)\beta q\big] + (n - 1) \ln\big[1 - \beta + \beta q\big] \\ &= \ln\Big[1 - \frac{n\beta q}{1 - \beta(1 - q)}\Big] + n \ln\big[1 - \beta(1 - q)\big] . \end{aligned} \qquad (19.43)$$

Now we should recall that the replica trick was introduced with the aim in mind of performing the limit $n \to 0$ at the end of the calculation, cf. (19.5,19.30). For the function obtained in (19.43) this step can indeed be

---

[2] It is by no means clear that such a state corresponds to the deepest minimum, if any, of the function Φ. We will come back to this problem at the end of this section.

carried out now without any problem. Using l'Hospital's rule, we obtain for the limit

$$\lim_{n \to 0} \frac{1}{n} \text{Tr} \ln \Lambda = \frac{-\beta q}{1 - \beta(1 - q)} + \ln\left[1 - \beta(1 - q)\right]. \tag{19.44}$$

Let us examine whether this also works for the other contributions to the function Φ defined in (19.28). To evaluate the sum over configurations s^a in the last term we use the identity

$$\sum_{ab}^{a<b} s^a s^b = \frac{1}{2}\left[\sum_{ab} s^a s^b - \sum_a (s^a)^2\right] = \frac{1}{2}\left[\left(\sum_a s^a\right)^2 - n\right]. \tag{19.45}$$

Then the contribution to the 'free energy per neuron' takes the form

$$\exp\left[-\frac{1}{2}\alpha\beta^2 rn\right] \sum_{s^1\ldots s^n} \exp\left[\beta m\sigma\left(\sum_a s^a\right) + \frac{1}{2}\alpha\beta^2 r\left(\sum_a s^a\right)^2\right]. \tag{19.46}$$

To evaluate such an expression with a mixed bilinear form in s^a by now has become a matter of routine to us. Again introducing the inevitable auxiliary integration we obtain with the help of the Gaussian formula (18.5)

$$\sum_{s^1\ldots s^n} (2\pi)^{-1/2} \int dz\, e^{-\frac{1}{2}z^2} \prod_a e^{\beta(z\sqrt{\alpha r} + m\sigma)s^a}$$

$$= (2\pi)^{-1/2} \int dz\, e^{-\frac{1}{2}z^2} \left[2\cosh\beta(z\sqrt{\alpha r} + m\sigma)\right]^n. \tag{19.47}$$

This can be used to evaluate the limit $n \to 0$ of the last term in (19.28):

$$-\lim_{n \to 0} \frac{1}{n\beta} \ln\left\langle\!\!\left\langle \sum_{s^1\ldots s^n} \exp\left[\beta \sum_a m_a \sigma s^a + \frac{i}{N} \sum_{ab}^{a<b} \bar{r}_{ab} s^a s^b\right]\right\rangle\!\!\right\rangle$$

$$= \lim_{n \to 0} \frac{1}{n\beta}\left(\frac{1}{2}\alpha\beta^2 rn\right.$$

$$\left. - \ln \int \frac{dz}{\sqrt{2\pi}} e^{-\frac{1}{2}z^2}\left[1 + n\left\langle\!\!\left\langle \ln[2\cosh\beta(z\sqrt{\alpha r} + m\sigma)]\right\rangle\!\!\right\rangle + \ldots\right]\right)$$

$$= \frac{1}{2}\alpha\beta r - \frac{1}{\beta}\frac{1}{\sqrt{2\pi}}\int dz\, e^{-\frac{1}{2}z^2}\left\langle\!\!\left\langle \ln[2\cosh\beta(z\sqrt{\alpha r} + m\sigma)]\right\rangle\!\!\right\rangle. \tag{19.48}$$

Using the relations

$$\sum_a (m_a)^2 = nm^2, \qquad \frac{i}{\beta N}\sum_{ab}^{a<b} \bar{r}_{ab} q_{ab} = \frac{n(n-1)}{2N} p\beta rq, \tag{19.49}$$

which are valid if replica symmetry holds, and with the help of (19.44,19.48) we arrive at the following result for the free energy per element of the network:

$$f = \frac{1}{2}m^2 + \frac{\alpha}{2}\left[1 + \beta r(1-q) - \frac{q}{1-\beta(1-q)} + \frac{1}{\beta}\ln\left[1 - \beta(1-q)\right]\right]$$

$$- \frac{T}{\sqrt{2\pi}} \int dz\, e^{-\frac{1}{2}z^2} \left\langle\!\left\langle \ln[2\cosh\beta(z\sqrt{\alpha r} + m\sigma)] \right\rangle\!\right\rangle. \tag{19.50}$$

Note that still an average has to be taken, as indicated by the angular double brackets, but this refers only to the single condensed pattern ($\mu = 1$). The fluctuations, which are caused by the whole set of $p-1$ additional patterns stored in the network, are contained implicitly in the Gaussian parameter integral.

Now the saddle-point conditions (19.29) can be given in closed form. Differentiation of the function f with respect to the order parameters m, r, and q leads to

$$\frac{\partial f}{\partial m} = 0 \quad \rightarrow \quad m = \int \frac{dz}{\sqrt{2\pi}} e^{-\frac{1}{2}z^2} \left\langle\!\left\langle \sigma \tanh\left[\beta(z\sqrt{\alpha r} + m\sigma)\right] \right\rangle\!\right\rangle, \tag{19.51}$$

$$\frac{\partial f}{\partial r} = 0 \quad \rightarrow \quad q = 1 - \frac{T}{\alpha} \int \frac{dz}{\sqrt{2\pi}} z\, e^{-\frac{1}{2}z^2} \left\langle\!\left\langle \sqrt{\frac{\alpha}{r}} \tanh\left[\beta(z\sqrt{\alpha r} + m\sigma)\right] \right\rangle\!\right\rangle$$

$$= \int \frac{dz}{\sqrt{2\pi}} e^{-\frac{1}{2}z^2} \left\langle\!\left\langle \tanh^2\left[\beta(z\sqrt{\alpha r} + m\sigma)\right] \right\rangle\!\right\rangle, \tag{19.52}$$

$$\frac{\partial f}{\partial q} = 0 \quad \rightarrow \quad r = q\left[1 - \beta(1-q)\right]^{-2}. \tag{19.53}$$

In the second equation we have performed an integration by parts to get rid of the factor z under the integral.

At this stage we should pause and look back at the results which were obtained in Chapt. 18 for the case of a finite number of stored patterns, i.e. $\alpha = 0$, for comparison. We note that the expression for the function f in the limit $\alpha \to 0$ reduces to (19.50). Accordingly (19.51) is a generalization of the old result (18.23). The main difference is the presence of an additional integration over the variable z which describes the influence of the large number of patterns ($p \to \infty$) stored simultaneously by the net. Also the equation determining q, (19.52), had its analog in Chapt. 18 where it appeared as the diagonal element of the stability matrix $Q_{\mu\nu}$ of (18.38).

19.2 Phase Diagram of the Hopfield Network

As the final step in determining the statistical properties of the network we have to average over the condensed pattern σ in (19.51,19.52). This operation, however, turns out to be completely trivial as we admitted only a single condensed pattern. An inspection of the integrals shows that the result is invariant under the simultaneous change of sign of σ and of the integration variable z. Therefore the distribution of probabilities for the values $\sigma = +1$ and $\sigma = -1$ does not matter and the angular brackets can simply be removed, leading to

$$m = \int \frac{dz}{\sqrt{2\pi}} e^{-\frac{1}{2}z^2} \tanh\left[\beta(z\sqrt{\alpha r} + m)\right] , \tag{19.54}$$

$$q = \int \frac{dz}{\sqrt{2\pi}} e^{-\frac{1}{2}z^2} \tanh^2\left[\beta(z\sqrt{\alpha r} + m)\right] . \tag{19.55}$$

The combination of these two integral equations together with (19.53) determine the values of the three order parameters m, q, and r.

What information can be extracted from these results? Clearly the most interesting quantity is the magnitude of the parameter m which is the overlap of the network configuration with the condensed pattern. The network 'remembers' this pattern, provided that there is a stable solution with $m \neq 0$. We will study under which conditions this happens, beginning with the special case of *zero temperature*, $T \to 0$ ($\beta \to \infty$). In this limit the hyperbolic tangent reduces to a step function

$$\tanh[\beta(z\sqrt{\alpha r} + m)] \to \mathrm{sgn}(z\sqrt{\alpha r} + m) , \tag{19.56}$$

and the equation for the order parameter m becomes

$$m = 2 \int_0^{m/\sqrt{\alpha r}} \frac{dz}{\sqrt{2\pi}} e^{-\frac{1}{2}z^2} = \mathrm{erf}(m/\sqrt{2\alpha r}) , \tag{19.57}$$

where $\mathrm{erf}(x)$ is the error function. The parameter q approaches $q \to 1$ in the zero temperature limit. A simple equation for this limit most readily obtained from the first version of (19.52), multiplied by β:

$$\begin{aligned}
\beta(1-q) &= \int \frac{dz}{\sqrt{2\pi}} \frac{z}{\sqrt{\alpha r}} e^{-\frac{1}{2}z^2} \mathrm{sgn}(z\sqrt{\alpha r} + m) \\
&= 2 \int_{m^2/2\alpha r}^{\infty} \frac{e^{-x} dx}{\sqrt{2\pi\alpha r}} = \left(\frac{2}{\pi\alpha r}\right)^{1/2} e^{-m^2/2\alpha r} .
\end{aligned} \tag{19.58}$$

The equation for the parameter r, (19.53), can be combined with (19.57, 19.58). Introducing the variable $x = m/\sqrt{2\alpha r}$ we are finally led to a simple transcendental equation

$$x\sqrt{2\alpha} = F_0(x) \equiv \mathrm{erf}(x) - \frac{2x}{\sqrt{\pi}} e^{-x^2} . \tag{19.59}$$

Let us study the solutions of this equation and their dependence on the parameter α, which can be qualitatively understood from the graph of the function $F_0(x)$, cf. Fig. 19.1. Only for small values of α where the slope of the straight line defining the left-hand side of (19.59) is small does there exist a nonvanishing solution. The resulting dependence of the overlap parameter m on α had been depicted in Fig. 4.5. For low pattern loading the fraction of errors $\frac{1}{2}(1-m)$ is exponentially small

$$m \simeq 1 - \sqrt{\frac{2\alpha}{\pi}} \exp\left(-\frac{1}{2\alpha}\right) \quad \text{for } \alpha \to 0 . \tag{19.60}$$

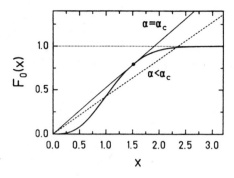

Fig. 19.1. Graphical solution of (19.59) leading to the critical storage capacity at zero temperature.

If α exceeds a critical value α_c the slope becomes so large that the straight line no longer intersects the curve $F_0(x)$, besides the solution $x = 0$. The numerical determination of the point of osculation between the straight line and the curve leads to the critical value $x_c \approx 1.513$, $\alpha_c \approx 0.138$. Here the overlap $m_c = \mathrm{erf}(x_c)$ has the remarkably large value $m_c \approx 0.967$, which discontinuously drops down to $m = 0$ at higher α. Thus the Hopfield network at zero temperature exhibits a *discontinuous phase transition* as a function of the pattern loading α, jumping from a phase which is almost completely 'magnetized' to one which has no 'magnetization' at all. In the language of memory function this means that at $\alpha = \alpha_c$ the network undergoes a sudden collapse from nearly perfect recollection to a state of amnesia or total confusion. This is a remarkable result which rewards us for the considerable effort devoted to its derivation.

With increasing temperature T the critical storage capacity α_c decreases continuously. It reaches the limit $\alpha_c = 0$ at $T = 1$. This result, of course, is consistent with the value of the critical temperature obtained in Sect. 18.2 for the case $\alpha = 0$. The complete phase diagram [Am87a] of the Hopfield network, without taking into account replica-symmetry breaking, is given in Fig. 19.2. Note that it is more complicated than we might have expected from our discussion up to now. Four different phase regions can be distinguished: the *paramagnetic phase* P where all order is destroyed by the high temperature T; the *spin-glass phase* SG, which does not allow the retrieval of stored patterns; a *mixed phase* F+SG, where the stored patterns are metastable, i.e. the global energy minima are spin-glass states, and a *"ferromagnetic" phase* F, where the stored patterns are global energy minima. The equilibrium states of the network have large overlap with a single key pattern.

The boundary between the regions P and SG represents a phase transition of second order and the boundary between F+SG and F a first-order (discontinuous) phase transition. At zero temperature the latter transition occurs at $\alpha \approx 0.051$. Figure 19.2 is a quantitative version of the phase diagram of the Sherrington–Kirkpatrick model of spin glasses introduced earlier (cf. Fig. 17.2).

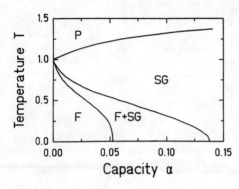

Fig. 19.2. The complete phase diagram of the Hopfield model exhibiting four different phases (after [Am87a]).

To end this section we have to confront the problem, whether the approximation (19.38), which assumes complete replica symmetry, is justified. Sherrington and Kirkpatrick in their early work on the spin-glass model [Sh75, Sh78] already noted that this assumption cannot be strictly correct. At small temperatures they found the unphysical result that the entropy of the system assumes negative values. Clearly this cannot be correct since the system has a finite number of degrees of freedom so that its entropy must be positive by definition. On the other hand, numerical simulations have led to values of the free energy per unit, f, which agrees quite well with the analytical results, i.e. with an accuracy of about 5%.

What is wrong with our treatment? As it turns out, the problem lies in the tacit assumption that the equations determining the saddle point (19.51–19.53), produce a solution which corresponds to a minimum of the free energy f. In fact, they lead to a maximum of f [Al78]! This can be quite easily seen by writing out the q-dependent terms contained in (19.50) before going to the limit $n \to 0$:

$$\frac{1}{n}\Phi = \frac{\alpha}{2}\left[1 + \beta r + (n-1)\beta rq + \frac{1}{\beta n}\ln\frac{1 - \beta - (n-1)\beta q}{1 - \beta + \beta q}\right.$$
$$\left. + \frac{1}{\beta}\ln\left[1 - \beta + \beta q)\right]\right] + \cdots . \tag{19.61}$$

The condition of a vanishing first derivative just reproduces (19.53), which determines the order parameter r. Now the second derivative with respect to q leads to

$$\frac{\partial^2(\Phi/n)}{\partial q^2} = \frac{(n-1)}{2}\beta\alpha\frac{(1-\beta)^2 + (n-1)\beta^2 q^2}{(1 - \beta + \beta q)^2[1 - \beta + \beta(1-n)q]^2} . \tag{19.62}$$

This expression is positive as long as $n > 1$. In the case $n < 1$, however, the second derivative takes on negative values, i.e. the saddle point corresponds to a local maximum of the free energy. This applies to the actual case of interest of the model, namely the limit $n \to 0$.

As a consequence there must exist values of the order parameters m_a, r_{ab}, and q_{ab} associated with a lower value of the energy. These values correspond to a complicated *breaking of the replica symmetry* characterized by the fact that the $(n \times n)$ matrix q_{ab} is invariant under subgroups of the full permutation group S_n. A typical example for this matrix in the case $n = 8$ might read

$$
(q_{ab}) = \begin{pmatrix}
0 & q_0 & q_1 & q_1 & q_2 & q_2 & q_2 & q_2 \\
  & 0   & q_1 & q_1 & q_2 & q_2 & q_2 & q_2 \\
  &   & 0   & q_0 & q_2 & q_2 & q_2 & q_2 \\
  &   &   & 0   & q_2 & q_2 & q_2 & q_2 \\
  &   &   &   & 0   & q_0 & q_1 & q_1 \\
  &   &   &   &   & 0   & q_1 & q_1 \\
  &   &   &   &   &   & 0   & q_0 \\
  &   &   &   &   &   &   & 0
\end{pmatrix} .
\tag{19.63}
$$

A detailed discussion of this phenomenon would exceed the scope of our presentation. More information can be found in [Pa80]; a selection of original papers is reprinted in [Me87]. While the breaking of replica symmetry may seem to be somewhat counterintuitive we should not be too surprised that the limit of $n \to 0$, which means that there is no copy of the network at all, has some peculiar mathematical properties.

As it turns out, however, replica-symmetry breaking seems to play only a minor role as far as the capability of the network for pattern recollection is concerned. The truly stable configurations turn out to differ from the learned patterns only at a few positions which are distributed irregularly over the network. This has little influence on the quality of pattern recognition, especially in the presence of thermal noise [Am87a]. Indeed the storage capacity is slightly enhanced [Cr86] to a value of $\alpha_c = 0.145$, while $m_c \simeq 0.983$.

19.3 Storage Capacity of Nonlinear Neural Networks

In the previous sections the Hopfield network has been analyzed using the methods of statistical mechanics borrowed from the study of spin glasses. There it was assumed that the network was "trained" using Hebb's rule, but the formalism can be applied also to modified learning rules.

1.   Let us quote the modifications which arise in the case of the *nonlinear extension of Hebb's rule*

$$
w_{ij} = \frac{\sqrt{p}}{N} \Phi\left(\frac{1}{\sqrt{p}} \sum_{\mu=1}^{p} \sigma_i^\mu \sigma_j^\mu \right) ,
\tag{19.64}
$$

with a general monotonously increasing function $\Phi(x)$ which was introduced in Sect. 10.2.2. The replica method can also be applied in this case [He87c] to find the storage capacity[3] $\alpha = p/N$. The equations which determine the saddle point (19.53–19.55) now read

---

[3] The case of a finite number of stored patterns is analyzed in detail in [He88a].

$$m = \int \frac{dz}{\sqrt{2\pi}} e^{-\frac{1}{2}z^2} \tanh\left[\beta(z\sqrt{\alpha r} + m)\right], \qquad (19.65)$$

$$q = \int \frac{dz}{\sqrt{2\pi}} e^{-\frac{1}{2}z^2} \tanh^2\left[\beta(z\sqrt{\alpha r} + m)\right], \qquad (19.66)$$

$$r = q\left[1 - \beta(1 - q)\right]^{-2} + C. \qquad (19.67)$$

The only difference to the earlier result is the presence of the term

$$C = \tilde{J}^2/J^2 - 1, \qquad (19.68)$$

where

$$J = \int \frac{dz}{\sqrt{2\pi}} e^{-\frac{1}{2}z^2} \left[z\Phi(z)\right], \qquad \tilde{J}^2 = \int \frac{dz}{\sqrt{2\pi}} e^{-\frac{1}{2}z^2} \left[\Phi(z)\right]^2. \qquad (19.69)$$

The numerical value of the constant C is determined by the function $\Phi(z)$. In the special case of the original Hebbian rule the function is linear, $\Phi(z) = z$, which yields $J = \tilde{J} = 1$, $C = 0$, so that (19.66–19.69) reduce to the old result, as they should. The *clipping of synapses* by means of the step function $\Phi(z) = \text{sgn}(z)$, leads to $J = \sqrt{2/\pi}$, $\tilde{J} = 1$, and thus $C = (\frac{\pi}{2} - 1) \simeq 0.571$.

2. As a further modification the network can be *diluted* by randomly switching off a fraction $1 - d$ of all synapses. The only effect is an increase of the constant C given by

$$C = \frac{1-d}{d} + \frac{1}{d}\left(\frac{\tilde{J}^2}{J^2} - 1\right). \qquad (19.70)$$

In the limit of complete "brain damage", $d \to 0$, the value of C diverges and there are no stable patterns.

3. As a final example of a learning rule which is amenable to the above analytical treatment we introduce the nonlinear learning function $\Phi(z) = \text{sgn}(z)\theta(|z| - a\sqrt{p})$. According to this rule the weak synapses are discarded and the strong ones are clipped to the value ± 1. This leads to the function

$$C(a) = \frac{\pi}{2} e^{a^2} \text{erfc}(a/\sqrt{2}) - 1, \qquad (19.71)$$

where $\text{erfc}(x) = 1 - \text{erf}(x)$ is the complementary error function. The function $C(a)$ is depicted in Fig. 19.3.

The critical storage capacity at temperature $T = 0$ again can be determined by solving a transcendental equation similar to (19.59). In the presence of a nonvanishing constant C this equation reads

$$x\sqrt{2\alpha} = F_C(x) \equiv \left[\left(\text{erf}(x) - \frac{2x}{\sqrt{\pi}}e^{-x^2}\right)^{-2} + \frac{C}{\text{erf}^2(x)}\right]^{-\frac{1}{2}}, \qquad (19.72)$$

which obviously is a generalization of (19.59). The function $F_C(x)$ is depicted in Fig. 19.4 for three values of the parameter C. The dashed line shows how the critical capacity is deduced from the condition of tangency between the

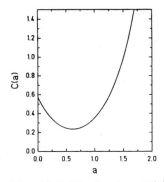

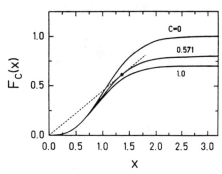

Fig. 19.3. The function $C(a)$ for various values of the cutoff parameter a according to (19.71).

Fig. 19.4. The function $F_C(x)$ which can be used to determine the storage capacity of modified networks. The parameters $C = 0$ and $C \simeq 0.571$ correspond to the linear model and the clipped model, respectively.

function $F_C(x)$ and the straight line $\sqrt{2\alpha}\,x$ for the case of clipped synapses, $C \simeq 0.571$.

Figure 19.5 shows the critical storage capacity α_c as a function of C. As one might have expected, α_c decreases with growing "nonlinearity parameter" C, the original linear Hebb rule leading to the largest storage capacity. The error rate $(1-m)/2$ of a condensed pattern is shown in Fig. 19.6 as a function of α. Obviously the value of the overlap m remains close to one under all conditions of practical interest. This means that the network retains nearly perfect recall of the stored patterns up to the critical point.

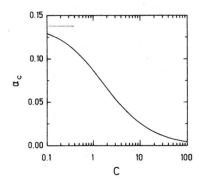

Fig. 19.5. The critical storage capacity α_c as a function of the parameter C.

The result of Fig. 19.5 has already been discussed in Sect. 10.2.3 for a network with randomly eliminated Hebbian synapses as a function of the dilution parameter d. The value of $\alpha_c(d)$ everywhere is larger than the linear function $d\,\alpha_c(d=1)$ which is proportional to the reduction of active synapses.

This indicates that in a way the efficiency of pattern storage can be increased by the process of dilution.

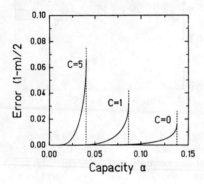

Fig. 19.6. The quality of pattern recall (fraction of wrong bits $(1 - m)/2$) plotted against the memory loading α for various values of the nonlinearity parameter C.

19.4 The Dynamics of Pattern Retrieval

The study of the thermodynamic properties of spin-glass-type neural networks has provided valuable information on the memory capacity and the nature of the retrieval states. This treatment, however, is applicable only to the *equilibrium properties* of the network. It does not address the question how the configuration $s_i(t)$ evolves in time starting from a given initial nonequilibrium state $s_i(0)$. While no general solution of this problem is known, apart from numerical simulations, several limiting cases and approximations have been studied.

19.4.1 Asymmetric Networks

The most interesting type of dynamic evolution is displayed by *asymmetric networks* with $w_{ij} \neq w_{ji}$. Unfortunately such systems cannot be treated with the methods of ordinary equilibrium statistical mechanics. Owing to the absence of an energy (or Lyapunov) function the network is not forced to approach a stable attractor state. For finite size N and assuming a deterministic updating rule ($T = 0$) the network, after an initial transient phase, will enter a *periodic cycle* (since there are only finitely many network configurations). In the limit $N \rightarrow \infty$ the length of these cycles may diverge and the time dependence may become *chaotic*.

Spitzner and Kinzel [Sp89] have studied the dynamics of asymmetric networks in numerical simulations. They used an asymmetric version of the Hopfield network with couplings

$$w_{ij} = \epsilon_{ij} \frac{1}{N} \sum_{\mu=1}^{p} \sigma_i^\mu \sigma_j^\nu , \qquad i \neq j \tag{19.73}$$

with random unbiased patterns σ_i^μ. Here the factors ϵ_{ij} are randomly chosen to be 0 or 1 subject to the condition $\epsilon_{ij}\epsilon_{ji} = 0$, $\epsilon_{ij} + \epsilon_{ji} = 1$, which serves to switch off either w_{ij} or w_{ji}, but not both of them.

This network shows some similarity with its symmetric counterpart, having a reduced critical storage capacity $\alpha_c \simeq 0.078$. Despite the strong asymmetry of the couplings many of the simulation runs of the network ended up in stable states. In the finite networks simulated numerically the average length of cycles $\langle t_c \rangle$ was quite small compared to the total number 2^N of possible configurations. However, the scaling with N (e.g. in the case $\alpha = 0.1$: $\langle t_c \rangle \simeq 2$ for $N = 200$ and $\langle t_c \rangle \simeq 60$ for $N = 800$) indicates that the cycle length $\langle t_c \rangle$ diverges exponentially for $N \to \infty$. It is expected that the dynamics becomes *chaotic* in the nonretrieval regime ($\alpha > \alpha_c$). Such behavior was also observed for networks with completely random weights w_{ij} [Kü88, Ba89].

19.4.2 Highly Diluted Networks

One interesting case for which the dynamics has been solved *exactly* is that of a *highly diluted network* with asymmetric couplings. As mentioned in Sect. 10.2.3 here one assumes that a large number of randomly selected Hebbian synapses are switched off according to a probability distribution

$$\rho(\epsilon_{ij}) = \frac{C}{N}\,\delta(\epsilon_{ij} - 1) + \left(1 - \frac{C}{N}\right)\delta(\epsilon_{ij}) \tag{19.74}$$

in (19.73), C being the average number of connections per neuron.

At zero temperature the relaxation of a network trained according to Hebb's rule is described by (cf. (3.5))

$$s_i(t+1) = \text{sgn} \sum_{j \neq i} \frac{\epsilon_{ij}}{N}\left(\sum_{\mu=1}^{p} \sigma_i^\mu \sigma_j^\mu\right) s_j(t) . \tag{19.75}$$

The treatment of the dynamics is facilitated if the *parallel-updating rule* according to Little is used. This stands in contrast to the sequential-updating rule which was the basis of the thermodynamic calculation of equilibrium properties in the previous sections. The overall characteristics of the network will not depend in an essential way on the updating rule [Fo88a].

The interesting observable quantity to be followed when studying the time-dependent evolution of the network is the set of *overlaps* m^μ with the stored patterns σ_i^μ:

$$m^\mu(t) = \frac{1}{N}\sum_{i=1}^{N} \sigma_i^\mu s_i(t) . \tag{19.76}$$

Derrida, Gardner, and Zippelius [De87c] have succeeded in obtaining an exact solution for the $N \to \infty$, $C \to \infty$ limit of the model (19.74) under the condition of *extreme dilution*

$$C \ll \ln N \,, \tag{19.77}$$

in which case the spins $s_i(t)$ at different sites remain uncorrelated. For $T = 0$ the overlap develops according to the discrete map

$$m^\mu(t+1) = \int_{-\infty}^\infty \frac{dy}{\sqrt{2\pi}} \, e^{-y^2} \, \mathrm{sgn}\left[m^\mu(t) - \sqrt{2\alpha}\,y\right] \,. \tag{19.78}$$

This iteration law has a stable nontrivial fixed point m^* if the pattern loading $\alpha = p/C$ is smaller than the critical capacity $\alpha_c = \frac{2}{\pi}$. Starting from a configuration which lies within the basin of attraction of pattern μ the overlap approaches $m^\mu(t) \to m^*$ monotonically.

However, this does not signal that the network becomes stationary! The individual spins $s_i(t)$ keep fluctuating with time, even if the global overlap observable $m^\mu(t)$ becomes constant. Moreover, the dynamics depends sensitively on the initial conditions. Let us assume that two runs of the same network are started with initial configurations $s_i(0) \neq \tilde{s}_i(0)$ which both converge to the same attractor m^*. One can show that their trajectories in phase space diverge with time, irrespective of how close to each other they have been initially. This is just the criterion for *chaotic motion* [Sc84]. Similar behavior was also found for an exactly solvable model of a layered feed-forward network [De88b, Me88].

19.4.3 The Fully Coupled Network

Tools from statistical physics and probability theory can also be applied to investigate the dynamics of a *fully connected network* although here no exact solution has been found. We will briefly review a calculation which deduces the storage capacity and the size of the basins of attraction in a Hopfield–Little network with finite pattern loading $\alpha = p/N$ from the retrieval dynamics [Za89].

The zero-temperature time evolution of these overlaps can be written as

$$\begin{aligned}
m^\mu(t+1) &= \frac{1}{N} \sum_{i=1}^N \sigma_i^\mu \mathrm{sgn}\left[\sum_{j \neq i}^N \frac{1}{N} \sum_{\nu=1}^N \sigma_i^\nu \sigma_j^\nu \, s_j(t)\right] \\
&\approx \frac{1}{N} \sum_{i=1}^N \mathrm{sgn}\left[m^\mu(t) + \sigma_i^\mu U_i^\mu[s(t)]\right]
\end{aligned} \tag{19.79}$$

with the fields

$$U_i^\mu[s] = \frac{1}{N} \sum_{\nu \neq \mu} \sum_{j \neq i} \sigma_i^\nu \sigma_j^\nu s_j \,. \tag{19.80}$$

The time evolution can be studied analytically in the limit of an infinite system, $N \to \infty, p \to \infty, \alpha = p/N = \mathrm{const}$. For this we have to assume that the pattern spins $\sigma_i^\mu = \pm 1$ are independent uniformly distributed *random variables*. By use of the central-limit theorem one can deduce that the fields $U_i^\mu[s]$

will be random variables described by a superposition of Gaussian probability distributions. To evaluate (19.79) one has to assume that the random variables σ_i^μ and U_i^μ are statistically independent. With this approximation the limit $N \to \infty$ can be performed using Birkhoff's ergodic theorem. We quote the result of Zagrebnov and Chvyrov [Za89]:

$$m^\mu(t+1) = F_\alpha(m^\mu(t)) \,, \tag{19.81}$$

where $F_\alpha(m)$ is an error function of a somewhat complicated argument

$$F_\alpha(m) = \operatorname{erf}\left(\frac{m}{\sqrt{2[\alpha + 2(1 - |m|)]}}\right) \,, \tag{19.82}$$

which depends on the memory loading parameter α. Thus a simple analytical expression has been deduced which can be iterated to obtain the full (discrete) time evolution of the overlaps $m^\mu(t)$. In the limit $t \to \infty$ the overlap will approach one of the fixed points of the transcendental equation

$$m^* = F_\alpha(m^*) \,. \tag{19.83}$$

Figure 19.7 shows a graphical construction of these solutions. If the pattern loading α exceeds a *critical capacity* α_c there is only one fixed point, namely $m_0^* = 0$, which is an attractor of the dynamical evolution law (19.82) for any starting overlap. Thus the network has completely lost any memory of the stored patterns. Below $\alpha = \alpha_c$ there are three fixed points m_0^*, m_1^*, m_2^*. From the figure it is immediately obvious that two of them are stable, their regions of attraction being separated by the intermediate, unstable fixed point m_1^*. Numerical solution of (19.83) leads to the critical capacity $\alpha_c \approx 0.1398$ corresponding to the overlap $m_c^* \approx 0.9698$, which means a fraction of 1.5% wrong bits in the retrieved pattern. The *basin of attraction* of a stored pattern is given by the interval $m_1^* \leq m(0) \leq 1$. This region is largest, of course, in the limit $\alpha = 0$, where the critical overlap is $m_1^* = 0.808$, i.e. a pattern is recognized if it has less than 10% wrong bits.

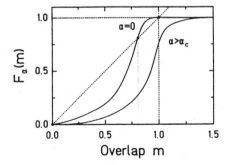

Fig. 19.7. Graphical solution of the equation determining the fixed points of the time evolution of the overlap parameter $m^\mu(t)$. The *dotted line* shows the border of the basin of attraction in the most favorable case $\alpha = 0$.

Figure 19.7 and the resulting values for the critical capacity and overlap are remarkably similar to the results obtained for the thermodynamic equilibrium properties of a Hopfield network deduced in Sect. 19.2, see Fig. 19.1. The values of the critical capacity and overlap obtained there, $\alpha_c \approx 0.138, m_c^* \approx 0.967$, are hardly distinguishable from the results quoted above. The underlying transcendental equations (19.82,19.83) and (19.57) with (19.53) are very similar (but not identical). As a bonus, however, the calculation presented here also reveals how the equilibrium state is approached in time.

These analytical results have been checked by *numerical simulations*. These have to be performed carefully since the properties of small networks are influenced by *finite-size effects*. To observe the correct thermodynamic phase structure several thousand neurons are needed in the simulation [Am87a]. Networks with sizes up to $N = 6000$ were studied in [Za89]. From the results a critical capacity of $\alpha_c \approx 0.145$ was deduced, somewhat larger than the analytical value $\alpha_c \approx 0.1398$. Further, more serious differences between theory and simulation were observed: (1) the basins of attraction appear to be considerably larger than predicted and (2) in the case of nonrecognition of a pattern with too small $m^\mu(0)$ the overlap does not vanish as expected for the trivial attractor m_0^*. Such a remanent magnetization also has been observed in spin-glass models [He87d]. Apparently these deviations result from neglecting correlations between the random variables σ_i^μ and U_i^μ, which was required to carry out the calculation.

20. The Space of Interactions in Neural Networks

In the previous chapters we have applied the tools of statistical mechanics to determine the phase structure and equilibrium properties of the Hopfield network. The theory was based on an energy function which contained the "neuron spins" s_i as dynamical variables with interactions based on Hebb's rule or its nonlinear modification. While this rule has an appealing simplicity it generates synaptic efficacies w_{ij} which are far from optimal. One would like to study the storage properties of the best possible network for a given task. The question thus is: Given a set of p patterns $\sigma_i^\mu, i = 1, \ldots, N; \mu = 1, \ldots, p$, is there a network which has these patterns as fixed points? This is satisfied if the local field $h_i = \sum_i w_{ij} s_j$ points in the same direction (i.e. has the same sign) as the spin at site i so that the network is in a stationary state under the deterministic updating rule (3.5). We postulate the *imbedding condition*

$$\gamma_i^\mu[w] \equiv \sigma_i^\mu \sum_{j=1}^N w_{ij} \sigma_j^\mu / \|w_i\| > \kappa \,, \tag{20.1}$$

where $\|w_i\| = \sqrt{\sum_j w_{ij}^2}$ is the Euclidian norm of the interactions coupling to neuron i. It is clear that for a growing number of patterns p it will be increasingly difficult to satisfy the set of Np equations (20.1) with a constraint $\kappa \geq 0$ needed for stability. A quantitative attack on this problem again can make use of the techniques of statistical mechanics [Ga88a, Ga88b]. Now, however, the dynamical variables of the theory are the synaptic couplings w_{ij}. The neuron spins s_i, on the other hand, are fixed to a selection of "quenched" patterns σ_i^μ, which we will to assume to be random and uncorrelated.[1] The problem can be cast into the language of statistical mechanics if we introduce a formal *energy function*

$$H[w] = \sum_{i=1}^N \sum_{\mu=1}^p \theta\left(\kappa - \sigma_i^\mu \sum_{j=1}^N {}' w_{ij}\sigma_j^\mu / \|w_i\|\right) = \sum_{i=1}^N \sum_{\mu=1}^p \theta\left(\kappa - \gamma_i^\mu[w]\right) \,, \tag{20.2}$$

---

[1] This "space of interactions" approach was pioneered by the late Elizabeth Jane Gardner who died on 18 June 1988 at the age of 30. Her work, containing important contributions to the theory of neural networks, in particular concerning the storage capacity of Little–Hopfield networks, has been honored in a special issue of the Journal of Physics (J. Phys. A: Math. Gen. 22, 1954 ff. (1989)).

which depends on the configuration $[w] \equiv \{w_{ij}\}$ in the space of interactions. The prime in the sum over j is meant to indicate that we assume the absence of self-couplings, i.e. the sum does not include the diagonal element w_{ii}. The function $H[w]$ simply counts the number of weakly imbedded pattern spins with stability γ_i^μ less than κ (which for $\kappa = 0$ is just the number of errors in the memorized patterns). Using this energy function the thermodynamics of the system can be studied. In the *canonical approach* the partition function for an ensemble of networks in thermal equilibrium at a fictitious temperature $T = 1/\beta$ (not related to the operating temperature of the network introduced earlier) reads

$$Z = \int \left(\prod_{i,j}' dw_{ij} \right) \rho[w]\, e^{-\beta H[w]} , \tag{20.3}$$

where $\rho[w]$ is the normalized density of states in the space of interactions. As usual the interesting quantity is the logarithm of the partition function (related to the free energy $F = -\beta^{-1} \ln Z$) from which, e.g., the "mean energy" can be calculated:

$$E = -\frac{\partial \ln Z}{\partial \beta} . \tag{20.4}$$

In the interesting zero-temperature limit, E is the minimum number of wrong pattern spins W that can be realized by any network. This quantity still depends in a complicated way on the specific structure of the patterns σ_i^μ to be stored. Assuming statistically independent random patterns a meaningful general statement on the storage properties will be obtained by performing the *average* over the distribution of spin values in the patterns. Denoting this average over the quenched random patterns by double angular brackets one has to calculate

$$W = -\lim_{\beta \to \infty} \frac{\partial}{\partial \beta} \langle\!\langle \ln Z \rangle\!\rangle . \tag{20.5}$$

This averaged logarithm of the partition function can be evaluated using the replica trick discussed in connection with the spin-glass model.[2]

While the calculation can be worked out for finite β [Ga88b] we will immediately consider the zero-temperature limit $\beta \to \infty$;

$$Z = \int \left(\prod_{i,j}^{N}{}' dw_{ij} \right) \rho[w] \prod_{i=1}^{N} \prod_{\mu=1}^{p} \left[e^{-\beta}\theta(\kappa - \gamma_i^\mu) + \theta(\gamma_i^\mu[w] - \kappa) \right]$$

$$\rightarrow \int \left(\prod_{i,j}^{N}{}' dw_{ij} \right) \rho[w] \prod_{i=1}^{N} \prod_{\mu=1}^{p} \theta(\gamma_i^\mu - \kappa) . \tag{20.6}$$

According to its general definition (17.10) the partition function approaches

---

[2] For an alternative treatment using the "cavity method" [Me87], which studies the reaction of the system if its size is increased from N to $N + 1$, see [Me89b].

$$Z = \sum_{E_n} \Omega(E_n)e^{-\beta E_n} \to \Omega(0) , \tag{20.7}$$

where $\Omega(E_n)$ represents the number of states with energy E_n. $\Omega(0)$ counts the number of possible states with energy $E = 0$. We recognize (20.6) as the fractional volume in the space of interaction where the stability condition (20.1) is satisfied. The thermodynamical quantity of interest is the *entropy*, which for the zero-energy state is given by the microcanonical formula

$$S = \ln \Omega(0) = \ln \int \left(\prod_{i,j}^{N}{}' dw_{ij} \right) \rho[w] \prod_{i=1}^{N} \prod_{\mu=1}^{p} \theta(\gamma_i^{\mu}[w] - \kappa) . \tag{20.8}$$

If there are no constraints which relate the rows of the w_{ij} matrix to each other (such as the symmetry condition $w_{ij} = w_{ji}$ of the Hebbian model) the expression for the entropy factorizes:

$$S = \sum_{i=1}^{N} \ln \Omega_i \tag{20.9}$$

with

$$\Omega_i = \int \left(\prod_{j}^{N}{}' dw_{ij} \right) \rho[w] \prod_{\mu=1}^{p} \theta(\gamma_i^{\mu}[w] - \kappa) . \tag{20.10}$$

After averaging over an ensemble of random patterns σ_i^{μ} the fractional volume will be the same at all sites so that it will be sufficient to perform the calculation of $\langle\!\langle \ln \Omega_i \rangle\!\rangle$ only once for an arbitrary fixed site i. Using the *replica trick*

$$\langle\!\langle \ln \Omega_i \rangle\!\rangle = \lim_{n \to 0} \frac{\langle\!\langle (\Omega_i)^n \rangle\!\rangle - 1}{n} , \tag{20.11}$$

we have to evaluate (for fixed i)

$$\langle\!\langle \ln \Omega_i \rangle\!\rangle = \left\langle\!\!\!\left\langle \prod_{a=1}^{n} \int \left(\prod_{j}^{N}{}' dw_{ij}^a \right) \rho[w^a] \prod_{\mu=1}^{p} \theta\left(\sigma_i^{\mu} \sum_{j=1}^{N}{}' w_{ij}^a \sigma_j^{\mu} / \|w_i\| - \kappa \right) \right\rangle\!\!\!\right\rangle . \tag{20.12}$$

To evaluate this expression we have to specify the geometry of the space of interactions. If the couplings w_{ij} can take on arbitrary real values there is an infinite class of equivalent configurations which differ only by scale transformations $w_{ij} \to \lambda_i w_{ij}$. (Observe that λ_i drops out of γ_i^{μ}, (20.1)). Thus without restricting generality we can impose the normalization condition

$$\|w_i\|^2 = \sum_{j=1}^{N}{}' w_{ij}^2 = N , \tag{20.13}$$

which defines the *spherical model*. The corresponding integration measure is

$$\rho[w] = \frac{\delta\left(\sum_j' w_{ij}^2 - N\right)}{\int \left(\prod_j' dw_{ij}\right) \delta\left(\sum_j' w_{ij}^2 - N\right)} \, . \tag{20.14}$$

In another model of considerable interest only discrete couplings $w_{ij} = \pm 1$ are allowed (*Ising bonds*):

$$\rho[w] = \frac{\delta(w_{ij} - 1) + \delta(w_{ij} + 1)}{\int \left(\prod_j' dw_{ij}\right) [\delta(w_{ij} - 1) + \delta(w_{ij} + 1)]} \, . \tag{20.15}$$

Since the Ising model is difficult to treat theoretically we will first concentrate on the spherical model.

20.1 Replica Solution of the Spherical Model

Following Gardner [Ga88a] we introduce the following integral representation for the step function in (20.12):

$$
\begin{aligned}
\theta\big(\gamma_i^\mu[w^a] - \kappa\big) &= \int_\kappa^\infty d\lambda_\mu^a \, \delta\big(\gamma_i^\mu[w^a] - \lambda_\mu^a\big) \\
&= \int_\kappa^\infty \frac{d\lambda_\mu^a}{2\pi} \int_{-\infty}^\infty dx_\mu^a \exp\Big[ix_\mu^a\Big(\lambda_\mu^a - \sigma_i^\mu \sum_j' w_{ij}^a \sigma_j^\mu / \sqrt{N}\Big)\Big].
\end{aligned}
\tag{20.16}
$$

The exponential dependence makes it easy to take the average over the pattern spins σ_i^μ which are assumed to follow a random distribution with $\Pr(\sigma_i^\mu = \pm 1) = \frac{1}{2}$. The relevant part of (20.12) is

$$
\begin{aligned}
\left\langle\!\!\left\langle \prod_{a=1}^n \prod_{\mu=1}^p e^{-ix_\mu^a \sum_j' w_{ij}^a \sigma_i^\mu \sigma_j^\mu / \sqrt{N}} \right\rangle\!\!\right\rangle &= \prod_{j=1}^N{}' \prod_{\mu=1}^p \left\langle\!\!\left\langle \prod_{a=1}^n e^{-ix_\mu^a w_{ij}^a \sigma_i^\mu \sigma_j^\mu / \sqrt{N}} \right\rangle\!\!\right\rangle \\
&= \prod_{j=1}^N{}' \prod_{\mu=1}^p \left(\frac{1}{2} \prod_{a=1}^n e^{-ix_\mu^a w_{ij}^a / \sqrt{N}} + \frac{1}{2} \prod_{a=1}^n e^{+ix_\mu^a w_{ij}^a / \sqrt{N}}\right) \\
&= \prod_{j=1}^N{}' \prod_{\mu=1}^p \cos\left(\sum_{a=1}^n x_\mu^a w_{ij}^a / \sqrt{N}\right) \\
&= \exp\left[\sum_{j=1}^N{}' \sum_{\mu=1}^p \ln\cos\left(\sum_{a=1}^n x_\mu^a w_{ij}^a / \sqrt{N}\right)\right] \\
&\simeq \exp\left(-\frac{1}{2} \sum_{j=1}^N{}' \sum_{\mu=1}^p \sum_{a,b=1}^n x_\mu^a x_\mu^b w_{ij}^a w_{ij}^b / N\right),
\end{aligned}
\tag{20.17}
$$

where in the last step the Taylor expansion $\ln\cos x = -\frac{x^2}{2} + \dots$ was used.

We introduce the set of $\frac{1}{2}n(n-1)$ parameters

$$q_{ab} = \frac{1}{N} \sum_{j=1}^{N}{}' w_{ij}^a w_{ij}^b , \qquad a < b , \tag{20.18}$$

which are the mutual overlaps between the couplings realized in the various replica copies (corresponding to the Edwards–Anderson parameters in the spin-glass case). Equation (20.12) becomes

$$
\begin{aligned}
\langle\!\langle \, \Omega_i^n \, \rangle\!\rangle \;=\;& \mathcal{N} \int \Big(\prod_a \prod_j{}' dw_{ij}^a \Big) \prod_a \delta \Big(\sum_j{}' (w_{ij}^a)^2 - N \Big) \\
&\times \int_\kappa^\infty \Big(\prod_{a,\mu} \frac{d\lambda_\mu^a}{2\pi} \Big) \int_{-\infty}^\infty \Big(\prod_{a,\mu} dx_\mu^a \Big) \\
&\times \exp\Big[\sum_\mu \Big(i \sum_a x_\mu^a \lambda_\mu^a - \frac{1}{2} \sum_a (x_\mu^a)^2 - \sum_{a<b} x_\mu^a x_\mu^b \, q_{ab} \Big) \Big] ,
\end{aligned}
\tag{20.19}
$$

where the normalization factor is

$$\mathcal{N}^{-1} = \int \Big(\prod_a \prod_j{}' dw_{ij}^a \Big) \prod_a \delta \Big(\sum_j{}' (w_{ij}^a)^2 - N \Big) . \tag{20.20}$$

The λ_μ^a and x_μ^a integrations clearly factorize with respect to the pattern index μ into $p = \alpha N$ identical factors. The result will be denoted by

$$
\begin{aligned}
e^{N\alpha G_1(q_{ab})} \;=\;& \Big[\int_\kappa^\infty \Big(\prod_a \frac{d\lambda^a}{2\pi} \Big) \int_{-\infty}^\infty \Big(\prod_a dx^a \Big) \\
&\times e^{\, i \sum_a x^a \lambda^a - \frac{1}{2} \sum_a (x^a)^2 - \sum_{a<b} x^a x^b \, q_{ab}} \Big]^p .
\end{aligned}
\tag{20.21}
$$

In order to integrate over the q_{ab} dependence we will introduce conjugate variables F_{ab} through the following representation of unity

$$
\begin{aligned}
1 \;=\;& \int_{-\infty}^\infty dq_{ab} \, \delta \Big(q_{ab} - \frac{1}{N} \sum_j{}' w_{ij}^a w_{ij}^b \Big) \\
=\;& \int_{-\infty}^\infty dq_{ab} \int_{-\infty}^\infty \frac{dF_{ab}}{2\pi/N} \, e^{\, iNF_{ab} \left(q_{ab} - \frac{1}{N} \sum'_j w_{ij}^a w_{ij}^b \right)} ,
\end{aligned}
\tag{20.22}
$$

which will be introduced for the $\frac{1}{2}n(n-1)$ combinations $a < b$. Also the normalization constraint will be expressed by an integral representation of the delta function

$$\delta \Big(\sum_j{}' (w_{ij}^a)^2 - N \Big) = \int_{-\infty}^\infty \frac{dE_a}{2\pi} \, e^{\, iE_a \left(\sum'_j (w_{ij}^a)^2 - N \right)} . \tag{20.23}$$

This leads to

$$\langle\!\langle \Omega_i^n \rangle\!\rangle = \mathcal{N} \int \left(\prod_a \frac{\mathrm{d}E_a}{2\pi} \right) \int \left(\prod_{a<b} \frac{\mathrm{d}q_{ab}\mathrm{d}F_{ab}}{2\pi/N} \right) \mathrm{e}^{\mathrm{i}N \sum_{a<b} F_{ab}q_{ab}} \, \mathrm{e}^{-\mathrm{i}N \sum_a E_a}$$

$$\times \, \mathrm{e}^{N\alpha G_1(q_{ab})} \int \left(\prod_a \prod_j{}' \mathrm{d}w_{ij}^a \right) \mathrm{e}^{\mathrm{i} \sum_a E_a \sum_j'(w_{ij}^a)^2 - \mathrm{i} \sum_{a<b} F_{ab} \sum_j' w_{ij}^a w_{ij}^b} \, .$$

$$(20.24)$$

The integrations over $\mathrm{d}w_{ij}^a$ factorize with respect to the site index j. We introduce the notation

$$\mathrm{e}^{NG_2(E_a,F_{ab})} = \left[\int \left(\prod_a \mathrm{d}w^a \right) \mathrm{e}^{-\mathrm{i} \sum_{a<b} F_{ab} w^a w^b + \mathrm{i} \sum_a E_a(w^{a\,2}-1)} \right]^N , \quad (20.25)$$

neglecting the difference between N and $N-1$ in the exponent. This finally leads to

$$\langle\!\langle \Omega_i^n \rangle\!\rangle = \mathcal{N} \int \left(\prod_a \frac{\mathrm{d}E_a}{2\pi} \right) \int \left(\prod_{a<b} \frac{\mathrm{d}q_{ab}\mathrm{d}F_{ab}}{2\pi/N} \right) \mathrm{e}^{NG(q_{ab},E_a,F_{ab})} , \qquad (20.26)$$

where

$$G(q_{ab}, E_a, F_{ab}) = \alpha G_1(q_{ab}) + G_2(E_a, F_{ab}) + \mathrm{i} \sum_{a<b} F_{ab}\, q_{ab} \, . \qquad (20.27)$$

The integrals in (20.26) are managable in the $N \to \infty$ limit by using the saddle-point method and imposing the ansatz of *replica symmetry*

$$q_{ab} = q, \qquad\qquad F_{ab} = F, \qquad\qquad E_a = E \qquad\qquad (20.28)$$

for all values of a and $b > a$. The saddle-point is defined by the three equations

$$\frac{\partial G}{\partial E} = \frac{\partial G}{\partial F} = \frac{\partial G}{\partial q} = 0 \, . \qquad\qquad (20.29)$$

In order to evaluate the derivatives we need explicit expressions for the functions $G_1(q)$ and $G_2(E, F)$. The latter function is given by an n-dimensional Gaussian integral

$$G_2(E, F) = \ln \int \left(\prod_a \mathrm{d}w^a \right) \mathrm{e}^{-\mathrm{i}\frac{1}{2}F \sum_{a\neq b} w^a w^b + \mathrm{i}E \sum_a (w^{a\,2}-1)} \qquad (20.30)$$

and can be evaluated with the methods of Sect. 19.1.2. The matrix Λ of the bilinear form in the exponent is of the type (19.40) so that the explicit expression for the determinant (19.43) can be employed in the Gaussian integration formula:

$$\begin{aligned} G_2(E, F) &= \ln \left((2\pi)^{n/2} (\det \Lambda)^{-1/2} \right) - \mathrm{i}nE \\ &= \frac{1}{2} \Big[n \ln(2\pi/\mathrm{i}) - \ln(-2E + (n-1)F) \\ &\qquad + (1-n)\ln(-2E - F) \Big] - \mathrm{i}nE. \end{aligned} \qquad (20.31)$$

Differentiating (20.27) we can evaluate the first two saddle-point conditions

$$0 = \frac{\partial G}{\partial E} = -in + \frac{\partial G_2}{\partial E} ,$$

$$0 = \frac{\partial G}{\partial F} = i\frac{1}{2}n(n-1)q + \frac{\partial G_2}{\partial F} . \tag{20.32}$$

In the limit $n \to 0$ the resulting coupled algebraic equations are solved by

$$E = i\frac{1-2q}{2(1-q)^2} ,$$

$$F = i\frac{q}{(1-q)^2} . \tag{20.33}$$

For the third saddle-point condition, which will be needed to fix the value of q, we have to evaluate the function $G_1(q)$ defined in (20.21) which is somewhat laborious. $G_1(q)$ again contains an n-dimensional Gaussian integration

$$G_1(q) = \ln \int_\kappa^\infty \left(\prod_a \frac{d\lambda^a}{2\pi} \right) \int_{-\infty}^\infty \left(\prod_a dx^a \right) e^{-\frac{1}{2}\sum_a x^{a^2} - \frac{q}{2}\sum_{a \neq b} x^a x^b + i\sum_a x^a \lambda^a}$$

$$= \ln \int_\kappa^\infty \left(\prod_a \frac{d\lambda^a}{2\pi} \right) (2\pi)^{n/2} (\det A)^{-1/2} e^{\frac{1}{2}\sum_{a,b}(i\lambda^a)(A^{-1})_{ab}(i\lambda^b)} , \tag{20.34}$$

using (19.15). The matrix A has 1 in the diagonal and all off-diagonal elements are equal to q. This leads to the determinant $\det A = [1+(n-1)q][1-q]^{n-1}$. The inverse matrix is given by

$$A^{-1} = \frac{1}{[1+(n-1)q](1-q)} \begin{pmatrix} 1+(n-2)q & -q & \cdots & -q \\ -q & 1+(n-2)q & \cdots & -q \\ \vdots & \vdots & \ddots & \vdots \\ -q & -q & \cdots & 1+(n-2)q \end{pmatrix} ,$$

which, taking the $n \to 0$ limit, leads to

$$G_1(q) = \ln \int_\kappa^\infty \left(\prod_a \frac{d\lambda^a}{2\pi} \right) \left(\frac{2\pi}{1-q} \right)^{n/2} e^{-\frac{1}{2(1-q)^2}\left[(1-q)\sum_a \lambda^{a^2} - q\left(\sum_a \lambda^a\right)^2 \right]} . \tag{20.35}$$

The remaining n-dimensional λ^a integration is somewhat unpleasant since its lower boundary does not extend to infinity and because, owing to the squared sum term in the exponent, the integrals do not decouple. However, with the familiar introduction of an auxiliary integration according to (18.5) we can get rid of the offending term and factorize the λ^a integals.

$$G_1(q) = \ln \left(\frac{2\pi}{1-q} \right)^{n/2} \int_\kappa^\infty \left(\prod_a \frac{d\lambda^a}{2\pi} \right)$$

$$\times \int_{-\infty}^\infty \frac{dz}{\sqrt{2\pi}} e^{-\frac{1}{2}\frac{1}{1-q}\sum_a \lambda^{a^2} - \frac{\sqrt{q}}{1-q}\left(\sum_a \lambda^a\right)z - \frac{1}{2}z^2}$$

$$\simeq \ln\left(\frac{2\pi}{1-q}\right)^{n/2}\int_{-\infty}^{\infty}\frac{dz}{\sqrt{2\pi}}\,e^{-\frac{1}{2}z^2}\left[\int_{\kappa}^{\infty}\frac{d\lambda}{2\pi}\,e^{-\frac{1}{2}\frac{1}{1-q}(\lambda+\sqrt{q}z)^2}\right]^n,$$

$$(20.36)$$

where we have dropped a term vanishing in the $n \to 0$ limit. Equation (20.36) can be written as

$$G_1(q) = \ln \int_{-\infty}^{\infty} Dz \left[\int_{\tau}^{\infty} Dt\right]^n \qquad (20.37)$$

with the abbreviations

$$\tau = \frac{\sqrt{q}z + \kappa}{\sqrt{1-q}} \qquad \text{and} \qquad Dz = \frac{dz}{\sqrt{2\pi}}e^{-z^2/2}. \qquad (20.38)$$

The integral over the Gaussian measure is $\int_{-\infty}^{\infty} Dz = 1$. In the limit $n \to 0$ (20.37) can be further simplified using

$$\ln \int_{-\infty}^{\infty} Dz\,\Phi^n(z) = \ln \int_{-\infty}^{\infty} Dz\,e^{n\ln\Phi} \simeq \ln \int_{-\infty}^{\infty} Dz\,(1 + n\ln\Phi)$$

$$= \ln\left(1 + n\int_{-\infty}^{\infty} Dz\ln\Phi\right) \simeq n\int_{-\infty}^{\infty} Dz\ln\Phi. \qquad (20.39)$$

Thus

$$G_1(q) = n\int_{-\infty}^{\infty} Dz\ln\int_{\tau}^{\infty} Dt \equiv n\int_{-\infty}^{\infty} Dz\ln H(\tau), \qquad (20.40)$$

where $H(x) = \frac{1}{2}\mathrm{erfc}(x/\sqrt{2})$ is related to the complementary error function $\mathrm{erfc}(x) = 1 - \mathrm{erf}(x)$. Now the third saddle-point condition (20.29) can be evaluated:

$$0 = \alpha\frac{\partial G_1}{\partial q} + i\frac{1}{2}n(n-1)F. \qquad (20.41)$$

After performing an integration by parts and using (20.33) this leads to the transcendental equation

$$q = (1-q)\frac{\alpha}{2\pi}\int_{-\infty}^{\infty} Dz\,\frac{e^{-\tau^2}}{H^2(\tau)}, \qquad (20.42)$$

which determines q as a function of α and κ. The only missing ingredient needed to evaluate the entropy is the normalization factor (20.20), which can be evaluated in an analogous manner.

$$\mathcal{N}^{-1} = \int\left(\prod_a\prod_j{}' dw_{ij}^a\right)\prod_a\delta\left(\sum_j{}'(w_{ij}^a)^2 - N\right)$$

$$= \int\left(\prod_a\frac{dE_a}{2\pi}\right)e^{N\left(-i\sum_a E_a + G_2(0,E_a)\right)}. \qquad (20.43)$$

Here replica symmetry is ensured trivially and the saddle-point integration leads to

$$\mathcal{N}^{-1} = \left[\sqrt{\frac{1}{(2\pi)(2N)}}\, e^{N/2}(2\pi)^{N/2}\right]^n \simeq e^{Nn\frac{1}{2}(1+\ln 2\pi)} , \qquad (20.44)$$

where the latter expression contains only the term of leading order in N. (Note that (20.44) can be derived from the general formula for the surface of a sphere of radius $R = N^{1/2}$ in n dimensions: $S_n = R^{n-1}\,\pi^{n/2}\,n/(n/2)!$.) Collecting the results (20.26, 20.31, 20.33, 20.40, 20.44), we find the asymptotic expression

$$\langle\!\langle\, \Omega_i^n \,\rangle\!\rangle = \exp\left[Nn\left(\alpha\int_{-\infty}^{\infty} Dz\,\ln H(\tau) + \frac{1}{2}\ln(1-q) + \frac{1}{2}\frac{q}{1-q}\right)\right], \quad (20.45)$$

where q is to be determined from the saddle-point condition (20.42). According to (20.10), (20.11) this finally yields the expectation value of the entropy

$$\langle\!\langle\, S \,\rangle\!\rangle = N \lim_{n\to 0} \frac{\langle\!\langle\, \Omega_i^n \,\rangle\!\rangle - 1}{n}$$

$$= N^2\left(\alpha\int_{-\infty}^{\infty} Dz\,\ln H(\tau) + \frac{1}{2}\ln(1-q) + \frac{1}{2}\frac{q}{1-q}\right). \quad (20.46)$$

The entropy of the synaptic system is an extensive quantity, being proportional to the number of dynamical degrees of freedom w_{ij}, N^2.

20.2 Results

The expression (20.39) for the entropy of the error-free state solves the problem of the optimal storage capacity of a Hopfield network comprising N units trained with $p = \alpha N$ patterns. With increasing α the entropy will fall since the number of states $\Omega(0)$, i.e. the volume in the space of interactions, shrinks. The capacity limit α_c is reached when error-free configurations no longer exist which formally means $\langle\!\langle\, S \,\rangle\!\rangle < 0$. An inspection of (20.46) reveals that this happens when the overlap parameter q approaches 1 and the integral goes to $-\infty$.

The corresponding value of α follows from the transcendental equation (20.39). The asymptotic behavior of the error function gives $H(x) \to \frac{1}{\sqrt{2\pi x}}e^{-x^2/2}$ for $x \to +\infty$ and $H(x) \to 1$ for $x \to -\infty$. Thus when $(1-q) \to 0$ the z-integral in (20.39) receives its main contribution from the region $z > -\kappa$ where it diverges like $(1-q)^{-1}$. The result for $\alpha_c = \alpha(q = 1, \kappa)$ is

$$\alpha_c^{-1} = \int_{-\kappa}^{\infty} Dz\,(z+\kappa)^2 . \qquad (20.47)$$

For the smallest acceptable stability parameter, $\kappa = 0$, the integral is analytically solvable with the result

$$\alpha_c^{\max} = 2 . \qquad (20.48)$$

Thus the Hopfield network with general continuous interactions can store up to $p = 2N$ uncorrelated patterns without errors, i.e. *2 bits of information per unit*. This result already had been found long ago by Cover [Co65] who derived it using an elegant geometric argument, and was rediscovered and elaborated by [Ve86, Mc87]. The capacity limit was confirmed by numerical simulations [Kr87a]. Note, however, that for $\kappa = 0$ the stability is marginal; a slight perturbation can render the stored pattern unstable. The *basins of attraction* will be increased for a positive stability threshold κ. Figure 20.1 shows that this goes together with a decrease of storage capacity.

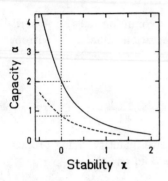

Fig. 20.1. Capacity α of a neural network as a function of the stability coefficient κ. *Full line*: continuous synapses (spherical model); *dashed line*: binary synapses (Ising model)

Various generalizations of the model have been studied.

1. *Biased patterns.* The analysis so far has assumed that the stored patterns are random with $\sigma_i^\mu = \pm 1$ having equal probability. To study the storage of *biased patterns* one has to introduce a "mean magnetization" m. This is achieved when the pattern spins are drawn from a biased random distribution

$$\Pr(\sigma_i^\mu) = \frac{1}{2}(1 + m)\,\delta_{\sigma_i^\mu,1} + \frac{1}{2}(1 - m)\,\delta_{\sigma_i^\mu,-1} \,. \tag{20.49}$$

Then the patterns will have a mean activation of $\langle \sigma_i^\mu \rangle = m$ and a mean correlation $\langle \sigma_i^\mu \sigma_i^\nu \rangle = m^2$ for $\mu \neq \nu$. (Note, however, that under this condition the correlation coefficient between two patterns defined by $N R_{\mu\nu} = \sum_i \langle \sigma_i^\mu \sigma_i^\nu \rangle - \sum_i \langle \sigma_i^\mu \rangle \langle \sigma_i^\nu \rangle$ is still zero.)

Taking the spin average $\langle\!\langle \dots \rangle\!\rangle$ with the distribution (20.49) will lead to a modification of (20.17). The replica calculation can be carried through [Ga88a] leading to a function $\alpha_c(m)$. Since the patterns get increasingly "diluted" the value of α_c will grow as m deviates from 0. For $m \to 1$ the result is

$$\alpha_c = -\frac{1}{(1 - m)\ln(1 - m)} \,. \tag{20.50}$$

However, when measuring not the number of stored patterns but their information content the capacity actually decreases. The information contained in all the "sparsely coded" patterns is determined by the probabilities p_k of having the various possible spin states s_k. According to Shannon's formula the information in bits is given in terms of the probability of occurrence of a state with probability p_k by the expression

$$I = -\frac{1}{\ln 2} \sum_k p_k \ln p_k , \tag{20.51}$$

summing over all sites and patterns. For biased random patterns this gives

$$\begin{aligned}
I &= -N^2 \alpha_c \frac{1}{\ln 2} \left[\frac{1-m}{2} \ln \frac{1-m}{2} + \frac{1+m}{2} \ln \frac{1+m}{2} \right] \\
&\rightarrow N^2 \frac{1}{2\ln 2} \simeq N^2 \times 0.72 \qquad \text{for} \quad m \rightarrow \pm 1 .
\end{aligned} \tag{20.52}$$

Thus each synapse can hold at most 0.72 bits of information in the limit of extremely sparse coding.

2. *Ising bonds.* For the case of Ising bonds $w_{ij} = \pm 1$ the calculation can be repeated with minor modifications. The discrete phase-space measure of (20.15), which replaces the integration over w_{ij} by a two-valued summation, has to be used. This will influence the function $G_2(E_a, F_{ab})$: the parameter E_a drops out since the normalization constraint $\sum_j w_{ij}^2 = N$ is satisfied automatically. As a result the formula for the storage capacity (20.47) is replaced by [Ga88b]

$$\alpha_c^{-1} = \frac{\pi}{2} \int_{-\infty}^{\infty} Dz \, (z + \kappa)^2 , \tag{20.53}$$

which implies

$$\alpha_c^{\mathrm{max}} = \frac{4}{\pi} . \tag{20.54}$$

But this result is too good to be true: as two-valued variables the couplings w_{ij} certainly will not be able to hold more than one bit of information, therefore (20.54) must be wrong! The source of the error can only lie in the approximation introduced by the replica method. Indeed, it has been shown [Ga88b] by studying the second derivatives of the function $G(q_a, E_{ab}, F_{ab})$ at the saddle point that the replica-symmetric integration is stable in the spherical model, but *unstable* for Ising bonds in the interesting region $q \rightarrow 1$. Therefore the calculation in the latter case cannot be trusted.

A closer investigation of this problem [Kr89] has led to a more realistic value of the critical storage capacity. While the "energy" (20.2) of the replica-symmetric solution, i.e. the number of wrong patterns, remained zero up to the capacity value quoted above ($\alpha = \alpha_c = 4/\pi$), the entropy was observed to become negative much earlier, at $\alpha = \alpha_S \simeq 0.83$ (taking $\kappa = 0$).[3]

---

[3] The situation is complicated further by the fact that the replica-symmetric solution, judged from the stability of the saddle-point condition, is stable up to $\alpha \simeq 1.015$.

The Ising-bond neural network seems to be characterized by a solution with first-order replica-symmetry breaking. This solution has been constructed explicitly, and the storage capacity was confirmed to be $\alpha_S \simeq 0.83$. This claim has also been substantiated by a heuristic argument concerning the geometry of configurations in the space of interactions [Kr89].

Numerical simulations are rather difficult because there is no learning rule with guaranteed convergence. Therefore the search for good combinations of the couplings w_{ij} has to be attacked as a combinatorial optimization problem. Applying a simulated annealing technique and also a heuristic search method to nets of sizes up to $N = 81$ the authors of [Am89b] found that the capacity $\alpha_c(\kappa)$ falls considerably below the replica-symmetric prediction. The capacity was estimated as $\alpha_c{}^{\max} \simeq 0.6 \ldots 0.9$, while [Ga89a] obtained $\alpha_c{}^{\max} \simeq 0.75$. The large uncertainty is caused by strong finite-size effects which make extrapolations to the limit $N \to \infty$ difficult. Thus the numerical studies are consistent with the analytical result $\alpha_c = \alpha_S \simeq 0.83$.

3. *Sign-constrained synapses.* It is a remarkable observation, and at first sight might seem counterintuitive, that it is not essential for the operation of a neural network to have synaptic couplings of varying signs. In fact the capacity of a network with sign-constrained synapses is simply reduced by a factor of two with respect to the unconstrained case. It is possible to impose a sign condition which classifies each neuron as either inhibitory or exhibitory,[4] characterized by a set of constants $g_j = \pm 1$ which can be chosen freely. Thus we demand that for all target neurons i the synaptic coefficients w_{ij} emerging from neuron j have the same sign g_j.

It is not too difficult to include this additional constraint in the calculation which gives the storage capacity [Am89c]. The constraint can be included into the function $\rho[w]$ of (20.14), which describes the density of states in the space of interactions. For the spherical model (continuous synapses) the volume of interaction space, cf. (20.10), then takes the form

$$\Omega_i \propto \int \left(\prod_j{}' \mathrm{d}w_{ij} \right) \delta \left(\sum_j{}' w_{ij}^2 - N \right) \prod_{\mu=1}^p \theta(\gamma_i^\mu[w] - \kappa) \prod_{j=1}^N \theta(w_{ij}g_j). \quad (20.55)$$

The last factor in this equation is responsible for the sign constraint. The calculations of Sect. 20.1 can be carried through with minor modifications. The optimal storage capacity is found to be [Am89c]

$$\alpha_c^{-1} = 2 \int_{-\kappa}^\infty \mathrm{D}z \, (z + \kappa)^2 \,, \quad (20.56)$$

which differs from (20.47) simply by a factor of 2. Thus a fully coupled neural network with sign constraints can store up to $p = N$ unbiased patterns, $\alpha_c{}^{\max} = 1$. By restricting the signs of the couplings the storage capacity has been cut in half.

---

[4] In neurobiology this is known as Dale's rule, cf. Sect. 1.1.

As the coefficients g_j can be chosen at will it is possible to have a functioning network with purely excitatory couplings! This can be characterized as a *ferromagnetic spin glass*. In fact, any sign-constrained network is equivalent to such a system owing to a "local gauge invariance" obeyed by the system. The state of the system remains essentially unchanged under the simultaneous local sign transformation of the couplings $w_{ij} \rightarrow g_j w_{ij}$ and of the spins $s_i \rightarrow g_i s_i$. Thus choosing, e.g., all couplings to be positive, $g_j = +1$, does not restrict generality.

It should be noted that the constrained matrix w_{ij} is not simply obtained by switching off the wrong-signed couplings (this would lead to a lower storage capacity [Sh87]) but follows a new distribution. There is one distinction between the operation of networks with and without sign constraint: in the former case the *ferromagnetic* state, where all neurons have the same activation (-1 or $+1$), is a very powerful attractor of the dynamics. Thus when a starting pattern is not in the basin of attraction of one of the stored patterns the network most likely will end up in the ferromagnetic state which can be taken as a sign of "nonrecognition". Unconstrained networks are more easily confused by an input pattern too noisy to be recognized correctly and converge to a wrong pattern. One sometimes refers to an associative memory with this property as being "opinionated".

4. *Symmetry constraint.* The storage capacity will be affected if the couplings exhibit a definite symmetry, i.e. if the values w_{ij} and w_{ji} are related to each other. This can be investigated by introducing a symmetry constraint $\delta\left(\sum'_j w_{ij} w_{ji} - \eta N\right)$ in addition to the spherical constraint $\delta\left(\sum'_j w_{ij}^2 - N\right)$ into the integration measure (20.14). Symmetry coefficients of $\eta = \pm 1$ describe fully symmetric/antisymmetric couplings while for $\eta = 0$ there is no correlation. Since the new constraint couples the rows of the synaptic matrix w_{ij} the problem no longer factorizes with respect to the site index i. An analytical treatment was found possible only in a simplified model: the network is assumed to be *diluted* so that each neuron is coupled only to $C \simeq \ln N$ other neurons, instead of $C \simeq N$ in the fully coupled case. For this model the function $\alpha_c(\kappa, \eta)$ was obtained and the following found [Ga89b]: (1) the optimum capacity $\alpha_c^{\max} = 2$ is reached at a finite positive value of η, i.e. the couplings w_{ij} and w_{ji} have a positive correlation in the optimal network; (2) the capacity of fully symmetric networks is $\alpha_c^{\max} = \alpha_c(0,1) \simeq 1.28$ (note that here $\alpha = p/C$).

5. *Higher-order couplings.* Microscopic studies of neural tissue have revealed that the picture of neurons interacting through simple synaptic contacts with efficacy w_{ij} is oversimplified. It is quite common that two or even more axonal branches jointly make contact to a dendrite, thus forming a *synaptic junction of higher order*, see Fig. 1.6. The post-synaptic potential will be influenced if the incoming activation signals are correlated. Such a correlation is described by multiplying the neuron variables s_i. The Hopfield model can be generalized

to this situation by including additional polynomial terms in the expression for the local field h_i [Pe86a].[5] In full generality we may write

$$h_i = \sum_{\lambda=1}^{\infty} \sum_{j_1 < \ldots < j_\lambda}^{N}{}' w_{ij_1\ldots j_\lambda}^{(\lambda)} s_{j_1} \cdots s_{j_\lambda} , \tag{20.57}$$

where λ denotes the order of the multineuron synapses characterized by the set of efficacies $w_{ij_1\ldots j_\lambda}^{(\lambda)}$. The ordering of the indices in the sum was introduced to avoid double counting. The prime indicates that self-couplings are excluded, $i \neq j_1, \ldots j_\lambda$. If the synaptic efficacies are fixed, the local field together with the usual updating rule (3.5) determines the time evolution of the spin variables $s_i(t)$.

How does the memory capacity of a fully coupled higher-order network compare to the ordinary case of simple synapses ($\lambda = 1$)? This question has been answered recently with the use of the space-of-interactions approach [Ko90]. Kohring for simplicity started from a set of pure couplings of order λ

$$h_i = \sum_{[J]} w_{ij_1\ldots j_\lambda} s_{j_1} \cdots s_{j_\lambda} \tag{20.58}$$

and studied the analog of the *spherical model* by normalizing the couplings to

$$\|w_i\|^2 = \sum_{[J]} w_{ij_1\ldots j_\lambda}^2 = \binom{N-1}{\lambda} \equiv M . \tag{20.59}$$

Here the index $[J]$ denotes summation over $j_1 < \ldots < j_\lambda \neq i$. The binomial coefficient M gives the number of λ synapses coupling to a given neuron (the use of N instead of $N-1$ in (20.15) is irrelevant in the large-N limit). The imbedding condition, which guarantees the local stability of pattern μ at site i, is generalized to

$$\gamma_i^\mu[w] \equiv \frac{1}{\|w_i\|} \sigma_i^\mu \sum_{[J]} w_{ij_1\ldots j_\lambda} \sigma_{j_1}^\mu \cdots \sigma_{j_\lambda}^\mu > \kappa . \tag{20.60}$$

The fractional volume of interaction space Ω_i is given by

$$\Omega_i \propto \int \left(\prod_{[J]} dw_{ij_1\ldots j_\lambda} \right) \delta \left(\sum_{[J]} w_{ij_1\ldots j_\lambda}^2 - M \right) \prod_{\mu=1}^{p} \theta \left(\gamma_i^\mu[w] - \kappa \right) , \tag{20.61}$$

which is an obvious generalization of (20.10).

The average $\langle\!\langle \ln \Omega_i \rangle\!\rangle$ can be evaluated with the replica trick. As it turns out, averaging over the pattern spins σ_j^μ again leads to (20.19) with the overlap parameter between different replicas replaced by

---

[5] The use of higher-order synapses for invariant pattern recognition has been discussed in Sect. 8.3.

$$q_{ab} = \frac{1}{M} \sum_{[J]} w^a_{ij_1\ldots j_\lambda} w^b_{ij_1\ldots j_\lambda} , \qquad a < b , \tag{20.62}$$

instead of (20.18). Thus the calculations of Sect. 18.1 apply to multineuron couplings of arbitrary order. The only dependence on λ enters through the replacement $N \rightarrow M$. The values of the capacity limit $\alpha_c(\kappa)$ given in (20.47) and $\alpha_c^{\mathrm{max}} = 2$ remain valid, but now the number p of stable patterns is related to α by

$$p = \alpha M = \alpha \binom{N-1}{\lambda} \tag{20.63}$$

instead of $p = \alpha N$. As a consequence, the number of patterns that can be stored in a multiconnected network with couplings of order λ grows like $p \simeq N^\lambda$ [Ba87b]. However, this enhanced capacity is bought at the expense of a larger number of required synapses; the storage density of information remains limited to two bits per synapse.

Kohring [Ko90] has also studied the *dynamics* of multiconnected networks, finding significant differences in the size of the basins of attraction between the cases $\lambda = 1$ and $\lambda > 1$.

6. *Feed-forward networks.* The capacity analysis given above is not limited to a Hopfield-type autoassociative memory. It applies as well to feed-forward networks without hidden layers (i.e. simple perceptrons) representing Boolean functions. Consider a network with input units $\sigma_j = \pm 1$, $j = 1, \ldots, N$ and output units $S_i = \pm 1, i = 1, \ldots, N'$. The output units are activated according to the usual discrete threshold rule

$$S_i = \mathrm{sgn}\left(\sum_{j=1}^N w_{ij} \sigma_j \right), \qquad i = 1, \ldots, N' . \tag{20.64}$$

This represents a Boolean function which yields certain N'-bit output vectors for each of the 2^N input combinations. One can ask the question: At how many different "grid points" or "examples" σ_j^μ can the function values S_i^μ be independently prescribed? Since (20.64) has the same structure as the fixed-point condition for the autoassociative memory both problems are equivalent. In particular, in the absence of correlations and for unconstrained weights w_{ij} there are at most $p = 2N$ such examples that can be learned. It will not be possible to enforce additional values of the Boolean function without introducing errors. [De89, Pa87b, Ca87b] discuss the relevance of this limitation of capacity to the ability of perceptrons for generalization and classification, cf. Sect. 9.2.2. A more recent review is [Wa93b].

Part III

Computer Codes

21. Numerical Demonstrations

21.1 How to Use the Computer Programs

In the first two parts of this book we have introduced various concepts for the architectural design and learning protocol of neural networks. In the first part we have tried to present a comprehensive overview, discussing the potential and also the limitations of many of the models which have been proposed. Subsequently the properties of some of these models which yield themselves to theoretical analysis have been studied in more detail using tools borrowed from statistical physics and probability theory. However, although such a treatment may be elegant and enlightening, it leaves something to be desired. It is generally true that the full grasp of a new concept is best obtained by practical applications and exercises. The subject of neural networks, especially when being viewed as a new paradigm of computer programming, is very well suited to such an approach. Since the principles of most of the network architectures can be studied by *numerical simulations* on a fairly small scale using a personal computer we have decided to append a selection of programs to this book which can be run on IBM-PC-compatible computers.

The purpose of these programs is to provide practical demonstrations of some of the main classes of neural networks. The reader is encouraged to 'play' with these programs: by varying the input data and the model parameters one can gain a familiarity with the models which is hard to acquire from theoretical ruminations alone. Of course, running prefabricated software is only of limited educational value. We therefore provide both the executable versions of the programs and the source codes and urge readers to their implement own modifications of the codes. Ambitious programmers will certainly find ample opportunity to improve and extend our demonstration programs. Also, because of space limitations (and the wish to get this book finished within a finite time) we have implemented only a few of the plethora of known neural-network architectures. The reader should feel encouraged to venture beyond this limited domain, e.g. by writing programs for unsupervised learning, Boltzmann machines, or any other model he is fond of.

In the following chapters we will discuss the programs one by one. For ease of reference each chapter gives a summary of the corresponding network model as it was introduced in earlier parts of the book. (Where necessary new theoretical concepts are also discussed.) This is followed by a brief description of the program and its input and output. For full details the program source code should be consulted. Section 21.2 provides some information on the software implementation.

When executing the programs the user has to interact with them by providing input parameters. These are, e.g., the size of the network, or model parameters determining the learning mode or convergence behavior of the algorithms. Most of these parameters can be entered through input panels where a set of default values are predefined. These values can be changed by moving the cursor to the respective position and typing in the chosen number (don't forget to press <Return> when finished!). The progam tries to check for illegal entries and hopefully rejects them. When you are satisfied with the parameters you can leave the input panel by pressing <Escape>. If the program requires additional input data often standard values are suggested, enclosed in square brackets. For the first encounter with a program the best strategy may be just to take over the predefined parameters.

Finally a word of caution: our programs are intended to illustrate basic operational principles of neural networks. One should not expect too much of them with regard to the ability to solve realistic problems, let alone to emulate operations of the brain. Neither the speed nor the memory capacity of presently available personal computers is sufficient for this task. For the more ambitious investigator a large number of general-purpose or dedicated software neurosimulators are on the market. Also dedicated hardware products like neural-network add-on cards and full-sized neurocomputers are commercially available to help bridge the gap in computing power between personal computer and large massively parallel systems.

It is instructive, and very sobering, to take a look at the performance of biological neural networks for comparison [DA88]. At the top of the hierarchy, the human brain contains at least 10^{14} synapses, which gives a measure of the storage capacity. The computing power is measured by the number of *connections per second* (CPS). In the human brain this number is of the order of 10^{16} CPS assuming an updating frequency of 100 cycles per second. Even the brain of a lowly fly has 10^8 synapses with perhaps 10^9 CPS. This number rivals the computing power of the most advanced supercomputers available at present, which is of the order of 10^8 CPS, and of dedicated electronic neurochips, see Table 12.1 in Chapter 12. (Those readers who practice catching house-flies by hand will appreciate this fact.) The capacity of a PC-based simulation program with perhaps 10^5 CPS appears very modest indeed when viewed from this perspective.

21.2 Notes on the Software Implementation

The programs in this book were developed on IBM AT and PS/2 personal computers under the DOS operating system. They should run without modifications on any personal computer compatible with this standard.

The accompanying software diskette contains the source code of the demonstration programs, the executable modules, and some input data files. In addition there are a few files containing preprocessor directives and a collection of auxiliary routines for graphics display and screen control. The software contained on the diskette is briefly described in the file READ.ME. This text file also explains how the software can be installed on the hard disk of your computer.

The code is written in the programming language C, which we chose for its versatility, portability, and widespread availability on personal computers. We expect that the programs can be adapted for other computer systems with relative ease (complete portability is, alas, a goal which is hard to achieve given the fragmented state of present-day hardware and software development). We followed the ANSI-C standard and tried to avoid any compiler-specific statements. The executable modules were generated using the BORLAND C++ compiler (version 4.0) (note, however, that the code is written in plain C, object-oriented features are not used). The programs can be compiled without modification also with Borland's TURBO-C++ compiler (version 1.0) and with the MICROSOFT-C compiler (version 6.0); some minor incompatibilities were overcome by including preprocessor directives of the type #ifdef __TURBOC__ (commands) #endif.

Special tricks and programming hacks have generally been avoided to make the construction of the code transparent even if this meant sacrificing a bit of execution speed or elegance. For example, multidimensional arrays are accessed via the index notation a[i][j] instead of using pointer manipulations. In general we do not make use of dynamic storage allocation.

Indices which can take on N different values according to the convention used in C run in the range $i = 0, \ldots, N - 1$ although in the program description we sometimes use the more 'natural' notation $i = 1, \ldots, N$. It is assumed that the data types char, int, and long can hold 1, 2, and 4 bytes, respectively. Arrays of real numbers often are dimensioned as 4-byte variables of type float instead of double to save memory space while sacrificing some numerical accuracy. This can be easily changed by the user.

A problem was posed by the necessity of displaying high-resolution graphics. Although many compilers offer quite versatile graphics commands, there is no generally agreed standard for this task. Since our programs require only a fairly simple graphics display we support it by a small set of 'self-knitted' auxiliary routines. These routines work with the CGA, EGA, and Hercules monochrome graphics adapters (and compatibles). We suggest using the EGA color mode which is available on most personal computers. The operation of the graphics routines is described in the file READ.ME. The reader should be able to replace calls to these routines with the graphics commands made available by his or her favorite C compiler or software package without too much effort. Access to the text screen is controlled by a public-domain version of the "CURSES" library.

To allow for convenient modification of the programs we have provided files supporting the MAKE utility. In this way each of the programs can be re-compiled and linked with a single command, e.g. invoking make -f asso.mak to generate the executable code asso.exe. The user may have to modify the path names in these make-files to specify in which directory the compiler is located. For details again consult READ.ME.

The floating-point emulation library is linked so that the programs can make use of a numerical coprocessor but do not require its presence. However, the execution speed of most of the programs without a coprocessor will be uncomfortably slow.

Nearly all of the programs at some stage make use of random numbers, e.g. for providing the initialization of the weights of a network, for generating patterns to be learned, etc. The (pseudo)random numbers are obtained through the standard function rand() which we initialize, by default, using the system time. In this way each execution run of a program will follow a different path. However, it is also possible to repeat a run with exactly the same random numbers. For this you just have to call the program with an optional command line parameter (for example entering asso 65779) which then is used as the seed number for the random-number generator.

Finally we mention that several of the programs offer the opportunity to slow down the execution speed (personal computers just have become too fast in recent years). By pressing the key s you can cycle through the modes 'normal speed', 'slow motion', and 'suspended operation'.

22. ASSO: Associative Memory

This program implements an associative-memory network of the type advanced by Hopfield [Ho82] and Little [Li74]. In accordance with the discussion of Chapt. 3 the network consists of a set of N units s_i which can take on the values ± 1. A set of p patterns σ_i^μ each containing N bits of information is to be memorized. This storage is distributed over the set of $(N \times N)$ real-valued synaptic coefficients w_{ij} mutually connecting the neurons. Memory recall proceeds by initially clamping the neurons to a starting pattern $s_i(0)$. Then the state of the network develops according to a relaxation dynamics in discrete time steps $s_i(0) \to s_i(1) \to \dots$. Whether a neuron flips its state depends on the strength of the 'local field'

$$h_i(t) = \sum_{j=1}^{N} w_{ij} s_j(t) . \tag{22.1}$$

Several varieties of updating rules can be introduced. In a deterministic network the new state of the neuron is uniquely controlled by the local field derived from the previous time step

$$s_i(t+1) = \mathrm{sgn}\big[h_i(t) - \vartheta\big] ; \tag{22.2}$$

ϑ is a global threshold which is normally set equal to zero for balanced patterns (equal numbers of "on" and "off" neurons). In contrast to this rule, in a stochastic network only the probability for the new state is known, namely

$$\Pr\{s_i(t+1) = \pm 1\} = f\big(\pm(h_i(t) - \vartheta)\big) , \tag{22.3}$$

where $f(-x) = 1 - f(x)$. The probability function $f(x)$ usually is taken to be

$$f(x) = \left(1 + e^{-2\beta x}\right)^{-1} , \tag{22.4}$$

where the parameter $\beta = 1/T$ has the interpretation of an inverse temperature.

Given the updating rule (22.2) or (22.3) one still has the choice of timing. The updating steps can be taken as follows.

- *In parallel*, i.e. all values s_i are updated 'simultaneously' depending on the state of the other neurons $s_j(t)$ before the common updating step.

- *Sequentially* in the sense that the updating proceeds in a given order (e.g. with increasing value of the label i). The local field is calculated as a function of $s_j(t), j < i$ and $s_j(t-1), j \geq i$.
- *Randomly*, so that the update proceeds without a fixed order and a given neuron is updated once per unit time interval only on the average.

More interesting than the relaxation dynamics are the *learning rules* which determine the synaptic coefficients w_{ij} from the patterns σ_i^μ which are to be stored. The following alternatives are implemented in the program ASSO.

(1) The original Hebb rule

$$w_{ij} = \frac{1}{N} \sum_\mu \sigma_i^\mu \sigma_j^\mu \tag{22.5}$$

which was discussed extensively in Parts I and II of this book.

The aim of the following more involved learning rules is to ensure the stability of the sample patterns by iteratively increasing the *stability coefficients*

$$\tilde{\gamma}_i^\mu = \sigma_i^\mu \sum_j w_{ij} \sigma_j^\mu = \sum_j w_{ij} \sigma_i^\mu \sigma_j^\mu . \tag{22.6}$$

The tilde serves to distinguish this coefficient from the normalized stability

$$\gamma_i^\mu = \frac{\tilde{\gamma}_i^\mu}{\|w_i\|} \tag{22.7}$$

introduced in Sect. 10.1.3 and Chapt. 20 Here $\|w_i\|$ is the Euclidian norm

$$\|w_i\| = \left(\sum_k w_{ik}^2 \right)^{\frac{1}{2}} . \tag{22.8}$$

(2) The iterative learning rule of Diederich and Opper [Di87], cf. Sect. 10.1.2, in the form

$$w_{ij}^{\text{new}} = w_{ij}^{\text{old}} + \sum_\mu \delta w_{ij}^\mu \tag{22.9}$$

with the synaptic modifications

$$\delta w_{ij}^\mu = \frac{1}{N} \theta \left(1 - \tilde{\gamma}_i^\mu \right) \sigma_i^\mu \sigma_j^\mu . \tag{22.10}$$

(3) The learning rule of Gardner [Ga88a] and Forrest [Fo88b], which is slightly different from (22.10) in that the threshold for synaptic modification is determined locally for each neuron:

$$\delta w_{ij}^\mu = \frac{1}{N} \theta \left(\kappa - \gamma_i^\mu \right) \sigma_i^\mu \sigma_j^\mu . \tag{22.11}$$

κ is a positive real constant which determines the 'depth of imbedding' of the patterns.

(4) The prescription of Abbott and Kepler [Ab89] results from (22.10) by scaling of the synaptic increment with a smoothly varying coefficient f_i

$$f_i = \frac{1}{N} \|w_i\| \left(\kappa - \gamma_i^\mu + \delta + \sqrt{(\kappa - \gamma_i^\mu + \delta)^2 - \delta^2} \right) . \qquad (22.12)$$

δ is a small parameter of the order $\delta \approx 0.01$.

Note: To speed up computations, the couplings w_{ij} are scaled by a factor of N and treated as *integer variables* in the program ASSO. Therefore only a restricted form of the learning rule (22.12) could be implemented where the $N\delta w_{ij}^\mu$ is rounded to the next higher integer. If many iteration steps are required, the accumulated values of w_{ij} can overflow the allowed range of integers, and the iteration fails.

(5) The learning rule of Krauth and Mézard [Kr87a]. The algorithm repeatedly visits all neurons and selectively enhances the pattern $\mu_0(i)$ which locally has the lowest stability. The updating rule is

$$\delta w_{ij} = \frac{1}{N} \sigma_i^{\mu_0(i)} \sigma_j^{\mu_0(i)} , \qquad (22.13)$$

where $\mu_0(i)$ is that pattern for which the stability coefficient takes its minimum value

$$\gamma_i^{\mu_0(i)} = \min_{\nu=1\ldots p} \gamma_i^\nu \equiv c_i . \qquad (22.14)$$

The iteration terminates when all coefficients c_i are larger than a predetermined positive real constant C. This learning rule in principle is able to exhaust the maximum storage capacity of a Hopfield network, i.e. $\alpha = 2$ [Kr87a].

These learning rules generate synaptic coefficients w_{ij} of both signs. However, it is also interesting to study the case where all elements in a column of the matrix w_{ij} (i.e. the outgoing synapses of neuron j) have a *predetermined sign* ϵ_j. According to Dale's law, see Sect. 1.1, this seems to be the case in biological neural systems. It is remarkably simple to introduce such a sign constraint [Am89d] for the iterative learning rules discussed above: one simply has to discard those learning steps which would produce a synapse of the wrong sign. Starting from a configuration having the correct signs, the constraint can be enforced in subsequent steps by the prescription

$$\delta w'_{ij} = \delta w_{ij} \, \theta \big[w_{ij}(w_{ij} + \delta w_{ij}) \big] , \qquad (22.15)$$

which ensures that the sign is not changed during the learning process. The increments δw_{ij} can be taken from any of the learning rules introduced above.

22.1 Program Description

Since the human brain is not particulary adept at analyzing long bit-strings the state of the network is graphically displayed as a two-dimensional rectangular field with dimensions (nx × ny) filled with dark or light squares.

(Keep in mind, however, that this representation of the data just serves to make a nice display. The Hopfield network knows nothing about geometrical properties or pattern shapes.) The maximum dimensions are nx=13 , ny=12, which leads to a network of $N = 156$ neurons. The network can be trained both with ordered and with random patterns. The former are read from a data file ASSO.PAT.

This is an ordinary text file which has to contain a first line having the format
    Columns: nx Rows: ny Patterns: pat
where nx, ny, and pat are integers characterizing the size and number of patterns. This is followed by nx scan lines for the patterns which are formed by blank spaces and asterisks (or any other nonblank character). The pat different patterns are separated from each other by a line of arbitrary content. We have provided a data file which contains renditions of the 26 capital letters $A \ldots Z$. In addition the file ASSO1.PAT contains a 'skinny' version of the same set of letters, i.e. the values $\sigma_i^\mu = -1$ are heavily favored against $\sigma_i^\mu = +1$. To use it, one has to rename this file ASSO.PAT (don't forget to save the old pattern file beforehand).

The user is asked to specify the total number of patterns patts and the number of predefined patterns plett among them (which may be set to zero). In the case patts>plett the patterns read in from the file are complemented by the necessary number of random patterns. These have no 'mean magnetization', i.e. the probabilities for $\sigma_i^\mu = \pm1$ are equal. Up to patts=312 patterns are allowed, which is useful when studying the saturation properties of the iterative learning rules. However, for large values of patts, learning times tend to become intolerably large.

Next the parameter wsign can be specified. It determines, whether the synaptic couplings can take on any real value or are forced to have a specified sign. This is a special case ($\epsilon_j = +1$ or -1 for all j) of the sign constraint mentioned in the last section. If we put wsign=1 or 2, the sign constraint is implemented by modifying the iterative learning rules according to (22.15).

Finally the variable rule is read in and this selects the learning rule for the synaptic coefficients. In addition to the five choices explained above also heteroassociation of patterns $(1 \to 2 \to 3, \ldots)$ can be learned using the Diederich-Opper rule. Since this will be the subject of Chapt. 23 we can skip the discussion here. We only mention that in the pattern-retrieval phase parallel updating has to be chosen if heteroassociation is to work.

After these parameters have been specified the learning procedure is started (by pressing <Escape>). The program displays in rapid succession the patterns to be learned. Some of the learning models ask for the entry of an additional parameter which determines the strength of the local embedding of the patterns. During the learning phase a marker indicates at which of the neurons the synapses are modified at a given instant. In the case of the non-iterative Hebbian rule this marker goes just once through the field of display. For the improved learning rules it moves patts times through the network and this is repeated in each of the iteration steps, also known as 'learning epochs'. However, in later steps the imbedding condition will be

satisfied at most locations and the number of synaptic modifications drops
sharply (if the learning rule leads to convergence). It is interesting to watch
how a few neurons (or rather the synapses attached to them) turn out to be
'stubborn' and have to be updated repeatedly. After each iteration step the
number of modified synapses change is displayed in an extra window to the
right of the screen. (In the case of the Krauth learning rule the lowest stability coefficient γ is shown instead.) The learning terminates if one of three
conditions is met. (1) Convergence is reached, i.e. change=0. (2) A timeout
is reached after 1000 iteration steps. (3) The user runs out of patience and
stops the iteration by pressing any key.

The learning process can also be temporarily interrupted by pressing the
key 'a' to analyze the synaptic matrix w_{ij}. The minimum, maximum, average, and mean square values of the coupling elements are given (note that
internally the program works with synaptic coefficients Nw_{ij} scaled to *integer* numbers). In addition the *symmetry parameter* η is shown; it is defined
by

$$\eta = \frac{\sum_{ij} w_{ij} w_{ji}}{\sum_{ij} w_{ij}^2} . \tag{22.16}$$

Subsequently the user is given the choice of inspecting a histogram of
the current distribution of coupling coefficients w_{ij}. In addition the stability coefficients can be analyzed. If this option is chosen, after a slight delay a table shows the distribution of values of the normalized local stability
coefficients γ_i^μ defined in (22.7) The sum over all entries in this table is
pN (=patts*ntot). Also the minimum, average, and maximum values are
shown. The distribution of the coefficients γ_i^μ contains valuable information
on the performance of the associative memory. In particular, the negative
coefficients (if any) correspond to those locations where a bit in one of the
patterns is stored incorrectly. The distribution of stability coefficients is also
displayed graphically as a histogram. The stability analysis can be performed
repeatedly during the learning phase, if desired.

After finishing the learning process the synaptic coefficients can be analyzed once again. Having studied this information one proceeds to a further
input panel asking for the following variables:

– update: The rule of updating the network, as discussed above. Random,
  sequential, or parallel dynamics can be chosen.
– sstep: Single-step mode.
– temp: The temperature T of the network.
– theta: The neural threshold parameter ϑ in (22.2). The same value is used
  for all neurons. For normal operation set $\vartheta = 0$.
– maim: This determines how the starting pattern is obtained. A correct
  pattern will be distorted either by random flipping of bits or by masking,
  i.e. blanking out, one or several lines at the bottom of the pattern.

After leaving this panel the network is ready to operate as an associative memory. The following control parameters can be chosen:

- s: Search for a pattern.
- p: Change the operating parameters as described above.
- m: Modify the synaptic coefficients.
- e: Exit the program.

Pressing the key 'm' leads to an input panel which allows one to change the synaptic matrix w_{ij} which previously had been determined by the learning process. Three distorting operations can be applied to the coupling matrix, characterized by the following variables.

- dead: This parameter determines the degree of *dilution*. It is the probability for a synapse w_{ij} to be disrupted, i.e. to have the value 0. Nonzero values of dead can be used to study the 'graceful degradation' of the network, i.e. the fact that its overall performance falls off smoothly as the components develop failures.
- cutoff: A nonzero positive value w_0 of this parameter leads to the *clipping* of the synaptic strength according to

$$w'_{ij} = \begin{cases} -w_0 & w_{ij} < -w_0 \\ w_{ij} & \text{for} \quad -w_0 \leq w_{ij} \leq w_0 \\ +w_0 & w_0 < w_{ij} \end{cases} . \tag{22.17}$$

- binary: If binary=1 is chosen, the synaptic matrix is projected to binary values (*Ising couplings*), $w'_{ij} = \text{sgn}(w_{ij})$. Couplings with value zero are not affected so that the dilution of the network is preserved.

Having changed the couplings the new matrix w'_{ij} can be analyzed again. After the original learning process a backup copy of the coupling matrix is stored. Therefore the "modify synapses" option can be selected repeatedly, alway operating on the original synaptic matrix.

If one presses the key 's', the program asks for the starting pattern, which for convenience is designated by a letter (even if you have defined your own type of pattern or if random patterns are learned). Thus at most the first 25 patterns can be chosen for searching, even if patts is larger. The last required input depends on the previous choice of maim. Either the number of blanked-out lines blank or the fraction of randomly flipped bits noise has to be entered. Note the definition of noise: it is not the fraction of bits which are inverted. Rather, the algorithm visits all neurons and selects each of them with the probability noise. The state of a selected neuron is then determined by 'tossing a coin'. In this way even the maximum value noise=1 will lead to a pattern in which about half of the bits are correct but there is no correlation between the original and the distorted pattern. Using noise directly to flip the neurons would lead to anticorrelated patterns. Since the Hopfield network is completely symmetric under spin flip $s_i \rightarrow -s_i$ no new information could be gained by allowing such starting configurations.

Now the pattern-retrieval phase can begin. Three frames are displayed which contain (1) the distorted starting pattern $s_i(0)$, (2) the correct pattern σ_i^μ which is to be recognized, (3) the actual state of the network. By sweeping through the network the values $s_i(t)$ are rapidly updated according to the rule specified before. Alternatively, in the single-step mode (sstep=1) after each updating sweep the program pauses until the return key is pressed. The updating cycles are stopped when the correct pattern has been recognized, i.e. $s_i(t)$ agrees with σ_i^μ. If the network fails to converge to the correct pattern the iteration can be interrupted manually.

To facilitate an analysis of the network performance two columns of numbers are displayed. These are the overlaps of the initial ($t = 0$) and of the actual state of the network with the set of stored patterns $m_\mu(t) = \frac{1}{N} \sum_{i=0}^N \sigma_i^\mu s_i(t)$. This overlap is related to the number of incorrect bits (the Hamming distance) by $H_\mu(t) = \frac{N}{2}(1 - m_\mu(t))$. The magnitude of this overlap will decide whether a configuration lies in the basin of attraction of a pattern μ.

Also displayed is the energy (or rather the Lyapunov function) $E(t) = -\frac{1}{2}\frac{1}{N}\sum_{i,j}^N w_{ij}s_i(t)s_j(t)$. In the case of the Hebbian learning rule, which leads to symmetric couplings w_{ij}, this quantity does not increase with time, i.e. the dynamics is characterized by a descent into a local energy minimum. For the other learning rules, however, the function $E(t)$ can move in both directions and has no special significance.

Having watched the iteration process you may again type s to confront the network with a new distorted starting pattern, type p to go back to the parameter entry panel, type m to modify the synapses, or type e to quit.

22.2 Numerical Experiments

To investigate the network in a systematic fashion one might start applying Hebb's rule (rule=1) to learn the shapes (more correctly: the bit patterns) of letters provided on the data file ASSO.PAT. Starting with the very modest task to memorize three letters (patt=3) the network performs nicely. The letters are stable fixed points having large basins of attraction. Distortions of up to 80% are tolerated and mostly lead to a correct recall. In fact, the network appears to have only five stable fixed points, i.e. two spurious states in addition to the memorized patterns (here and in the following we do not count the inverse patterns $s_i \rightarrow -s_i$ which of course are also stable). However the attempt to learn one additional letter leads to a sobering experience: only the letter A is recognized correctly, the patterns B,C,D are *unstable*. In fact the network has combined them to a single funny object which has features of all the three letters. This new pattern is a very powerful attractor; all starting configurations except those close to the pattern A appear to end up in this energy minimum. Adding the letter E to the training set has the curious side effect of stabilizing the pattern C, but in general the network performance declines further as patts is increased and quite soon the network converges only to superimposed patterns of its own design.

A system which becomes confused so easily is of no use in pattern recognition. Clearly the origin of the problem lies in the fact that our patterns are strongly correlated. (The overlap between, e.g., B and C is $m = 0.56$. The pattern A is nearly uncorrelated with the letters following it which explains its relatively robust stability.) To check this hypothesis we look at the performance when learning random patterns, which means setting plett=0. Indeed the performance is much improved and proves to be consistent with the theoretical predictions discussed in Chapts. 3 and 19. When the pattern loading $\alpha = p/N$ is low, all patterns are stable attractors. With increasing α errors creep in and around $\alpha = \alpha_c \approx 0.14$ (which is $p = 22$ for $N = 156$) most of the patterns are unstable, although many attractors still have a large overlap with a particular learned pattern.

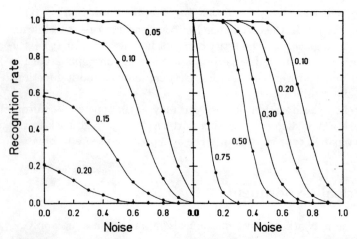

Fig. 22.1. The fraction of correctly recognized patterns as a function of the random noise in the starting configuration. The network of $N = 156$ units has been trained with $p = \alpha N$ random patterns using Hebb's rule (*left part*) or the learning rule of Krauth and Mézard with the parameter $C = 1$ (*right part*). Curves for various values of α are shown.

The results of a statistical analysis are shown in Fig. 22.1. The fraction of correctly recognized patterns is drawn as a function of the magnitude of initial distortion (the variable noise) for various values of the loading parameter α. Note that the basins of attraction shrink with the number of learned patterns. As α approaches α_c the recognition rate drops even in the absence of noise. This behavior is reminiscent of the phase transition expected in the thermodynamic limit $N \to \infty$. Of course, with a network as small as our toy system *finite-size effects* lead to a considerable smoothing of the phase transition. This is discussed in [Fo88b].

As we know from Sect. 10.1.2 the solution to the problem posed by correlated patterns, which at the same time gives a boost to the memory capacity, lies in improved learning schemes. E.g. the schemes of Diederich and Opper [Di87] and Gardner[Ga88a] have no problems in providing synaptic couplings which stabilize the 26 letters in a comparatively short learning time. If we choose the stability threshold $\kappa = 1$ in (22.10), 16 training epochs are needed. The scheme of Abbott and Kepler [Ab89] does the job within 9 epochs. The algorithm of Krauth and Mézard (22.13) with $C = 1$ needs 112 sweeps, each of which, however, involves fewer computations. Increasing the number of patterns or of the stability threshold will lead to prolonged learning times and eventually to the failure of convergence.

Figure 22.2 shows two histograms of the distribution of the normalized stability coefficients γ_i^μ for a network trained with 26 ordered patterns (i.e. letters). The full curve refers to Hebbian learning. Its broad shape with a long tail reaching into the region of instability ($\gamma < 0$) shows the detrimental effect of correlations. The corresponding distribution for 26 random patterns would be a Gaussian with a much narrower shape. The dashed histogram was obtained using the Krauth–Mézard learning rule with a threshold $C = 1$.

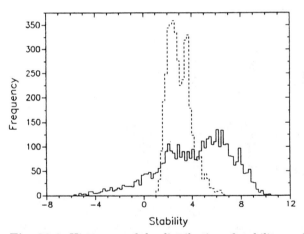

Fig. 22.2. Histogram of the distribution of stability coefficients γ_i^μ for an $N = 156$ network trained with 26 correlated patterns (capital letters). *Full line*: Hebb's learning rule. *Dashed line*: Krauth–Mézard learning rule using the stability threshold $C = 1$.

Figure 22.3 shows a few typical time evolutions of a network trained with the full set of 26 letters. Using sequential updating the state of the network is displayed after each full sweep. Parts (a) and (b) of the figure show the fate of the letter A distorted with 50% random noise. While the second example is successfully recognized, the first starting pattern, which has a slightly smaller overlap with the correct pattern, quite surprisingly is transformed into the inverse letter O.

While the network easily beats human competitors if the initial distortion is random, it is less successful with partially masked letters. Quite often the masking of two or three lines (out of 12) causes the recognition to fail. Parts (c) and (d) of Fig. 22.3 show an example of two highly correlated patterns where the basins of attraction seem to be shaped quite oddly. The letter I is recognized when it is strongly masked but the network wrongly decides in favor of the letter T with less masking.

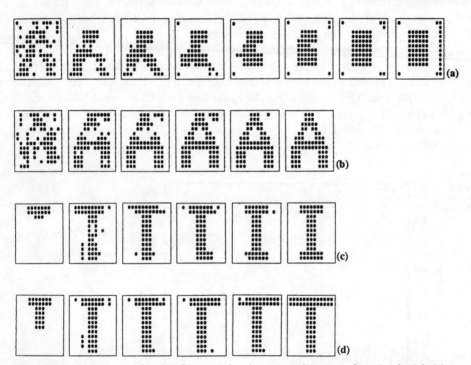

Fig. 22.3. Some examples for the time development of a network trained with 26 letters. **(a,b)** The first two runs both start with the letter A distorted with 50% random noise. While in (*b*) the letter is recognized correctly, the run (*a*) eventually leads to the inverse of the letter O. **(c,d)** The partially blocked letter I can either (*c*) be recognized correctly or (*d*) evolve into the highly correlated letter T.

When enforcing the *constraint of positive weights* (`wsign=1`) it is not surprising that the learning process converges more slowly. Nevertheless it is possible to learn the full set of 26 letters within a finite time when a small value of the stability threshold, e.g. $\kappa = 0.1$, is chosen. The basins of attraction are smaller than in a corresponding unconstrained network. As a characteristic difference the network often converges to the *ferromagnetic state* where the spins point in the same direction, i.e. all $s_i = +1$ or all $s_i = -1$.

We leave it to the reader to explore the influence of variations of additional parameters, like the temperature[1], the updating dynamics, and the threshold potential, on the network performance.

Finally note that the task performed by our associative memory is essentially rather trivial. It usually selects the pattern which is closest in Hamming distance to the bitstring it is asked to analyze. Incidentally the corresponding explicit computation of the Hamming distance is done as a side task in our program since it displays the vector of overlaps $m_\mu(t)$ after each sweep of the network. The problems of shape recognition and perception invariant under distortion are not addressed by this architecture.

Exercise: Modify the program ASSO.C so that it performs a statistical analysis of the network performance. Study the dependence on network size, pattern loading, and the learning rule.

---

[1] Note that the effect of the temperature depends on the strength of the local field. Thus without doing a rescaling of the coupling strengths the absolute value of T does not have a unique meaning when using the iterative learning laws.

23. ASSCOUNT: Associative Memory for Time Sequences

An obvious way to teach a Hopfield-type network to reproduce time sequences instead of stationary patterns is to introduce synaptic coefficients with an intrinsic time dependence [Am88, Gu88] as discussed in Sect. 3.4. Thus instead of the synaptic matrix w_{ij}, which acts instantaneously to generate the local field defined in (22.1), one may introduce a collection of synapses w_{ij}^τ acting with a distribution of characteristic time delays of magnitude τ. The local field, which according to (22.2) determines the probability for the updated neuron state $s_i(t+1)$, can be replaced by a generalized convolution in time

$$h_i(t) = \sum_{\tau=0}^{\tau_{\max}} \sum_{j=1}^{N} \lambda^\tau w_{ij}^\tau s_j(t-\tau) \,. \tag{23.1}$$

Here the coefficients λ^τ determine the relative strength of the various types of synaptic couplings. The ordinary Hopfield model corresponds to the case $\lambda^0 = 1$, $\lambda^\tau = 0$ for $\tau \geq 1$. The couplings in (23.1) can give rise to a complex time evolution of the network. In the simplest nontrivial case just two types of synapse are present. We study a network connected through prompt synapses w_{ij} and slow synapses w_{ij}' of relative strength λ with the fixed time delay τ. The local field reads (cf. Sect. 3.4)

$$h_i(t) = \sum_{j=1}^{N} w_{ij} s_j(t) + \lambda \sum_{j=1}^{N} w_{ij}' s_j(t-\tau) \,. \tag{23.2}$$

Such a network can be made to go through an ordered sequence of patterns $1 \to 2 \to 3 \to \dots$ [Gu88, Am88]. To this end the couplings w_{ij}' have to provoke the transition between sucessive patterns in the series. This is achieved by a very simple modification of the learning rules. For example, the Hebb rule (22.5) for heteroassociation of the pattern $\sigma^{\mu+1}$ given the pattern σ^μ as an input reads

$$w_{ij}' = \frac{1}{N} \sum_{\mu} \sigma_i^{(\mu+1)} \sigma_j^\mu \,. \tag{23.3}$$

The improved learning rules can be modified in the same way by replacing σ_i^μ by $\sigma_i^{\mu+1}$ in the updating rule for δw_{ij}^μ. The instantaneous synapses w_{ij} are used to stabilize the patterns; thus they are trained in the ordinary way for the task of autoassociation. If the coefficient λ is large enough, the network will act as a freely running counter. The state of the network will switch from one pattern to the next after τ clock ticks when the delayed coupling starts to 'pull in the direction' of the new pattern.

A second interesting mode of operation of this network arises if the parameter λ is a little too small to induce transitions by itself. Then it takes an external disturbance to destabilize the old pattern. Once the network is driven out of the basin of attraction of the last pattern into some unstable region the delayed coupling w'_{ij} (which does not yet feel the disturbance) provides a driving force toward the associated next pattern. Thus the network performs the task of counting the number of externally entered signal pulses. Such a pulse can be implemented by adding a spike of random noise with amplitude ρ to the local field:

$$h_i(t) = \sum_{j=1}^{N} w_{ij} s_j(t) + \lambda \sum_{j=1}^{N} w'_{ij} s_j(t - \tau) + \rho \xi_i \delta_{t,t_0} , \qquad (23.4)$$

where ξ_i is a random variable taking on values ± 1.

23.1 Program Description

The program ASSCOUNT.C is closely related to ASSO.C so that much of the discussion of the last section applies here also. We assume that the reader is already familiar with ASSO.C. Again the total number of patterns patts and the number of predefined patterns plett to be learned must be specified. For a difference, the ordered patterns take the shape of the ten numerals $0 \ldots 9$ which are provided on the input file ASSCOUNT.PAT. The data format is the same as in the file ASSO.PAT so that by copying this file to the new name it is possible also to use the sequence of 26 letters instead of the numerals as patterns for counting.

Since our interest now is no longer focused on the learning process, only two choices are provided: Hebb's rule and the iterative scheme of Diederich and Opper. Learning takes about twice as much time since the synaptic matrices w_{ij} and w'_{ij} have to be learned separately. The learned association of patterns is cyclical, i.e. the last pattern in the series, σ_i^p, is followed again by σ_i^1.

After learning is finished you are asked to specify the following set of parameters:

- update: The rule of updating the network which can be random or sequential.
- sstep: Single-step mode.
- temp: The operating temperature T of the network.
- tau: The time delay of the couplings w', an integer in the range $1 \ldots 10$.

- lambda: The relative strength of the delayed compared to the prompt couplings.
- rho: The amplitude of the counting signal as defined in (23.4).

The iteration of the network is started by providing the starting pattern for which a pixel representation of one of the numbers $0 \ldots 9$ can be chosen (of course, if you have changed the input file ASSCOUNT.PAT the patterns may look very different). You can start with a distorted version of this pattern, specifying a value noise $\neq 0$. The network is updated for 10 000 sweeps unless the iteration is interrupted manually. As in ASSO.C the running parameters can be modified and the iteration restarted. The external signal for triggering the counter is entered by pressing <SPACE>.

To have the delayed network configuration $s_i(t - \tau)$ available at any instant t, all the intermediate values $s_i(t-1) \ldots s_i(t-\tau)$ have to be stored. To avoid unnecessary data shuffling these values are stored in a cyclical buffer s0[t][i] of length tau. Two pointers t_in and t_out specify the locations where a configuration is stored and read out. These pointers are incremented after each sweep of the network using modulo τ arithmetic to account for the cyclical nature of the buffer.

23.2 Numerical Experiments

You might start with the set of 10 numerals (which corresponds to a modest memory loading factor $\alpha = 0.064$), using the learning rule for correlated patterns. At temperature $T = 0$, experiment with the parameter λ. Essentially, different operating modes of the network can be observed in the three distinct regions. $\lambda > \lambda_2$: freely running counter; $\lambda_1 < \lambda < \lambda_2$: counter for input signals. $\lambda < \lambda_1$: stable associative network. However, the transition between these parameter regions is soft. For $\lambda > 2$ the network evolves like a perfect clock, i.e. each counting step has a length equal to the synaptic delay time τ, since a jump of the action of the second term in (23.4) immediately induces a transition to the next pattern. When reducing the value of λ the transition from one pattern to the next often needs several sweeps and the clock runs slow.

Figure 23.1 shows an example of the spontaneously counting network with a marginal value of λ. For $\tau = 1$ the network has not enough time to settle at the attractors; it will operate with distorted and superimposed patterns. A further reduction of λ will eventually cause the network to 'get stuck' at one of the attractors with high stability (for our patterns the first of them happens to be the number '8'). At $\lambda = 0.4$ all the patterns are stable against spontaneous transitions. Now we are in the region where the network can serve as a perfect counter for external signals provided that their amplitude ρ is large enough. The theoretical expectation [Am88] for the required amplitude is $\rho > 1 - \lambda$ which is very roughly confirmed by our numerical experiment. Finally, below $\lambda = 0.2$ the strength of the delayed synapses is too small to induce the correct transition even for counting pulses of arbitrary strength.

The discussion so far referred to the fully deterministic case, $T = 0$. Introducing a finite temperature has a profound influence on the network in the 'chime counting' mode. Since the temperature-induced noise tends to destabilize a condensed pattern it aids the action of the delayed synapses. As soon as a sufficiently large fluctuation occurs, the state of the network will make a transition to the next pattern. E.g., take the parameters $\lambda = 0.4, \tau = 5$, which at zero temperature lead to a stationary network. At finite temperatures the network starts to count spontaneously! This effect can even serve as a 'thermometer' since the cycle time is sensitive to the temperature (a rough estimate: T=0.5 gives a cycle length of 65; T=0.15 gives a cycle length of 120). Of course, at very high temperatures the network no longer reproduces any of the learned patterns while near $T = 0$ the counting is frozen in long-lived metastable states.

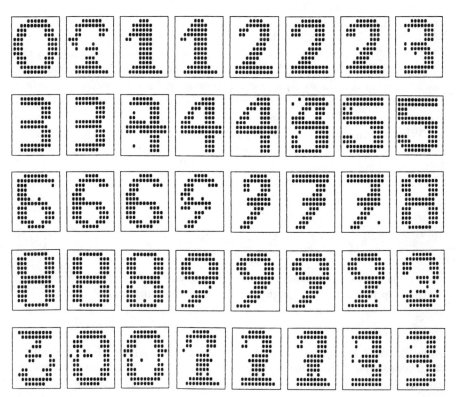

Fig. 23.1. The time development of a spontaneously counting network trained with the numeral patterns $0 \ldots 9$. The parameters are $\lambda = 0.7, \tau = 2$.

Exercise: Study in more detail the dependence on temperature and on the pattern loading α. Try to estimate the phase boundaries. Use random patterns to eliminate the influence of correlations.

24. PERBOOL: Learning Boolean Functions with Back-Prop

This program illustrates the function of a feed-forward layered neural network as described in Chapts. 5–7. Its task is to learn an arbitrary Boolean function of a small number of Boolean variables. This is achieved by a three-layer feed-forward network which is trained by a gradient-descent method, i.e. by the rule of error back-propagation. The network consists of an input layer $\sigma_k, 1 \leq k \leq n_{\text{in}}$, a hidden layer $s_j, 1 \leq j \leq n_{\text{hid}}$, and an output layer $S_i, 1 \leq i \leq n_{\text{out}}$. The synaptic connections are $\overline{w}_{jk}$ from the input to the hidden layer, and w_{ij} from the hidden to the output layer. In addition there are activation thresholds $\overline{\vartheta}_j$ and ϑ_i. The neurons have two activation values chosen as -1 and $+1$ (the transformation to the values 0 and 1 commonly used to represent binary numbers is trivial).

When an input pattern is fed into the network by clamping the neurons σ_k to represent an input pattern, the activation of the intermediate and output layers is determined according to the rule (6.4), namely

$$S_i = f(h_i), \qquad h_i = \sum_j w_{ij} s_j - \vartheta_i, \tag{24.1}$$

$$s_j = f(\overline{h}_j), \qquad \overline{h}_j = \sum_k \overline{w}_{jk} \sigma_k - \overline{\vartheta}_j. \tag{24.2}$$

The activation function $f(x)$ for discrete two-state neurons corresponds to the step function $f(x) = \text{sgn}(x)$. During the learning phase, however, this is replaced by the continuous "sigmoidal" function

$$f(x) = \tanh(\beta x), \tag{24.3}$$

since gradient learning is possible only if $f(x)$ is a smooth function with a finite derivative, which in our case is $f'(x) = \beta(1 - f^2(x))$. At the end of the training phase the limit $\beta \to \infty$ leading to a discontinuous activation function is taken[1].

The rule for error back-propagation minimizing the squared deviation of the network output from the target values according to (6.7–6.12) leads to the synaptic corrections

---

[1] If one wishes to work with two-state neurons throughout the training process, one can simply replace the derivative function $f'(x)$ by an averaged constant, which can be absorbed in the factor ϵ in (24.4, 24.5).

$$\delta w_{ij} = \epsilon \sum_\mu \Delta_i^\mu s_j^\mu, \qquad \text{where} \qquad \Delta_i^\mu = \left[\zeta_i^\mu - f(h_i^\mu) \right] f'(h_i^\mu) . \qquad (24.4)$$

$$\delta \overline{w}_{jk} = \epsilon \sum_\mu \overline{\Delta}_j{}^\mu \sigma_k^\mu, \qquad \text{where} \qquad \overline{\Delta}_j^\mu = \left(\sum_i \Delta_i^\mu w_{ij} \right) f'(\overline{h}_j{}^\mu) . \qquad (24.5)$$

The changes of the threshold values $\delta \vartheta_i, \delta \overline{\vartheta}_j$ are determined by nearly identical formulae. These are obtained by simply replacing the input signal (s_j^μ or σ_k^μ) by the fixed value -1.

A simple modification of the learning procedure [So88] consists in the replacement of (24.4) by

$$\Delta_i^\mu = \beta \left[\zeta_i^\mu - f(h_i^\mu) \right] . \qquad (24.6)$$

This change has the potential to speed up the convergence: the weight adjustments get increased in the "saturation region" where the value of h_i^μ is large and the factor $f'(h_i^\mu)$ approaches zero. For small values of h_i^μ (24.6) agrees with (24.4).

The learning consists of consecutive "training epochs" in which the full set of patterns σ_k^μ to be learned is presented to the network. The synaptic and threshold corrections are accumulated and applied at the end of an epoch. As a variation of this "batch learning" algorithm one can also update the network parameters after each single presentation of a pattern. This 'online learning' may, but is not guaranteed to, improve the learning speed.

An important modification which has the potential to improve the stability of the learning process consists in the introduction of a kind of hysteresis effect or "momentum". In regions of the parameter space where the error surface is strongly curved the gradient terms Δ can become very large. Unless the parameter ϵ is chosen to be inordinately small, the synaptic corrections δw_{ij} will then tend to overshoot the true position of the minimum. This will lead to oscillations which can slow down or even foil the convergence of the algorithm. To avoid this problem the correction term can be given a memory in such a way that it will no longer be subject to abrupt changes. This is achieved by the prescription[2]

$$\delta w_{ij}^{(n)} = \epsilon \sum_\mu \Delta_i^\mu s_j^\mu + \alpha \delta w_{ij}^{(n-1)} , \qquad (24.7)$$

where the index n denotes the number of the training epoch. If α is chosen fairly large, the search in the parameter space will be determined by the gradient accumulated over several epochs, which has a stabilizing effect. Unfortunately there is no general criterion how the gradient parameter ϵ and the momentum parameter α are to be chosen. The optimal values depend on the problem to be learned.

---

[2] In the language of numerical analysis this is called a relaxation procedure.

24.1 Program Description

The progam PERBOOL.C implements a three-layer network with dimensions
$\text{nin} \leq 5$, $\text{nhid} \leq 100$, $\text{nout} \leq 5$. The desired values can be entered through an
input panel which comes up immediately after starting the program.

The nin Boolean input variables can take on $\text{nbin} = 2^{\text{nin}}$ different values.
The network is trained with a set of $\text{patts} = \text{nbin}$ output patterns, i.e. the
output is completely prescribed for each possible input combination[3]. Even
in the "scalar" case, i.e. for a single output variable, $\text{nout} = 1$, there is an
embarrassing number of $2^{2^{\text{nin}}}$ possible Boolean functions. To define an arbitrary Boolean function we have to select a set of nout bit strings having the
length nin.

These bit strings are represented in the program by the long integer
variables key[i], which conveniently can hold just 32 bits. Each bit specifies the desired response of the corresponding output neuron to that combination of input values which is encoded by the binary value of the position of the bit within the variable key. Since this may sound confus-
ing, let us look at the familiar example of the XOR function. There are
$2^2 = 4$ combinations of the two input variables and the output is defined by
$00 \rightarrow 0, 01 \rightarrow 1, 10 \rightarrow 1, 11 \rightarrow 0$. The input can be interpreted as a 2-bit
binary number. Stepping through its value in descending order, the output
is represented by the bit string 0110. Thus the XOR function is encoded by
specifying the number $\text{key} = 0110_{\text{bin}} = 6_{\text{hex}} = 6_{\text{dec}}$.

The Boolean function to be learned can be defined for the program by giving
the code number(s) key[i] for $i = 1 \ldots \text{nout}$ in hexadecimal representation. Since
in general it is somewhat tedious to work out this binary code, the program allows
for an alternative and more direct way of specifying the Boolean function. Having
answered 'm' to the prompt, a menu is displayed in which the desired output bits
can be directly entered by moving the cursor to the appropriate place, typing 0
or 1, and pressing <Return>. (Note that for the case nout>1 the complete output
string ζ_i^μ, $i = 1 \ldots \text{nout}$, has to be entered: it is not possible to modify single bits.)
The program displays the hexadecimal value of the resulting coding variable(s) key
which will facilitate the data entry if the same function is to be used repeatedly. The
decoding of the input and output patterns takes place in the subroutine patterns()
and produces the arrays $\text{zetp} = \zeta_i^\mu$ and $\text{sigp} = \sigma_k^\mu$.

Having been given the Boolean function to be learned, the program displays a second input panel which allows to specify details of the learning
procedure.

- epsilon: The learning rate ϵ.
- alpha: The "momentum" constant α.
- batch: Selects the learning protocol (online=0, batch=1).
- mod_cost: Use of the modified cost function (see eq. (24.6)).
- beta: The inverse steepness β of the activation function $f(x)$.

---

[3] To study the issue of generalization one would train the network with a subset
of all possible inputs, i.e. use a smaller value of patts.

Subsequently the program enters the subroutine `learn()`. The synaptic coefficients and thresholds are initialised with random numbers in the range $-1 < \xi < +1$ Then one by one the input patterns are presented and a forward sweep is performed leading to output activations `sout[i]`. The difference of these values from the correct output `zout[i]` allows the calculation of modified network parameters according to the back-propagation algorithm. After each learning epoch the accumulated error $\sum_\mu \sum_i |\zeta_i^\mu - S_i^\mu|$ is shown. In addition the actual values of the network parameters are displayed on screen. Each line in the display shows the connections attached to one of the hidden neurons according to the following format:

Table 24.1. Format used for output of network configuration.

$\overline{w}_{1,1}$	$\cdots$	$\overline{w}_{1,nin}$	$\overline{\vartheta}_1$	$w_{1,1}$	$\cdots$	$w_{nout,1}$
$\vdots$	$\ddots$	$\vdots$	$\vdots$	$\vdots$	$\ddots$	$\vdots$
$\overline{w}_{nhid,1}$	$\cdots$	$\overline{w}_{nhid,nin}$	$\overline{\vartheta}_{nhid}$	$w_{1,nhid}$	$\cdots$	$w_{nout,nhid}$
				ϑ_1	$\cdots$	ϑ_{nout}

If the number of hidden neurons is large, only the first 20 lines of this table are displayed in order not to overflow the screen.

After any learning epoch the user can interrupt the learning by pressing the key 'r'. This serves to evaluate the network performance using the current coupling coefficients. Displayed are the states of the input, intermediate, and output neurons, with a flagging of the wrong output bits. Note that this display employs the discrete activation function, so that the number of wrong bits may deviate from that contained in the `error` variable.

By pressing 'p' it is possible to view and modify the values of the parameters used by the back-propagation algorithm during the learning process.

Perhaps more illuminating – and at least more entertaining – than the numerical output is a graphical representation of the learning process. This mode is entered if the key 'g' is pressed during learning. Then as a function of the learning epoch a collection of `patts=` 2^{nin} curves is shown. Each curve represents the activation of the (first) output neuron in response to one specific input pattern. Since in the learning phase the network operates with continuous functions $f(x)$, these curves move smoothly in the interval $-1 \ldots + 1$. The abscissa extends over a range of 200 training epochs. If this number is exceeded, the display is erased and the curves reappear at the left border. By repeatedly pressing 'g' one can switch between graphics and text display.

If the learning converges to the correct solution, the program tries to make a smooth transition to the discrete (i.e. theta-function) operating mode by slowly increasing the value of β by a factor of 1.05 after each epoch. This sets in if the error variable falls below the value $1/\beta$ and continues until `error` reaches 10^{-3}.

After convergence has been reached or when the learning process is stopped by pressing the key 'e' the complete set of synaptic coefficients and thresholds is displayed and the final network performance is shown.

We should mention that initially the program operates in the 'slow-motion' mode, inserting a delay of 0.5 second in each training epoch. By pressing the key s it is possible to switch to full speed.

24.2 Numerical Experiments

The simplest example of a Boolean function with a nontrivial representation is the exclusive-OR problem (XOR) discussed in Sect.5.2.3. This can be investigated by setting `nin=2`, `nout=1`, and `key=6`. The back-propagation algorithm can solve this problem without much effort provided that the network configuration contains at least two hidden units, `nhid=2`. Over a fairly wide range of parameters ϵ and α most learning runs lead to convergence to a valid solution. One typically might choose a gradient parameter $\epsilon = 1$ and a momentum factor $\alpha = 0.9$. The steepness parameter β in principle is redundant and may be kept fixed at the value $\beta = 1$. Owing to the special form of the activation function $f(x)$, this parameter only enters as a multiplicative factor attached to the local fields h_i. Therefore a change of β can be offset by inversely rescaling all synaptic coefficients and thresholds. If also the gradient parameter ϵ is rescaled (quadratically), the behavior of the network should be unchanged. In our program implementation this is not quite true since the initialization is performed with random numbers in the fixed interval $-1 \ldots + 1$. If a very small value of β is chosen, this has the same effect as an initialization with near-zero values, which adversely affects the convergence behavior.

Examining the performance of the network, one can notice that it finds a considerable variety of solutions to the XOR problem. Most solutions are essentially equivalent, differing by simultaneous sign changes of activation values s and coupling coefficients w which leave invariant the equations which describe the network. This is a reflection of the fact that with the help of negation operations the XOR function can be represented in various ways by using the deMorgan identity of Boolean algebra $\overline{A \wedge B} = \overline{A} \vee \overline{B}$. There are two distinct classes of representation of XOR where the output neuron acts either as an OR gate or as an AND gate, according to

$$A \not\simeq B = (A \vee B) \wedge \overline{A \wedge B},$$
$$A \not\simeq B = (A \wedge \overline{B}) \vee (\overline{A} \wedge B)$$

where $\not\simeq$ is used to denote the exclusive-OR or XOR function. An example of the former solution type was shown in Fig. 6.1. The following two tables show typical results of couplings and thresholds generated by the back-propagation algorithm. They are displayed in the format introduced in the last subsection.

+1.657	+1.615	−1.968	+3.105	+2.059	−2.034	+2.368	+3.117
+2.536	+2.929	+3.115	−2.494	−2.432	+2.624	+2.518	+2.623
			+2.306				−2.153

Starting from randomly initialized couplings, the learning algorithm often converges within about 20 iteration steps. However, it does not always find valid solutions of the XOR problem. In some instances it gets stuck at a local minimum, or the local fields are driven to large values where the derivative of the activation function $f'(h)$ is nearly zero, leading to exceedingly slow learning. Figure 24.1 shows an example where the network has just managed to excape from such a situation after about 100 training epochs.

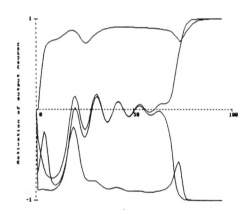

Fig. 24.1. A 2–2–1 perceptron learns the function XOR. The curves depict the activation of the output neuron S when the network is presented with the four possible input patterns as a function of training epochs. Parameters: $\epsilon = 0.1, \alpha = 0.9, \beta = 1$.

Going to problems of a higher dimension provides further insights. As the 'benchmark' problem we have predefined in the program PERBOOL the generalization of the XOR function to more than two variables. This function takes on the value 1 if there is an odd and 0 if there is an even number of 1s in the input. Thus it can also be called the 'parity' function; still another interpretation is 'addition modulo 2'. Let us take the case nin=5. Using the binary encoding of the Boolean function leads to key= $10010110011010010110100110010110_{bin}$ = 96696996_{hex}. The parity problem can be learned if the number of hidden neurons is at least equal to the input dimension, nhid=nin. However, even for nhid=nin-1 the network can produce solutions while operating with continuous activation functions. It is not too surprising that continuous variables s_i can encode more information than discrete ones. This solution is destroyed in the final stage of the learning process when the program tries to perform the limit $\beta \to \infty$.

Other interesting examples with nin=5 the reader might wish to study are the mirror symmetry function (key= 88224411_{hex}) and the shift register (key$_1$ = $AAAAAAAA_{hex}$, key$_2$ = $FFFF0000_{hex}$, key$_3$ = $FF00FF00_{hex}$, key$_4$ = $F0F0F0F0_{hex}$, key$_5$ = $CCCCCCCC_{hex}$).

The *symmetry function* signals whether the input bit pattern has mirror symmetry. As an example the following table shows the weights and thresholds of a 5-2-1 network which can recognize this property[4] .

+2.248	−1.058	+0.029	+1.020	−2.196	−1.149	+3.113
−2.659	+1.289	+0.021	−1.284	+2.628	−1.066	+2.759
						+2.627

We note the antisymmetric structure of the couplings in the input layer. The mirror symmetry property of a 5-bit pattern does not depend on the bit in the center. We observe that the third input unit is indeed decoupled and has no influence on the state of the network. The output layer performs a logical conjunction of the states of the two hidden neurons. The solution found above was obtained using the parameters $\beta = 1, \epsilon = 0.01, \alpha = 0.9$. However, very often the back-propagation algorithm does not converge to a valid solution.

Exercise 1: Study the distribution of learning times for the parity problem. Investigate the cases where the convergence fails. Are there true local minima which the algorithm can never escape?

Exercise 2: Modify the program PERBOOL in such a way that it solves the *encoder-decoder problem*. The task is to transmit information from an input layer with n_{in} units to an output layer of equal size through a "bottleneck". The network is required to map all input patterns of the form $\sigma = (0, \ldots, 1, \ldots, 0)$, i.e. with a single unit activated, to identical output patterns, $S = \sigma$. This task is nontrivial when the intermediate layer has small size, $n_{hid} \leq n_{in}$. Then the information contained in the input patterns has to be encoded in compact form. Obviously, the most compact encoding is achieved when the activation of the intermediate layer represents a binary number counting the input patterns. Thus at least $n_{hid} \geq \log_2 n_{in}$ hidden neurons are needed. Investigate the performance of the back-propagation algorithm for the encoder-decoder problem of size 4-2-4 or larger.

---

[4] Incidentally, two hidden neurons are sufficient to solve the mirror-symmetry problem of arbitrary dimension!

25. PERFUNC: Learning Continuous Functions with Back-Prop

As discussed in Sect. 6.4, feed-forward networks with hidden layers can represent any smooth continuous function $\mathrm{I\!R}^n \to \mathrm{I\!R}^m$. As an illustration the program PERFUNC is designed to solve the following task: when presented with the values of a specimen $g(x)$ out of a class of one-dimensional functions at a set of points x_i it predicts the function values at the output points x_i'. This is achieved by training the network with the error back-propagation algorithm by presenting various 'patterns' σ_i^μ where μ refers to specific values of the continuous parameters on which the function $g(x)$ depends. The training method is essentially the same as was used for Boolean-function learning, so that its description need not be repeated. As a slight variation, the couplings w_{ij} to the output layer are realized by *linear* transfer functions $\tilde{f}(x) = x$. In contrast to the bounded sigmoidal function $f(x) = \tanh(\beta x)$ used for the interior layers, this enables the output signals S_i to take on arbitrarily large values.

25.1 Program Description

PERFUNC is closely related to the program PERBOOL described in Chapt. 24. Differences arise from the fact that *0, 1, or 2 hidden layers* are allowed and from the way training patterns are provided. Since learning can take a long time, the synaptic coefficients and thresholds can be read in from a data file to provide for a fast demonstration of the network's capabilities. Alternatively, to operate in learning mode the following parameters can be specified via the input panel of the program.

- nlayers: The number of hidden layers which may take the values 0, 1, or 2.
- nin, nout, nhid: The number of neurons in the input, output, and hidden layer. If there are two hidden layers, these will have equal numbers of neurons.
- ntyp: The type of function $g(x)$ to be learned. The three choices are:

$$g_1(x) = c_1 x + c_2 ,$$
$$g_2(x) = \sin\left[(c_1 x + \tfrac{1}{2}c_2)\, 2\pi\right] ,$$
$$g_3(x) = c_2\, x^{1+c_1} .$$

– nparm: The number of function parameters, e.g. the frequency and phase of the sine function $g_2(x)$. If nparm=1, the second parameter is absent, $c_2 = 0$, or $c_2 = 1$ in the case of $g_3(x)$.
– patts: The number of different patterns to be used in training. A pattern consists of the input and output values σ_l, $l = 1 \ldots n_{\text{in}}$, S_i, $i = 1 \ldots n_{\text{out}}$ produced by a representative of the class of functions $g(x)$. It is generated by randomly selecting values for the parameters c_1 and c_2 in the range $-1 \ldots + 1$.
– errmax: Target value for the mean error per output neuron. Learning is stopped when this accuracy is reached.

Subsequently the program asks for the set of input and output coordinates x_i and x'_i.

A second input panel asks for the following parameters which determine the learning phase:

– epsilon: The learning rate ϵ.
– alpha: The "momentum" constant α.
– multi: The multiplicity of pattern presentations. If multi> 1, the same pattern is presented repeatedly to the back-propagation algorithm before going to the next pattern. (Note: the momentum parameter α has no effect in this mode.)
– fixpat: This variable determines whether a fixed set of patterns is used for training, or whether at the beginning of each training epoch new patterns are chosen.
– beta: The inverse steepness β of the activation function $f(x)$.

The learning process is started by initializing all couplings and thresholds to random numbers in the range $-1 \ldots 1$. After each learning epoch the current coupling and threshold values are displayed according to the format introduced in Sect. 24.1. (For nlayers=2 only the parameters of the first hidden layer and the output layers are displayed; for nlayers=1 the lines (columns) correspond to the input (output) units of the network.) In addition a listing of the error value as a function of the training epoch number is maintained. During the training phase the user can interact with the program in various ways. After pressing the key 'p' the parameters which influence the learning process may be changed: epsilon,alpha,multi, fixpat, beta. With the key 'r' the current network performance may be investigated. Finally, by pressing 'e' the learning is stopped.

Three different choices are offered for examining the current performance:

– n: A numerical listing of the errors for an equidistant mesh of parameter values c_1, c_2 in the interval $-1 \ldots +1$. If nparm=1, the values $S_1^{\mu=1}$, ζ_1^μ, and the mean square error $D^\mu = \left(\sum_{i=1}^{n_{\text{out}}} (S_i^\mu - \zeta_i^\mu)^2 \right)^{1/2}$ are listed. For nparm=2 a table of values of D^μ is displayed, where the columns correspond to the variation of the second parameter c_2 with an increment of 0.2, while the rows are calculated for different values of c_1 (increment 0.1). In addition the mean error for all patterns is printed.

- p: With this option the program shows a graphical plot of the deviations $S_i^\mu - \zeta_i^\mu$ $i = 1, \ldots n_{out}$ as a function of the parameter c_1. The range of display and the value of c_2, which is kept fixed, can be entered.
- f: The program prompts for the entry of the constants c_1 and c_2. Then a plot of the function $g(x)$ is drawn together with the input values $g(x_i)$ (marked by + symbols) and the output values $g(x_i')$ produced by the network (marked by x).

After a final examination of the network performance the couplings can be stored on a data file.

25.2 Numerical Experiments

Since the learning of continuous functions is rather time consuming, it might be wise to start with very simple examples. There is only one case where an exact solution is possible: the linear network trained to represent a linear function. Indeed, for a $1 - 1$ network trained with the general linear function $g_2(x)$ (nlayers=0, nin=2, nout=1, nparm=1) the learning algorithm quickly finds the solution, which is $w_{11} = \frac{x_2 - x'}{x_2 - x_1}$, $w_{12} = \frac{x' - x_1}{x_2 - x_1}$, $\vartheta_1 = 0$. This is quite insensitive to the details of the training rule as long as the gradient parameter \epsilon is not too large. Otherwise the algorithm becomes unstable and the weights 'explode', leading to an exponentially growing error. Learning is achieved both with a fixed set of 'patterns' (i.e. combinations of c_1 and c_2) and with patterns which are selected at random in each step, although in the latter case the convergence is markedly slower. If there are more input values than necessary, the problem is underdetermined. This does not impede the learning process, but different sets of weights w_{1i} are found in each run, because of the random initial conditions.

For a network with a *nonlinear* transfer function $f(x)$ the representation of a linear function will not be exact. For the trivial $1 - 1 - 1$ architecture (nlayers=1, nin=1, nhid=1, nout=1) trained with the one-parameter linear function the network represents the general function $S = w \tanh(\bar{w}\sigma - \bar{\vartheta}) - \vartheta$. This has to provide a fit to the function $\zeta = (x'/x)\sigma \equiv c\sigma$. In principle this can be accomplished with arbitrary precision by scaling the weights so that the linear region of the tanh function is probed: $\bar{\vartheta} = \vartheta = 0, \bar{w} = \epsilon, w = c/\epsilon$ with $\epsilon \to 0$. Experiments with the program show that a fairly good representation of the linear function can be achieved, but the relative size of the weights w and $\bar{w}$ remains of the order of one instead of diverging.

To study the learning of more challenging tasks, i.e. nonlinear functions depending on two parameters, the reader has to muster some patience. It can take several thousand learning epochs until a satisfactory representation of the function is reached. Provided that the gradient factor epsilon has a sufficiently small value, some degree of convergence can be achieved for various combinations of the parameters which determine the training process. However, it seems to be necessary that the same patterns are presented repeatedly in subsequent training epochs. If in each epoch new values of c_1 and c_2 are chosen at random (fixpat=0), the errors remain at a high level.

The data files PERFUNC1.DAT and PERFUNC2.DAT contain some typical results. Networks comprising one (or two) hidden layers with the topology $5 - 20 - 5$ (or $5 - 10 - 10 - 5$) have been trained to represent the two-parameter sine function $g_2(x)$. The networks provide a mapping from the function values at the input grid $x_i = 0.1, 0.2, 0.3, 0.4, 0.5$ to the output grid $x_i' = 0.6, 0.7, 0.8, 0.9, 1.0$. The mean deviation is of the order of about $|S_i - \zeta_i| \simeq 0.02$. The error starts to grow at the edges of the range of parameter values for which training examples have been provided. The network cannot be expected to (and indeed shows no tendency to) "generalize" in the sense of, e.g., extrapolating to sine functions with higher frequency. This is easily observed by plotting the error as a function of c_1 (using the option p in the 'Display network performance' panel) in the range of, say, $-2 \ldots + 2$. As shown in Fig. 25.1 the error rapidly grows outside the training interval $-1 \ldots + 1$. Pictorially speaking, the effect of the learning is to "dig a hole" into the error surface in the region where training patterns are provided.

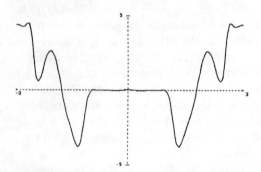

Fig. 25.1. Values of the error of the output signals of a network trained to represent the sine function $g_2(x)$ at five grid points, drawn as a function of the frequency parameter c_1. Outside the region used for training, $-1 < c_1 < +1$, the error rapidly increases.

Experience with the back-propagation algorithm shows that learning becomes very slow when many hidden layers of some complexity are involved. To train networks with a large number of hidden layers the use of alternative strategies, such as the genetic algorithms (see Chapt. 16) seems to be advantageous.

26. Solution of the Traveling-Salesman Problem

26.1 The Hopfield–Tank Model

As we know from the spin-glass analogy neural networks can model a complicated 'energy landscape' and hence can be employed for solving optimization problems by searching for minima in this energy surface. In Sect. 11.3.1 we have discussed an approach for solving the paragon combinatorial optimization task, the traveling salesman problem (TSP), using a neural network. Hopfield and Tank [Ho85] have mapped the N-city TSP onto a network with $N \times N$ formal neurons $n_{i\alpha}$. These are assumed to have a graded response with a continuous output value in the range $0 \ldots 1$ determined by the local field $u_{i\alpha}$ through the Fermi function

$$n_{i\alpha} = \frac{1}{1 + e^{-2u_{i\alpha}/T}} \qquad (26.1)$$

with formal temperature T. The neurons are arranged in a square array, the row i denoting the number of the city and the column α denoting the station of the tour on which this city is visited. A valid tour is therefore characterized by an activation pattern with exactly N neurons 'firing' and $N(N-1)$ 'quiescent'. There must be exactly one entry of 1 in each row and each column of the neural matrix $n_{i\alpha}$.

The task is to find that tour which has the shortest total length among the many admissible solutions. Using the Hopfield–Tank representation the TSP amounts to a minimization problem with constraints. The 'energy' function to be minimized can be written as

$$E = E_0 + \lambda_1 E_1 + \lambda_2 E_2 + \lambda_3 E_3 . \qquad (26.2)$$

Here E_0 measures the total length of a tour and the additional terms are intended to ensure constraint satisfaction, the constants λ_n being Lagrange parameters. Specifically we have

$$E_0 = \frac{1}{2} \sum_{i,j} \sum_{\alpha} d_{ij} n_{i\alpha} (n_{j,\alpha+1} + n_{j,\alpha-1}) , \qquad (26.3)$$

where d_{ij} denotes the distance between the cities i and j. To get a closed tour cyclic boundary conditions are imposed, i.e. one identifies $\alpha = 0$ with

$\alpha = N$ and $\alpha = N + 1$ with $\alpha = 1$. Normally the matrix d_{ij} contains the Euclidean distance in a two-dimensional plane[1]

$$d_{ij} = \sqrt{(x_i - x_j)^2 + (y_i - y_j)^2} \, . \tag{26.4}$$

The additional contributions to (26.3) are chosen such that they vanish if each city is visited at most once,

$$E_1 = \frac{1}{2} \sum_i \sum_{\alpha,\beta}^{\alpha \neq \beta} n_{i\alpha} n_{i\beta} \, , \tag{26.5}$$

and if at each stop at most one city is visited,

$$E_2 = \frac{1}{2} \sum_{i,j}^{i \neq j} \sum_\alpha n_{i\alpha} n_{j\alpha} \, . \tag{26.6}$$

These constraints are not yet sufficient; they would be solved, e.g., by the trivial solution $n_{i\alpha} = 0$. Therefore we have to enforce the presence of N entries of magnitude 1 through the constraint

$$E_3 = \frac{1}{2} \left(\sum_i \sum_\alpha n_{i\alpha} - N \right)^2 \, . \tag{26.7}$$

The approach to a solution of minimal energy $E[n]$ can be described by a differential equation in time t, as explained in Chapt. 12. As independent variables we use the values of the local field $u_{i\alpha}$ which are related in a unique way to the neural activation according to (26.1). Then we have to solve

$$\frac{du_{i\alpha}}{dt} = -\frac{u_{i\alpha}}{\tau} - \frac{\partial E_0}{\partial n_{i\alpha}} - \sum_{n=1}^{3} \lambda_n \frac{\partial E_n}{\partial n_{i\alpha}} \, , \tag{26.8}$$

which is a set of N^2 coupled first-order differential equations. The first term on the r.h.s. corresponds to a damping with τ representing the typical time constant of the neurons; the additional terms provide a gradient-descent driving force to an energy minimum.

The differential equation (26.8) has to be solved numerically. To keep the computation as simple as possible one can employ Euler's method for this task, which is obtained by replacing the differential quotient by the quotient of forward differences. Explicitly, the following neural updating rule results

$$u_{i\alpha}^{\text{new}} = u_{i\alpha}^{\text{old}}$$
$$+ \Delta t \left[-\frac{1}{\tau} u_{i\alpha} - \sum_j d_{ij} (n_{j,\alpha+1} + n_{j,\alpha-1}) \right.$$

---

[1] Note that the minimization algorithm is in no way restricted to this choice. Any distance matrix d_{ij} is admissible: it does not even have to satisfy the metric axioms!

$$-\lambda_1 \sum_{\beta \neq \alpha} n_{i\beta} - \lambda_2 \sum_{j \neq i} n_{j\alpha} - \lambda_3 \left(\sum_{j,\beta} n_{j\beta} - N \right) \Bigg] . \tag{26.9}$$

A modification of this algorithm is described in [Wa89a]. In this approach not only the neuron activations $n_{i\alpha}$ but also the Lagrange multipliers λ_n are treated as dynamical variables. Their value is taken to increase with time according to the value of the corresponding energy E_n. This is called the basic differential multiplier method (BDMM). Now (26.8) has to be solved in conjunction with the three additional differential equations [Pl88]

$$\frac{d\lambda_n}{dt} = +E_n \quad , \text{ where } \quad n = 1, 2, 3 . \tag{26.10}$$

This modification was found to improve the convergence of the algorithm since it gradually strengthens the constraints which enforce a valid solution. A rigorous justification, however, appears to be lacking.

26.1.1 TSPHOP: **Program Description**

The program TSPHOP.C attempts to solve the TSP using the Hopfield–Tank approach. The coordinates x_i, y_i of the cities are distributed over the two-dimensional unit square.

As usual the program starts with an input panel where the operating parameters can be specified. These parameters are:

- ncity: The size N of the problem, i.e. the number of cities to be visited. If ncity=0 is chosen the program reads the size and the city coordinates from a data file.
- ordered: This determines whether the coordinates are chosen randomly (0) or in a regular fashion (1). In the latter case the number of cities N is truncated to the next square of an integer. The coordinates are the interior points of a square lattice, i.e. they take on the values $k/(1 + \sqrt{N})$ where $k = 1, \ldots, \sqrt{N}$.
- deltat: The time increment Δt used in the numerical integration of Eq. (26.9).
- temp: The inverse steepness parameter T of the neuron activation function (26.1).
- bdmm: If this parameter is set to 1 the Lagrange multipliers λ_i are treated as variables to be integrated according to the differential equation (26.10) (the "basic differential-multiplier method").
- tau: The time constant τ of the neural evolution equation (26.8).
- lambda: Three numbers characterizing the (starting values of) the Lagrange multipliers $\lambda_1, \lambda_2, \lambda_3$ for the three constraints enforcing the convergence to valid solutions (exactly one nonvanishing entry in each row and column of the neuron activation matrix).

If `ncity=0` was chosen, the program reads in the coordinates from a data file named `INCITY.DAT`. This is a text file which in the first line contains the value of N. The subsequent lines specify the coordinate pairs x_i, y_i of the cities. Such a file is created automatically if new coordinates are to be generated by the program (the old content is destroyed in this case). In this way the same problem can be attacked repeatedly by the same program or the performance of the alternative algorithms to be discussed in the next sections can be compared.

After the city coordinates and the matrix of Euclidean distances d_{ij} have been computed the program chooses initial values for the neural activations, $n_i(0)$. For a completely unbiased start one might choose the same value $n_i(0) = \frac{1}{N}$ for all neurons i. However, to avoid getting stuck at a spurious fully symmetric state and to allow for a nondeterministic operation of the program some random noise is added. This is achieved by starting with the local fields

$$u_i(0) = -\frac{1}{2} \ln N \left(1 + 0.1\xi\right),\qquad(26.11)$$

where ξ is a random variable in the range $-1 \ldots + 1$.

Subsequently the numerical integration of the neural activations is started. Ten integration steps are performed in sequence, then an intermediate result is shown. In the text-display mode the energy values E_0, E_1, E_2, E_3 and the total energy E are given, together with the Lagrange multipliers $\lambda_1, \lambda_2, \lambda_3$ (printed below the corresponding energy). In addition the quantity

$$\eta = \frac{1}{N} \sum_{i\alpha} n_{i\alpha}^2 \qquad(26.12)$$

is given. η can be interpreted as the *saturation* of the network since, starting from $\eta = 1/N$, it approaches the value $\eta = 1$ if a valid solution is reached. This has no bearing on the length of the tour.

The main part of the text screen is filled by the neural activation matrix $n_{i\alpha}$ (row=city number, column=station). For large problems the display is truncated to 20 entries. To enhance readability the values $n_{i\alpha}$ are multiplied by a factor of 1000 and displayed as integer numbers.

As an alternative to the numerical display of the state of the network the currently chosen tour of the traveling salesman can also be traced on the graphics screen. This graphics mode is switched on and off by pressing the key 'g'. Finally, the integration can be stopped by pressing 'e'.

The display of the current tour is faced with a characteristic difficulty. Only after successful completion of the integration, if at all, can we expect to find an ideal valid solution where exactly one neuron in each row and column is 'on' and the others are 'off'. In the intermediate steps of the calculation the activations will be distributed in the range $0 < n_{i\alpha} < 1$. If one wants to deduce a tour despite of this imperfection a more or less arbitrary mapping of the continuous activation matrix to a discrete and valid path has to be chosen. This is achieved in the program using a heuristic prescription. For each city i the station α is marked which has the maximum activation $n_{i\alpha}$. If there is exactly one such marked entry for each station then a unique solution has been found. Otherwise the program attempts to allocate that city out of the set of simultaneously visited cities having lowest activations to that station out of the set of omitted stations which has the highest activation for the given city. The procedure is repeated until a valid tour is reached, although this does not always succeed.

In the graphics mode the length of the tour constructed by this prescription is displayed, together with the variable cheat which counts the number of reallocations that were required to construct a valid tour out of the activation matrix $n_{i\alpha}$. Of course, finally the value cheat=0 should be reached, otherwise the calculation has failed to solve the TSP.

26.2   The Potts-Glass Model

The original Hopfield–Tank approach does not provide a practicable way to solve the traveling salesman problem. Very frequently the solutions found do not represent a valid tour and if they do they often are far from optimal [Wi88]. At least part of the problem originates from the chosen representation of the tour by a set of N^2 independent Ising spin variables $n_{i\alpha} = 0$ or 1, $i, \alpha = 1, \ldots, N^2$. The dimension of the space of spin configurations is thus 2^{N^2}, which is very much larger than the number of valid tours $N!$ (which includes permutations of the same tour). This led Peterson and Söderberg [Pe89d] to suggest a more economical representation which makes use of "graded neurons". Here each of the N cities is described by a single neuron $\mathbf{s}_i$, $i = 1, \ldots, N$. The output of this neuron, however, is described by a unit vector in an N-dimensional vector space $\mathbf{s}_i = (0, \ldots, 1, \ldots, 0)$. Writing out the component $s_{i\alpha} = 0$ or 1 this looks quite similar to the previous case of two-valued variables $n_{i\alpha}$. However, the graded-neuron representation automatically satisfies the constraint

$$\sum_\alpha s_{i\alpha} = 1 \,, \tag{26.13}$$

which guarantees that a city is visited exactly once. In this way the space of configurations has dimension N^N, which still is larger than $N!$ but contains much less redundancy than the representation with independent two-valued variables. The vector-valued variables $\mathbf{s}$ are not a new invention: spin systems described by such variables are familiar to solid-state physicists as the N-state *Potts model*. The physics of the Potts model is reviewed in [Wu83].

In this way the TSP is mapped not onto an Ising spin glass but onto a Potts spin glass. The main idea, however, remains the same, namely to construct an energy function which has a global minimum when the spin variables take on values representing a valid tour with minimum length and then to search for the ground state of the spin system. The choice of such an energy function is not unique. One may use

$$E = \sum_{i,j} \sum_{\alpha} d_{ij} s_{i\alpha} s_{i,\alpha+1} + \frac{A}{2} \sum_{\alpha,\beta}^{\alpha \neq \beta} s_{i\alpha} s_{i\beta} + \frac{B}{2} \sum_{\alpha} \left(\sum_i s_{i\alpha} - 1 \right)^2 . \quad (26.14)$$

As in (26.3)–(26.7) the first term measures the length of the tour, and the remaining terms are Lagrange multipliers enforcing the constraints. The term containing the constant A agrees with 26.5 and guarantees that the same city is not visited twice. The third contribution enforces the balance of the solution, i.e. each station is allocated to exactly one city. The constraint terms can be rewritten in the following way

$$\begin{aligned} E &= \sum_{i,j} \sum_{\alpha} d_{ij} s_{i\alpha} s_{i,\alpha+1} \\ &+ \frac{A}{2} \sum_i \sum_{\alpha} s_{i\alpha}^2 + \frac{B}{2} \sum_{\alpha} \left(\sum_i s_{i\alpha} \right)^2 + \frac{1}{2}(A - B)N , \quad (26.15) \end{aligned}$$

where the constraint (26.13) has been used. The last term in (26.15) is an irrelevant constant.[2]

In the search for the ground state (or at least a state with energy as low as possible) of the spin glass the *mean-field approximation* of the Potts glass can be invoked. This has already be discussed for Ising spin systems, cf. Sect. 4.1. Here we give a derivation of the mean field equations following [Pe89d].

26.2.1 The Mean-Field Approximation of the Potts Model

In the following calculation we evaluate the thermodynamic expectation value of the spin variables s_i using the saddle-point approximation. We start from the canonical partition function (see Sect. 17.1)

$$Z = \sum_{[s]} e^{-E(s_1,\dots,s_N)/T} . \quad (26.16)$$

The $[s]$ summation runs over all possible spin configurations $s_1,\dots,s_N$ where each vector s_i can be any of the cartesian unit vectors $e_1,\dots,e_N$. With the introduction of delta functions the partition function can be expressed in terms of an integration over a set of conjugate auxiliary variables $u_i, v_i; i = 1,\dots,N$

$$\begin{aligned} Z &= \sum_{[s]} \int \left(\prod_i dv_i \right) e^{-E(s_1,\dots,s_N)/T} \prod_i \delta(s_i - v_i) \\ &= \sum_{[s]} \int \left(\prod_i dv_i \right) \int \left(\prod_i \frac{du_i}{2\pi} \right) e^{-E(v_1,\dots,v_N)/T} \prod_i e^{iu_i \cdot (s_i - v_i)} . \quad (26.17) \end{aligned}$$

---

[2] Note that also the A-dependent constraint in principle is dispensable. According to $s_{i\alpha}^2 = s_{i\alpha}$ and (26.13) it is automatically satisfied. However, for the numerical solution where $s_{i\alpha}$ is represented by continuous variables it was found useful to retain the term in the energy function.

Now an analytic continuation to imaginary values of the $\mathbf{u}_i$ variables is performed: $i\mathbf{u}_i \rightarrow \mathbf{u}_i$.

$$Z = \int \left(\prod_i d\mathbf{v}_i\right) \int \left(\prod_i \frac{d\mathbf{u}_i}{2\pi}\right) e^{-E(\mathbf{v}_1,\ldots,\mathbf{v}_N)/T - \sum_i \mathbf{u}_i \cdot \mathbf{v}_i} \sum_{[s]} \prod_i e^{\mathbf{u}_i \cdot \mathbf{s}_i}$$

$$= \int \left(\prod_i d\mathbf{v}_i\right) \int \left(\prod_i \frac{d\mathbf{u}_i}{2\pi}\right) e^{-E(\mathbf{v}_1,\ldots,\mathbf{v}_N)/T - \sum_i \mathbf{u}_i \cdot \mathbf{v}_i + \sum_i \ln z(\mathbf{u}_i)}. \quad (26.18)$$

In the last step use was made of the factorization property of the $[s]$ summation $\sum_{[s]} \prod_i \exp \mathbf{u}_i \cdot \mathbf{s}_i = \prod_i \left(\sum_{\mathbf{s}_i} \exp \mathbf{u}_i \cdot \mathbf{s}_i\right)$ and a "local partition function" $z(\mathbf{u}_i)$ was introduced by

$$z(\mathbf{u}_i) = \sum_{\mathbf{s}} e^{\mathbf{u}_i \cdot \mathbf{s}} = \sum_\alpha e^{u_{i\alpha}} . \quad (26.19)$$

In view of the fact that the function in the exponent of the integrand, let us call it $g(\mathbf{u}_1,\ldots,\mathbf{u}_N;\mathbf{v}_1,\ldots,\mathbf{v}_N)$, will be large for systems with $N \gg 1$ the integral can be solved using the *saddle-point method*. The saddle point is characterized by the simultaneous vanishing of the set of partial derivatives:

$$0 = \frac{\partial g}{\partial \mathbf{v}_i} = -\frac{1}{T}\frac{\partial E}{\partial \mathbf{v}_i} - \mathbf{u}_i ,$$

$$0 = \frac{\partial g}{\partial \mathbf{u}_i} = -\mathbf{v}_i + \frac{1}{z(\mathbf{u}_i)}\frac{\partial z}{\partial \mathbf{u}_i} = -\mathbf{v}_i + \frac{1}{z(\mathbf{u}_i)}\sum_{\mathbf{s}} \mathbf{s} e^{\mathbf{u}_i \cdot \mathbf{s}} . \quad (26.20)$$

Thus the saddle point is determined by

$$\mathbf{v}_i = \mathbf{F}_N(\mathbf{u}_i) ,$$
$$\mathbf{u}_i = -\frac{1}{T}\frac{\partial E}{\partial \mathbf{v}_i} \quad (26.21)$$

with a vector-valued function $\mathbf{F}_N(\mathbf{u})$ having the components

$$F_N^\alpha(\mathbf{u}) = \left(\frac{\sum_{\mathbf{s}} \mathbf{s} e^{\mathbf{u} \cdot \mathbf{s}}}{\sum_{\mathbf{s}} e^{\mathbf{u} \cdot \mathbf{s}}}\right)_\alpha = \frac{e^{u_\alpha}}{\sum_\beta e^{u_\beta}} . \quad (26.22)$$

Inspection of this relation shows that the functions $\mathbf{F}_N(\mathbf{u})$ are generalizations of the sigmoidal function $f(x) = \frac{1}{2}(1+\tanh x) = e^x/(e^x + e^{-x})$ familiar from the mean-field treatment of the Ising-spin model. It is immediately obvious that they satisfy the constraint

$$\sum_\alpha F_N^\alpha(\mathbf{u}) = 1 . \quad (26.23)$$

The variables $v_{i\alpha}$ have a direct physical interpretation: they give the probability that the spin number i points in the direction α. This is so because $\mathbf{v}_i$ is the thermodynamic expectation value of the spin variable $\mathbf{s}_i$:

$$\mathbf{v}_i = \langle \mathbf{s}_i \rangle_T = \frac{1}{Z} \sum_{[s]} \mathbf{s}_i e^{-E(\mathbf{s}_1,\ldots,\mathbf{s}_N)} . \quad (26.24)$$

Thus by solving the saddle-point equations (26.21) the optimization problem is solved, at least as far as the mean-field approximation is valid.

Equation (26.21) constitutes a set of coupled transcendental equation which, of course, cannot be solved in closed form. However, we can resort to the familiar strategy of *successive iteration* to obtain a fixed point of the equations. Experience has to show whether this is an economical way to find 'good' solutions of the problem.

26.2.2 TSPOTTS: Program Description

Using the explicit expression for the energy (26.14) the mean-field equations (26.21) read

$$v_{i\alpha} = F_N^\alpha(\mathbf{u}_i),$$

$$u_{i\alpha} = \frac{1}{T}\left[-\sum_j d_{ij}(v_{j,\alpha+1} + v_{j,\alpha-1}) + Av_{i\alpha} - B\sum_j v_{j\alpha}\right], \qquad (26.25)$$

with the function $F_N(\mathbf{u})$ defined in (26.22). The program TSPOTTS iterates these equations at a fixed temperature T until a stable solution is reached. This is used to begin the next iteration cycle which is performed at a reduced temperature. These steps are repeated until a satisfactory solution is found.

The input panel of the program asks for the following parameters:

- ncity: The number N of cities.
- ordered: Random or ordered distribution of city coordinates.
- A,B: The Lagrange multipliers A and B for the constraints enforcing a valid tour as formulated in (26.14).
- delta: A parameter δ used to decide when the successive iteration is to be ended, cf. (26.27).
- anneal: After a cycle of iterations at a constant temperature T the temperature is lowered by multiplication with the factor anneal < 1.

Given these parameters the temperature T has to be specified. Ideally this should be quite low so that the system can occupy a unique valley in the "energy landscape". On the other hand, at $T \to 0$ it will get stuck in the next available local minimum, which may be well off the desired global minimum. This conflict can be resolved by the strategy of simulated annealing discussed in Sect. 11.2. One starts with a high value of the temperature and slowly cools the system down. This prescription is implemented in the program. As the temperature is lowered the system undergoes a phase transition in which the system is drawn from the trivial high-temperature fixed point $v_{i\alpha} = \frac{1}{N}$ to a nontrivial solution. Of course, the annealing schedule should start at a temperature somewhat above the critical value T_c.

Peterson and Söderberg [Pe89d] have estimated this critical temperature. They started from the trivial solution and investigated its stability against perturbations by linearizing the set of coupled equations. By studying the distribution of eigenvalues of the resulting linear equations they obtained the following estimate for the critical temperature (assuming serial updating of the spins):

$$T_c \simeq \frac{1}{N} \max\left(B - A, B + \bar{d} + 1.3 N \sigma_d\right), \tag{26.26}$$

where $\bar{d}$ and σ_d are the average value and the standard deviation of the set of city-distances d_{ij}. This estimate for T_c is displayed by the program, which then asks for a starting temperature value. This should be chosen in the vicinity of T_c.

Subsequently the spin variables are initialized to uniform values $\frac{1}{N}$ (which is the trivial solution valid in the $T \to \infty$ limit) slightly disturbed by random noise $s_{i\alpha} = \frac{1}{N}(1 + 0.1\xi)$ as was done in the Hopfield–Tank program. Then the program starts iterating the set of equations (26.25), up to 40 times at a given temperature. The iteration is stopped if the accumulated change of the spin values in the updating procedure is smaller then a predetermined constant, i.e.

$$\frac{1}{N^2} \sum_{i\alpha} |v_{i\alpha}^{\text{new}} - v_{i\alpha}^{\text{old}}| < \delta. \tag{26.27}$$

After each annealing step the temperature gets lowered and the network configuration is displayed numerically. Alternatively by pressing the key 'g' a graphical display of the tour can be chosen. The display format and the heuristic algorithm to map the network configuration to a valid tour have been discussed in Sect. 26.1.1. The temperature T is successively lowered until the "saturation" η of the solution

$$\eta = \frac{1}{N} \sum_{i\alpha} v_{i\alpha}^2 \tag{26.28}$$

gets larger than 0.9. This indicates that the network is close to a uniquely specified valid solution (where η should be 1) and that the temperature is well below T_c.

In most cases the network then represents a valid tour. Even if this is not fulfilled a good solution can usually be obtained by a few reallocations of cities.

26.3 TSANNEAL: **Simulated Annealing**

The standard techniques used to solve combinatorial optimization problems do not rely on neural networks. Rather, they employ heuristic procedures which are based on the discrete nature of the problem.

For the case of the traveling-salesman problem good working strategies employ a set of elementary moves introduced by Lin and Kernighan [Li73]. These moves transform one configuration of the system (i.e. a valid tour connecting all the cities, let us denote it by C) to a different configuration C'. The most important of these moves are: (1) *Transport*: a segment of the tour is excised from its old position and inserted at a new place. (2) *Reversal*: a segment of the tour is selected and the order of cities in this segment is reversed. Then a suitable strategy has to be devised which, by applying a sequence of these elementary moves, transforms an arbitrarily chosen initial tour to an optimal or at least acceptable solution. A systematic (but not the only and not necessarily the fastest) way to do this is based on the thermodynamic concept of simulated annealing [Ki83] as discussed in Sect. 11.2.

The objective function to be minimized, i.e. the path length, is interpreted as the energy E of a physical system. An ensemble of such systems in thermal equilibrium is simulated. By carefully decreasing the temperature T the system is frozen out to a state of low energy (short path length). To establish thermal equilibrium the Metropolis method is used. An elementary move $C \to C'$ is chosen at random and the corresponding change in energy $\Delta E = E(C') - E(C)$ is calculated. Depending on the sign and magnitude of ΔE the move is accepted with probability

$$\Pr(C \to C') = \begin{cases} 1 & \text{for } \Delta E \leq 0 \\ e^{-\Delta E/T} & \text{for } \Delta E > 0 \end{cases}. \tag{26.29}$$

By accepting changes which enhance the energy the algorithm introduces thermal fluctuations and avoids being trapped in metastable states, i.e. local energy minima.

The program TSANNEAL.C, which is adapted from a routine published in the book *Numerical Recipes* [Pr86], implements this strategy of simulated strategy. It is meant to provide "benchmark" solutions against which the performance of the neural-network-inspired programs can be compared:

The program has only three input parameters:

- ncity: The number N of cities.
- ordered: The random or ordered distribution of city coordinates.
- anneal_fact: The reduction factor of the temperature T.

The program starts with an estimate for the temperature $T = L N^{-3/2}$, where L is the initial paht length. This value of T should be above the "melting point". Random elementary moves are chosen repeatedly (Segment Transport and Reversal having equal probabilities) and accepted or rejected according to the Metropolis condition (26.29). At a given temperature the iteration extends over at most $100 N$ moves. It will be stopped earlier, however, as soon as a number of $10 N$ moves has been accepted. Then the temperature is lowered according to a simple annealing schedule which reduces T by a certain fraction, typically by 5%. The program performance might be enhanced by the use of a more elaborate cooling schedule [La87a]. The problem is declared solved if during the last annealing cycle none of the attempted moves has been accepted. After each cycle the program graphically displays the current tour and the corresponding length.[3]

26.4 Numerical Experiments

Comparative experiments[4] with the three programs TSPHOP, TSPOTTS, and TSANNEAL should begin with a small number of cities, say $N = 10$ or less.

Experiments with the *simulated-annealing algorithm* show that it finds good solutions in virtually all attempts. Even for the 100-city problem (which is the largest dimension provided for in the program) it does not take longer than the duration of a coffee break.

The *Hopfield–Tank algorithm* will be found to converge very slowly. What is worse, very often the solution does not converge to a tour which is valid, let alone short. The situation is improved only slightly when the Lagrange parameters are treated as dynamical variables (bdmm=1). Of course, it is a toilsome task to search for an optimal combination of the parameter values λ_i fixing the constraints. In accordance with [Wi88] experiments with the program TSPHOP seem to indicate, however, that there is no set of parameters leading to consistently good solutions even for a small number of cities.

The *Potts-model algorithm* works much better. Usually it generates valid tours which are only slightly longer than the simulated annealing solutions. For large problem size N the computation requires more time; the computing time scales with $O(N^4)$. Here one has to take into account, however, that this would change if the algorithm were implemented as a parallel process instead of being simulated on a serial computer.

Figure 26.1 shows a typical result for the $N = 64$-city problem. The cities are arranged on a regular square grid so that it is easy for the human eye to recognize the optimal path. If $N = K^2$ is the square of an even number K the many tours of optimal length consist of N horizontal and vertical segments connecting neighboring lattice sites, the total tour length being $L = K^2/(1+K)$. For odd K one of the segments must be diagonal so that the expression for L contains an additional term $\sqrt{2}-1$ in the numerator. We did not succeed in producing a valid solution with the Hopfield–Tank algorithm. The results produced by the programs TSPOTTS and TSANNEAL are shown in Fig. 26.1. While the annealing algorithm has found one of the optimal solutions (for this regular distribution of cities there is a large number of tours of equal length) the Potts-model solution contains a number of localized "faults", i.e. diagonal instead of straight segments. The overall path length

---

[3] In contrast to the neural-network algorithms the display of intermediate results here poses no problem since a unique and valid tour is defined at any instant.

[4] We remark in passing that in the $N \to \infty$ limit the optimal tour length for cities distributed randomly on the unit square is known with high precision. It grows like $L \simeq 0.749 \sqrt{N}$.

in this way is increased by about 6% compared to the optimal tour, which still constitutes a quite respectable solution.

Fig. 26.1. Typical solutions of the traveling salesman problem with $N = 64$ cities arranged on a square lattice. *Left*: An optimal solution found with simulated annealing. *Right*: Approximate solution found with the Potts-model neural-network algorithm.

Experiments with the programs TSPHOP and TSPOTTS will probably convince the reader that neural networks do not constitute viable competitors with more standard methods of solving the traveling-salesman problem.[5] Nevertheless it is intriguing that combinatorial optimization problems can be attacked at all by this approach.

---

[5] Incidentally, the simulated-annealing method also performs poorly when compared to the best available heuristic methods based on a clever combination of Lin–Kernighan optimization moves. For moderately large numbers of cities there even exist programs [Pa87a] which are able to find *exact solutions* to the TSP using an affordable amount of computer time.

27. KOHOMAP: The Kohonen Self-organizing Map

As discussed in Sect. 15.2.3 Kohonen's topology-preserving map creates a representation of a multidimensional continuous "sensory space" on a grid of neurons. Here we will denote the sensory input by $\mathbf{x} = (x_1, \ldots, x_d)$. This is assumed to be a d-dimensional vector with real-valued components, taking on values in a subspace $V \in \mathbb{R}^d$. Exemplars of such vectors are repeatedly presented to the network, simulating the varying stimuli experienced by a sensory organ. The values of $\mathbf{x}$ are drawn randomly according to a given probability distribution which may be nonuniform over the range of possible values V. The $N = N_x \times N_y$ neurons of the network are thought to be organized in a two-dimensional grid so that each neuron can be labeled by a vector $\mathbf{n} = (n_x, n_y)$, where n_x, n_y are integer counting indices in the range $1 \leq n_i \leq N_i$.

The task is to determine a d-dimensional vector of "synaptic coefficients" $\mathbf{w_n}$ so that each neuron $\mathbf{n}$ is "responsible" for a range of input vectors for which the distance $\|\mathbf{x} - \mathbf{w_n}\|$ takes on the smallest value. The density of the associations of neurons should be such that regions with frequently occurring input signals are mapped more closely and no neurons are idle. In addition neighboring areas in the sensory space should be mapped to neighboring sites on the neural grid as far as topological constraints permit.

One might try to achieve this by an updating equation modeled after the perceptron learning rule

$$\mathbf{w_n}(t+1) = \mathbf{w_n}(t) + \epsilon f(\mathbf{w_n} \cdot \mathbf{x})\left(\mathbf{x} - \mathbf{w_n}(t)\right), \tag{27.1}$$

where $f(h)$ is a sigmoidal function of the local field $h = \mathbf{w_n} \cdot \mathbf{x}$ representing the post-synaptic excitation and the two terms in the brackets on the right correspond to learning of the new value $\mathbf{x}$ and unlearning of the old one, $\mathbf{w_n}(t)$. ϵ is a small positive coefficient. However, this rule does not lead to a mapping satisfying the criteria stated above. No mechanism is built in that forces a neuron to cooperate with its neighbors, allowing it to specialize on a certain subset of input stimuli. This is achieved by the following algorithm due to Kohonen [Ko82]:

1. The weights $\mathbf{w_n}$ are assigned arbitrary starting values $\mathbf{w_n}(0)$.
2. A sensory input vector $\mathbf{x} \in V$ is chosen randomly.

3. The neuron $\mathbf{n}_0$ is determined which has weights closest to the input vector:

$$\|\mathbf{x} - \mathbf{w}_{\mathbf{n}_0}\| = \min_{\mathbf{n}} \|\mathbf{x} - \mathbf{w}_{\mathbf{n}}\| . \tag{27.2}$$

4. The weights are updated according to the rule

$$\mathbf{w}_{\mathbf{n}}(t+1) = \mathbf{w}_{\mathbf{n}}(t) + \epsilon(t)g(\mathbf{n} - \mathbf{n}_0, t)(\mathbf{x} - \mathbf{w}_{\mathbf{n}}(t)) \quad \text{for all } \mathbf{n} . \tag{27.3}$$

The steps 2 to 4 are repeated until a satisfactory mapping is found.

The function $g(\mathbf{n} - \mathbf{n}_0, t)$ in step 4 is essential for the success of the algorithm. This function has its maximum size (normalized to unity) when $\mathbf{n}$ coincides with $\mathbf{n}_0$ and it decays to zero at large distances. The steepness of the decay is characterized by a width parameter Δ. The function g has the interpretation of inducing *lateral inhibition* among the neurons. The neuron $\mathbf{n}_0$ whose weights $\mathbf{w}_{\mathbf{n}_0}$ already are closest to the desired value $\mathbf{x}$ is most strongly affected by the input. The weights of the neighboring neurons are modified to a lesser extent, the influence decreasing with the distance $\|\mathbf{n} - \mathbf{n}_0\|$. Typical choices for the lateral inhibition are a Gaussian or a cutoff function

$$g(\mathbf{n} - \mathbf{n}_0) = e^{-\|\mathbf{n}-\mathbf{n}_0\|^2/2\Delta^2} , \tag{27.4}$$

$$g(\mathbf{n} - \mathbf{n}_0) = \theta(\Delta - \|\mathbf{n} - \mathbf{n}_0\|) . \tag{27.5}$$

Taking finite and constant values for the parameters ϵ and Δ, the learning algorithm leads to highly volatile weights which are modified after each new presentation of an input vector $\mathbf{x}$. A stable mapping can be achieved by making the parameters dependent on the updating time t. The values of $\epsilon(t)$ and $\Delta(t)$ are chosen to decrease gradually, i.e. the "range of influence" of a stimulus shrinks and also the magnitude of the weight change is reduced. In the limit $\epsilon \to 0$ and $\Delta \to 0$ the algorithm leads to a stable mapping which has the desired properties [Ko84].

27.1 Program Description

The program KOHOMAP.C implements the Kohonen mapping algorithm described above for an input space of dimension two. The sensory input stimuli are provided by vectors $\mathbf{x} = (x_1, x_2)$ with components distributed in a chosen subset of the square $[-1, +1]^2$. As illustrated in Fig. 27.1 two models are implemented: (1) a circular ring with inner radius 0.5 and outer radius 1; (2) two disjoint checkerboard squares of unit length.

The program starts by asking whether the network weights are to be read from a data file. If this option is chosen, the program only serves to redisplay a network configuration which was obtained earlier. If a full calculation is to be done (no input-data file) the program asks for the following input parameters:

Fig. 27.1. The input vectors **x** for the Kohonen map are drawn from the dashed areas of one of these geometrical figures.

- nx,ny: The x and y dimensions of the neuron grid.
- delta0: The initial value of the lateral-inhibition parameter $\Delta(0)$.
- eps0: The initial value of the updating-strength parameter $\epsilon(0)$.
- facdelta: A factor η_Δ which determines the rate of decay of the inhibition range $\Delta(t)$.
- faceps: A factor η_ϵ which plays the same role for the learning strength $\epsilon(t)$.
- model: This determines the geometry of the distribution of input vectors (=1: annulus, =2: squares).
- biased: If the value 1 is chosen, the distribution of **x** values is biased in favor of vectors with a positive first component ($x_1 > 0$) by a margin of $3:1$, otherwise the distribution is uniform.
- ordered: For a value of 1 the initial weights $\mathbf{w_n}$ are prepared on a regular rectangular grid in the space $[-1, +1]^2$. Otherwise all components are drawn from a uniform random distribution.

After these parameters have been specified the program generates ordered or random input weights $\mathbf{w_n}$ and displays the network configuration. The square $[-1, +1]^2$ is marked on the graphics screen and lines connecting the rows (full lines) and the columns (dotted lines) of the neural weight matrix $\mathbf{w_n}$ are drawn. The intersection points of this net thus correspond to the **x** values to which a neuron with labels (n_x, n_y) responds most strongly.

Following this display of the initial configuration, the program starts the learning process. From the previously specified probability distribution an input vector **x** is selected. The neuron $\mathbf{n_0}$ with weights closest to **x** is determined according to (27.2), using the norm $\|\mathbf{v}\| = |v_1| + |v_2|$. Then for all neurons **n** the correction to the weights $\mathbf{w_n}$ is calculated according to (27.3). The Gaussian lateral inhibition function (27.4) is used in conjunction with the Euclidean distance $\|\mathbf{n} - \mathbf{n_0}\| = \sqrt{(n - n_0)_x^2 + (n - n_0)_y^2}$. After a cycle of $N = N_x \times N_y$ updates the learning-strength and lateral-correlation-range parameters are reduced according to the prescription

$$\Delta^{\mathrm{new}} = \eta_\Delta \Delta^{\mathrm{old}} ,$$
$$\epsilon^{\mathrm{new}} = \eta_\epsilon \epsilon^{\mathrm{old}} . \tag{27.6}$$

This leads to an exponential decay of the parameters with learning time.[1]

---

[1] To obtain optimal solutions in the limit $t \to \infty$ a prescription leading to slower decay, like $\epsilon(t) \propto t^{-1}$, seems to be preferable [Ri88b].

At the end of an updating cycle the program graphically displays the current network configuration $\mathbf{w_n}$. The number of the last learning cycle and the current values of $\Delta(t)$ and $\epsilon(t)$ are printed. In addition the average spread $\|\mathbf{x} - \mathbf{w_{n_0}}\|$ between the input vectors $\mathbf{x}$ encountered in the learning cycle and the weights of the best adapted neuron $\mathbf{n_0}$ is given.

No stopping criterion (except of a timeout after 10^5 updates) is built into the program so that the learning process has to be interrupted manually by pressing the key e. The final configuration can be written on a data file.

27.2 Numerical Experiments

The performance of the Kohonen algorithm depends on a judicious choice of the parameters $\Delta(t)$ and $\epsilon(t)$. When the initial values of these parameters are too small, the motion in the space of weights $\mathbf{w_n}$ is slow. Furthermore, if the decay law is very rapid (i.e. the decay constants η_Δ and η_ϵ are considerably smaller than 1), the configuration is frozen in before the optimal equilibrium solution has been reached. This is very similar to the problem encountered with the temperature decrease in the simulated-annealing procedure. Thus the reader should experiment with various combinations of the parameter values in order to gain experience.

If the range of lateral inhibition $\Delta(t)$ decays faster than the learning rate $\epsilon(t)$ or is too small from the outset, the network may reach a configuration which adequately represents the distribution density of $\mathbf{x}$ values. However, the goal to get a topologically faithful mapping which preserves neighborhood relations will be violated since adjacent neurons do not experience similar "forces". On the other hand, keeping $\Delta(t)$ too large will lead to a poor covering of the sensory space since adjacent neurons are not able to develop their specific "area of competence" different from that of their neighbors.

The optimal choice of parameters also depends on the size of the network, i.e. the number of neurons N. When changing N the parameter ϵ should be scaled inversely to the number of neurons.

In the program KOHOMAP the dimension of the sensory input space is fixed to the value $d = 2$ which agrees with the natural dimension of the neuron grid. Nevertheless the effect of "automatic selection of feature dimensions" mentioned in Sect. 15.2.3 can be studied by setting $N_y = 1$. In this way the neural space is restricted to be one-dimensional. Then the Kohonen algorithm generates a mapping of a two-dimensional area onto a one-dimensional "string of beads". The transition from two dimensions to one dimension is nicely displayed in Fig. 27.2 for the example of a circular ring as sensory input space. The one-dimensional case is reminiscent of the well-known mathematical construction of "Peano" curves which completely fill a given space, recently made popular by the theory of fractals.

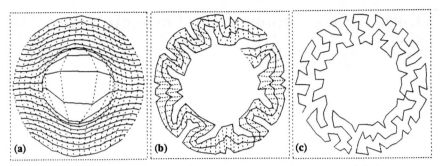

Fig. 27.2. Examples of Kohonen mappings of a circular ring on **(a)** a square net with $N_x \times N_y = 20 \times 20$, **(b)** a rectangular net 100×3, **(c)** a one-dimensional string 200×1.

Exercise: Construct Kohonen maps using the biased probability distribution (`biased= 1`). Try to establish how the "magnification rate" of the mapping (the density of the grid established by the algorithm) scales with the probability density of the sensory input vectors.

Exercise: Extend the program to the case of sensory input vectors **x** in a space of dimension higher than two.

28. BTT: Back-Propagation Through Time

As explained in Chapt. 14 an extension of the classical back-propagation algorithm can be used to train recurrent neural networks. Not only is it possible to construct networks which possess a set of predetermined stable states (attractors) but one can even impose time-dependent trajectories which are traced out in the state space of the network. The program RECURR implements the algorithm of back-propagation through time (BTT) derived in Sect. 14.3. The network consists of N fully coupled neurons which can take on continuous values in the interval $s_i = -1 \ldots + 1$. Either one or two these neurons represent the output of the network. The goal of the algorithm is to find a set of weights w_{ij} which guarantee that the output signals $s_i(t), i \in \Omega$, follow some predetermined functions $\zeta_i(t)$ as closely as possible over a time interval $0 \le t \le \tau$. This is achieved by minimizing an error functional, cf. (14.21), using the gradient descent method.

Here we collect the equations from Sect. 14.3 which are required to understand the operation of the program. The time evolution of the network activations is determined by the differential equation

$$\dot{s}_i = -\frac{1}{T_i}\left(s_i - f(h_i)\right) \tag{28.1}$$

where the local field is defined as

$$h_i(t) = \sum_k w_{ik} s_k(t) \, , \tag{28.2}$$

i.e. there are *no external inputs*. The summation runs over one extra neuron the value of which is kept to $s_{N+1} = -1$, which is the familiar way to implement a threshold ϑ_i. For the activation function we use a sigmoidal function with unit slope

$$f(x) = \tanh x \, . \tag{28.3}$$

The set of coupled first-order differential equations (28.1) is integrated forward in time from $t = 0$ to $t = \tau$. A set of conjugate variables z_i is generated by integrating the inhomogeneous linear differential equations

$$\dot{z}_i = \frac{1}{T_i} z_i - \sum_k \frac{1}{T_k} w_{ki} f'(h_k) z_k + (\zeta_i(t) - s_i)\delta_{i \in \Omega} \tag{28.4}$$

backward in time from $t = \tau$ to $t = 0$, starting with the initial condition $z_i(\tau) = 0$. The gradient learning rule for updating the weights reads

$$\delta w_{ik} = -\epsilon \frac{1}{T_i} \frac{1}{\tau} \int_0^\tau dt\, z_i f'(h_i) s_k \ . \tag{28.5}$$

The program also incorporates the option to adjust the time constants T_i. The learning rule for T_i can be obtained as in Sect. 14.3 by expressing the derivative of the error functional dE/dT_i in terms of the back-propagation signals z_i. The result is

$$\delta T_i = +\epsilon' \frac{1}{T_i} \frac{1}{\tau} \int_0^\tau dt\, z_i \dot{s}_i \ . \tag{28.6}$$

The learning constants for the weights and the time constants can be chosen separately. In the program we have put $\epsilon' = 0.02\epsilon$ since too fast variations in the time constants were found to have a detrimental effect on the learning process. The time constants are prevented from dropping below a minimum value $T_{\min} = 0.01$.

The program also implements the *teacher forcing* algorithm. Here during the forward integration of the network activations the output signals are forced to follow the target function, $s_i(t) = \zeta_i(t), i \in \Omega$. There are no conjugate variables z_i for the output units $i \in \Omega$. The equations (28.4), (28.5), and (28.6) essentially remain valid also in the case of teacher forcing if one makes the replacements

$$z_i \rightarrow \frac{1}{T_i} \left(f(h_i) - s_i \right) - \dot{\zeta}_i \quad \text{for} \quad i \in \Omega \ . \tag{28.7}$$

28.1 Program Description

As the first step after starting the program BTT the following parameters can be specified.

- ndim: The total number N of (fully interconnected) neurons.
- nout: The number of output neurons (either 1 or 2).
- tau: The upper time limit τ.
- tstep: The number of intermediate time steps used in the integration.
- tconst: The (initial) value of the time constants T_i, taken equal for all units.
- ftype: The type of function (1, 2, or 3) to be learned (see below).
- frequency: The frequency parameter ν.
- amplitude: The amplitude parameter A.

Three types of functions $\zeta_i(t)$ can be chosen, all of them being variants of Lissajous curves (in the two-dimensional case).

1. A circle

$$\zeta_1(t) = A \sin(2\pi\nu t) ,$$
$$\zeta_2(t) = A \cos(2\pi\nu t) .$$ (28.8)

2. The figure eight

$$\zeta_1(t) = A \sin(2\pi\nu t) ,$$
$$\zeta_2(t) = \frac{1}{2} A \cos(4\pi\nu t) .$$ (28.9)

3. A square

$$\zeta_1(t) = A \operatorname{triang}(2\pi\nu t) ,$$
$$\zeta_2(t) = A \operatorname{triang}(2\pi\nu t + \pi/2) .$$ (28.10)

The triangle function $\operatorname{triang}(x)$ in the last case is defined as a straight-line approximation to $\sin(x)$, i.e. it is consists of straight-line sections with slopes $\pm 2/\pi$ which share the extrema and zeroes with the sine function.

For the default values $\tau = 1$ and $\nu = 1$ each of the three functions traces out one cycle of a closed curve in the two-dimensional plane. In the one-dimensional case, nout=1, only $\zeta_1(t)$ is used, so that options 1 and 2 are the same.

After the network architecture and the target function have been defined the program presents a second input panel asking for parameters which are specific to the learning process.

- epsilon: The gradient parameter ϵ for the learning of the weights. Remember that for the time constants the much smaller learning constant $\epsilon' = 0.02\epsilon$ is used.
- alpha: The momentum factor α. This has been discussed several times; see, e.g. Chapt. 24, equation (24.7).
- teach_force: If this is set to 1 the teacher-forcing algorithm will be used.
- learn_tconst: If this is set to 1 the time constants T_i will be adjusted during the learning phase. Otherwise they are kept fixed to their initial values which were entered in the first input panel.
- linear: This option allows replacement the sigmoidal activation function (28.3) by the linear function $f(x) = x$.

The program run starts by initializing the weights w_{ij} with random numbers distributed uniformly over the range $-0.5 \ldots + 0.5$. Then the differential equations (28.1) are integrated forward in time using a simple Euler algorithm.[1] The initial condition is $s_i(0) = 0$ for hidden ($i \notin \Omega$) and $s_i(0) = \zeta_i(0)$ for output ($i \in \Omega$) units.

---

[1] Of course one could do better here, using, e.g., Runge–Kutta or another sophisticated integration method. However, it is not obvious, whether much accuracy would be gained in this way as long as the same discretization scheme is used during the training and the operating phase. The program can adapt to the properties of the integration algorithm.

If nout=2 the result of the first forward integration is presented graphically. The display shows the two-dimensional target curve $(\zeta_1(t), \zeta_2(t))$ (yellow) and the network trajectory $(s_1(t), s_2(t))$ (blue). The light-blue section corresponds to the time interval for which learning takes place ($0 \le t \le \tau$). Both curves contain some markers at equidistant time intervals $\tau/10$. The curve for the network trajectory has a dark-blue extension which corresponds to the time interval ($\tau \le t \le 2\tau$). This shows how the learned trajectory extrapolates beyond the training interval. One can see, for example, whether the network runs into a fixed point or has learned to represent a periodic orbit (limit cycle).

After inspecting the starting trajectory (which will have little resemblance with the target curve) the user has to press the key t to switch to the text screen. This starts the learning phase which consists of repeated cycles of forward integration (generating $s_i(t)$), backward integration (generating $z_i(t)$) and evaluation of the weight changes δw_{ij} (and optionally time-constant changes δT_i).

The text screen displays various information on the progress of the learning phase. The blue window in the lower left keeps track of the iteration steps, showing the value of the error functional E (given either by (14.21, 14.22) or by (14.39) in the case of teacher-forcing). Also shown is the sum of the absolute values of all weight changes $\sum_{ij} |\delta w_{ij}|$.

The upper part of the screen contains a listing of the updated weights w_{ij}, i counting the rows and j counting the columns. The last (two) row(s), printed in green, correspond to the weights feeding into the (two) output neuron(s). The extra column (no. $N+1$) contains the threshold ϑ_i of neuron number i.

Finally the display shows the values of the time constants T_i, the activations $s_i(\tau)$ reached at the upper boundary of the forward integration, and the error variables $z_i(0)$ reached at the lower boundary of the backward integration for each of the N units.

By pressing the key t the user can switch from the text screen to the graphical display of the trajectory described above. The learning continues and the screen is repeatedly updated after 20 learning steps. An alternative graphical display can be activated by pressing the key f. This shows all the activations $s_i(t)$ (hidden and output units) as functions of time t, together with the target functions $\zeta_i(t)$. The latter are drawn in white and yellow and the corresponding signals of the output neurons carry markers of matching color.

In addition to viewing its progress the user also can interact with the learning process by pressing p which brings up again the second input panel. This allows the user to modify the learning parameters and then to resume the learning process.

28.2 Numerical Experiments

Experiments with the program BTT perphaps should in the beginning use ftype=1, i.e. train the network to run through a *circle*. The randomly chosen initial weights w_{ij} in most cases lead to a rather inconspicuous starting trajectory which soon runs into a fixed point. However, using the default parameters ($N = 6$, $\tau = 1$, $T_i = 0.2$, $\nu = 1$, $A = 0.6$, $\epsilon = 1.$, $\alpha = 0.9$) a few learning steps are sufficient to make the trajectory resemble a circle. Often a few hundred steps produce a rather satisfactory result, although it will take much longer until the deviation settles to its final value.

The success depends on the value of the time constants T_i which should match the characteristic frequency of the target trajectory (typical value $1/2\pi\nu$). If the time constants are chosen too large the response of the neurons is too sluggish to follow the target function while very small values of T_i may lead to unstable behavior. The option to learn also the time constants according to equation (28.6) can help in this situation. Setting learn_tconst = 1 and starting with time constants of, e.g., $T_i = 0.05$ or $T_i = 1$, one observes that the learning algorithm tends to shift the values to more "reasonable" values. (Note that the optimal learning constant ϵ has to be adjusted in inverse proportion to T_i.) Success is not always guaranteed, however. Here one might experiment to find an optimal value for the ratio ϵ'/ϵ.

The same experiments can be repeated with the teacher-forcing algorithm. The results are quite similar if one takes into account that much smaller values of the learning constant have to be used, e.g. $\epsilon = 0.01$.

Encouraged with the success achieved with four hidden units (ndim = 6) one might wonder about the size of the minimal network which can represent the circular trajectory. As it turns out, this can be achieved without any hidden units! Using ndim = 2 the weight matrix very soon converges to a unique solution $w_{ij} = \begin{pmatrix} 1.345 & 1.773 & -0.004 \\ -1.761 & 1.291 & 0.014 \end{pmatrix}$. The matrix is nearly antisymmetric, with vanishing thresholds ϑ_i (third column). It is not difficult to understand why such a simple solution is possible. If we *linearize* the equation of motion (28.1) putting $f(x) \simeq x$ we have

$$T_1 \dot{s}_1 = (w_{11} - 1)s_1 + w_{12}s_2 - w_{13} ,$$
$$T_2 \dot{s}_2 = w_{21}s_1 + (w_{22} - 1)s_2 - w_{23} .$$

The circular trajectory (28.8) is a solution of these differential equations provided that $w_{11} = w_{22} = 1$, $w_{12} = 2\pi\nu T_1$, $w_{21} = 2\pi\nu T_2$, $w_{13} = w_{23} = 0$. The deviations of the learned weight matrix from these theoretical values mainly results from the nonlinearity of the activation function. To check this the program BTT also can be run in the *linear mode*[2], $f(x) = x$. In the linear

---

[2] While the activations s_i of the nonlinear network are constrained to the interval $(-1, +1)$, in the linear case their values are not guaranteed to remain bounded. Therefore occasionally the program may fail with an overflow message.

mode learning proceeds with great speed and the deviation of the network trajectory from circular target function drops down practically to zero. The weight matrix converges to $w_{ij} = \begin{pmatrix} 0.961 & 1.256 & 0.000 \\ -1.256 & 0.961 & 0.000 \end{pmatrix}$. The deviation from the theoretical values of $w_{11} = w_{22} = 1$, $w_{12} = -w_{21} = 2\pi\nu T = 1.257$ is due to the discretization error of the Euler integration. This is easily verified by changing the number of integration steps from its default value $\texttt{tstep} = 100$. In this way we also can confirm the earlier remark that the quality of the integration scheme does not really affect the operation of the network. If we use a small number of integration steps, say $\texttt{tstep} = 10$, the Euler algorithm should produce large discretization errors. Nevertheless the performance is not affected at all since the learning algorithm does not care whether an exact integration of the differential equations or some finite-difference scheme is implemented.[3]

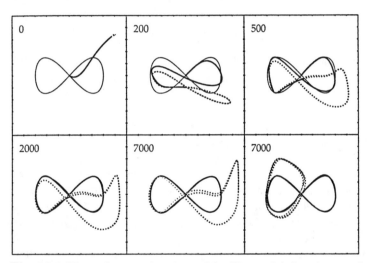

Fig. 28.1. Network approximations of the "figure eight" trajectory shown at various stages of learning with the BTT algorithm. Thin lines: target curve. Full lines: learned network output ($t = 0, \ldots, 1$). Dotted lines: extrapolated network output ($t = 1, \ldots, 3$). The lower right panel shows the result of a second learning run that differed only in the random initialization of the weights.

The ease with which the network is able learn the circular trajectory of course is somewhat deceptive. The function (28.8) emerges in a rather trivial manner as a stable limit circle of the linearized equations of motion $\dot{s} = F(s)$. A somewhat more demanding task is the *figure eight* (28.9). Here it is immediately obvious that hidden units are needed: The point $(0,0)$ is

---

[3] For the teacher-forcing algorithm, however, the achieved accuracy is found to depend on the step size since here the analytical derivatives $\dot{\zeta}_i(t)$ are used instead of their discretized counterparts.

passed by two branches of the trajectory moving in different directions. This requires some sort of memory which has to be represented by the hidden units. Using the default parameters (four hidden units) in most cases a satisfactory solution is found within a few thousand iteration steps. It turns out, however, that already two hidden units are sufficient to represent the figure eight. In the linear mode the output error of this four-neuron network can be made arbitrarily small.

A typical example obtained with the default parameters is shown in Fig. 28.1. The network output quite soon shows some resemblance to the target trajectory but it take much longer until the curves get congruent. It is interesting to follow the trajectory beyond the upper border τ of the learning interval. Although occasionally it will happen there is no guarantee that the network output is continued periodically. Usually the trajectory runs into some limit cycle which may or may not resemble the target curve. The two last panels of Fig. 28.1 illustrate this point: the dotted lines, which cover the time interval $\tau \leq t \leq 3\tau$, in one example trace out a distorted version of the figure eight while the limit cycle of the second run more closely resembles a circle. It is easy, however, to ensure that the limit cycle reached by the network agrees with the target curve, simply by extending the training interval over more than one period (e.g. by doubling the value either of τ or of ν).

As a final example the *square* function (28.10) can be used for training. In contrast to the two prior examples this function is not based on a simple superposition of sines, and it is harder to learn because it has discontinuous derivatives. The network output first will resemble a circle having the size of the square. Subsequently the edges have to be learned which is quite time consuming. Increasing the size of the network and allowing for variable time constants here is helpful. It is interesting to inspect the time-dependent activations $s_i(t)$ of the hidden neurons. At the edge points some of them undergo rapid variations, thus allowing the time derivatives $\dot{s}_i$ of the output units to become (nearly) discontinuous.

The reader should be aware that the program BTT does not yet exploit the full capacity of the recurrent neural network paradigm. As an extension the network could be made to respond to external inputs. Furthermore, as we have seen, the sample trajectories are learned better in the linear mode. Thus the strength of neural networks which lies in their use of nonlinear activation functions has not been exploited.

Exercise 1: Teach the network to follow different trajectories depending on the initial condition $s_i(t = 0)$.

Exercise 2: Construct a network which reacts to one or several *input variables* I_i. Consult [Pe90] for examples. The network might be taught to solve the XOR problem, or an input variable might be interpreted as a rotation angle which determines the orientation of the output trajectories.

29. NEUROGEN: Using Genetic Algorithms to Train Networks

This program uses a genetic algorithm to train a feed-forward neural network to predict time series. The network can have one or two hidden layers, adjacent layers being fully connected in the forward direction, and the nodes can take on any floating point value. The activation function is the familiar $\tanh(\beta x)$, but can be changed if desired. The training is accomplished by randomly choosing a population of genes – a gene is the list of weights and biases for the network – computing the performance of each member of the population and reproducing the population using standard genetic operators. The operators used are: *asexual reproduction* – simply copying the entire gene to the next population, *crossover* – cutting two genes at complementary locations and splicing them together, and *mutation* – each floating point copy is associated with a probability of mutation (adding a small random number to the weight or bias during copying) and a mutation size distribution.

One of the problems with genetic algorithms is getting them to converge to a final answer. This is accomplished here by decreasing the mutation size as the generations progress, much like simulated annealing. This concentrates the population into one area in the configuration space so that this region is explored fully. The drawback to using such a scheme is the loss of population diversity. This could possibly be remedied by doing detailed searches, collecting the best genes from these, and then using them as the starting population for another genetic algorithm.

29.1 Program Description

After starting NEUROGEN the program first asks whether a network is to be read in from file. This allows one to view the performance of a previously generated network which was saved to a data file. No further learning or modification steps can be applied to this network. If the answer is 'no' the user then is prompted to specify the type of time series that is to be learned. Four different choices have been prepared, but the reader is encouraged to add to this selection. The time-series data are read in from files which can be easily modified.

- *The logistic map*: This is defined by the recursive equation

$$x_{n+1} = rx_n(1 - x_n)$$

which despite its simplicity for suitable values of r leads to a series of iterates which exhibits deterministic chaos. The logistic map has been discussed in Sect. 7.1.1.
- *The Hénon map*: The recursive equation can be written in the form

$$x_{n+1} = 1 - rx_n^2 + bx_{n-1} \qquad r > 0 \quad , \quad |b| < 1 \quad .$$

In contrast to the logistic map two preceding values must be known to generate a new data point. The Hénon map too exhibits chaos in some regions of the r–b parameter space.
- *Sunspot numbers*: As an example of a time series based on data observed in nature the number of sunspots recorded at the Swiss Federal Observatory in the years $1700 \ldots 1987$ are provided. At first sight the sunspot activity appears to shows periodic oscillation. However, the period length is found to vary between 7 and 16 years (average value 11.3 years) and also the amplitudes look rather irregular. This makes the sunspot time series difficult to predict and one should not expect too much from our neural-network approach. More details on the sunspot data as a benchmark for time-series predictions have been given in Chapt. 16.
- *Sine function*: For those frustrated by experiments with the sunspot data we also provide a very regular and easy-to-predict time series. This is a simple harmonic function with a period of 11, which can be viewed as a mock-up of the sunspot numbers.

After the type of time series has been chosen an input panel opens which allows one to specify the layout of the network and various parameters affecting the learning process.

- nlayers: The number of hidden layers (1 or 2).
- nin: The number of input neurons.
- nhid: The number of neurons per hidden layer(s).
- pop: The number of networks in the population.
- t_learn: The number of time steps on which to train the networks.
- t_max: The length of the entire data set.
- extrap_mode: Determines the way in which the time series is extrapolated beyond t_learn. 0: The network looks only one step ahead and always gets presented the true data as input. 1: The network does a repeated extrapolation, using its own predictions (in the region beyond t_learn) as input.
- cross: The number of crossovers per generation.
- mutate: The probability of mutation for each floating point copy.
- s: The maximum probability of a gene being crossed over during the reproduction. The formula used to compute probabilities is $P_i = s \exp(-\text{alpha} * \text{rank}_i/\text{pop})$.

- `alpha`: This variable controls the selectivity of the crossover reproduction. A large value of alpha means that only the very best of each generation have a high probability of crossover.
- `copy`: Maximum number of asexual copies of each gene that can be placed in the next generation.
- `mut_range`: The program decreases the mutation size for the best members of the population by a factor $fact_i = 1 - exp[-rank_i/mut\_range]$. Thus the best member (rank = 0) is not subject to mutation at all and the value of `mut_range` controls how far into the ranking the suppression extends. If `mut_range` is set to zero the suppression factor is switched off entirely.
- `past`: The number of past generations over which to average the total population fitness. This average is used to test convergence.
- `mult`: This is the factor by which the mutation strength `stddev` gets reduced whenever the population is "stagnant".
- `trigger`: The measure of how "stagnant" a population has to be before the factor `mult` gets applied.

Once the parameters have been set the program is ready to run. The program reads the specified input data file which contains the time series. The first `nin` data points are presented to the input neurons and the network predicts the next point in the series. The input window slides along the data until the end of the series at `t_max` is reached.

Once the entire time series has been gone through, the performance of each of the `pop` genes is assessed. This is done by summing the absolute values of the difference between the predictions and the correct answers, averaged over all predicted values. (One could experiment with other error measures here). The genes are then ranked according to their score with the lowest score being the best, and the genetic operators described above are then applied to the population.

Mutations are controlled by the variables `mutate` and `stddev`. The fixed parameter `mutate` determines the probability for a mutation to happen, evaluated separately for each of the weights. If chance decides in favor of a mutation a Gaussian random number is drawn, gets multiplied by the factor $fact_i$ as described above, and is added to the weight. The standard deviation `stddev` of the Gaussian normal distribution initially has the value `stddev` = 1 and gets decreased in the course of the learning procedure. A kind of simulated annealing has been implemented in order to ensure convergence. The program averages the total population fitness over `past` generations and compares this with that same average from one generation earlier. If the difference between these is less than `trigger` then the mutation standard deviation `stddev` is decreased by the multiplicative factor `mult`. Convergence is tested by comparing the average population fitness with the best member's fitness. If these two numbers are very close, then the population has converged.

During the training phase the progress of the algorithm can be monitored. A large window shows a running list of the fitness (i.e. the mean prediction error) for the best and for the worst performer in the gene pool. Also the extrapolation error of the "champion" is displayed (this is relevant if t_max > t_learn). Also shown is a logarithmic measure for the diversity of the population. In the course of learning this number drops off to the value 1 as the population tends to become homogeneous. The last two columns show a sliding average of past and present performances of the whole population and the ratio of this average to the best performance. Occasionaly a red letter R appears in the display which indicates that according to the "annealing" prescription the value of stddev has been reduced.

A second window displays information on the fitness distribution of the population. Each line refers to one member of the population, showing its current rank and the rank its ancestor occupied in the previous generation. The following two columns give the performance (i.e. the mean error) achieved on the training data set, t ≤ t_learn and in the extrapolation region t > t_learn.

The learning process can be interrupted by pressing any of the keys s, p, w, e. Key w causes the program to wait until the next keypress and e ends the learning process. Pressing p leads back to the input panel for parameter selection which allows one to modify the learning parameters. Of course those of the entries in this panel which determine the network layout should be kept fixed, otherwise unpredictable behavior will result.

The performance of the population of networks can be inspected by pressing s, which leads to a choice of several alternatives. The time series can be graphically displayed by choosing the option t. The program then asks for the number (rank) of the network to be inspected and for a range of times, which allows one to take a look at a magnified part of the time series. A plot is generated which shows the "true" values of the time series (yellow dots) and the predictions of the network (blue dots). The data points are connected by straight lines to guide the eye. A vertical line marks the position of t_learn, i.e. the border between the training and the extrapolation data. For the special case of the logistic map, pressing the key f produces a display of x_{n+1} as a function of x_n which should be an inverted parabola, cf. Fig. 7.1. Entering l leads produces a numerical listing of the network weights and biases using the format described in Table 24.1. Finally, if the key w is pressed the configuration of the best network is saved to a data file which can be kept for further inspection.

References

[Ab89] L.F. Abbott and T.B. Kepler: Optimal Learning in Neural Network Memories, *J. Phys.* A **14**, L711 (1989)

[Ab92] P. Abreu et al.: Classification of the Hadronic Decays of the Z^0 into b and c Quark Pairs using a Neural Network, *Physics Letters B* **295**, 383 (1992)

[Ac85] D.H. Ackley, G.E. Hinton, and T.J. Sejnowski: A Learning Algorithm for Boltzmann Machines, *Cognitive Science* **9**, 147 (1985)

[Al78] J. de Almeida and D. Thouless: Stability of the Sherrington–Kirkpatrick Solution of a Spin-Glass Model, *J. Phys.* A **11**, 983 (1978)

[Al87] L.B. Almeida: A Learning Rule for Asynchronous Perceptrons with Feedback in a Combinatorial Environment, Proc. IEEE First Ann. Int. Conf. on Neural Networks, San Diego, June 1987, IEEE, New York (1987), p. 199

[Al88] J. Alspector, R.B. Allen, V. Hu, S. Satyanarayana: Stochastic Learning Networks and their Electronic Implementation, in: [An88a], p. 9

[Al89a] I. Aleksander (ed.): *Neural Computing Architectures. The Design of Brainlike Machines*, North Oxford Academic, London (1989)

[Al89b] L.B. Almeida: Backpropagation in Non-Feedforward Networks, in: [Al89a], p. 74

[Am72] S. Amari: Learning Patterns and Pattern Sequences by Self-Organizing Nets of Threshold Elements, *IEEE Trans. Electr. Comp.* **21**, 1197 (1972)

[Am80] S. Amari: Topographic Organization of Nerve Fields, *Bull. of Math. Biology* **42**, 339 (1980)

[Am85a] D.J. Amit, H. Gutfreund, and H. Sompolinski: Storing Infinite Numbers of Patterns in a Spin-Glass Model of Neural Networks, *Phys. Rev. Lett.* **55**, 1530 (1985)

[Am85b] D.J. Amit, H. Gutfreund, and H. Sompolinski: Spin-glass Models of Neural Networks, *Phys. Rev.* A **32**, 1007 (1985)

[Am87a] D.J. Amit, H. Gutfreund, and H. Sompolinski: Statistical Mechanics of Neural Networks Near Saturation, *Ann. Phys. (NY)* **173**, 30 (1987)

[Am87b] D.J. Amit, H. Gutfreund, and H. Sompolinski: Information Storage in Neural Networks with Low Levels of Activity, *Phys. Rev.* A **35**, 2283 (1987)

[Am88] D.J. Amit: Neural Networks Counting Chimes, *Proc. Natl. Acad. Sci. USA* **85**, 2141 (1988)

[Am89a] D. Amit: *Modelling Brain Functions*, Cambridge University Press, Cambridge (1989)

[Am89b] E. Amaldi and S. Nicolis: Stability–Capacity Diagram of a Neural Network with Ising Bonds, *J. Phys. France* **50**, 1333 (1989)

[Am89c] D.J. Amit, C. Campbell, and K.Y.M. Wong: The Interaction Space of Neural Networks with Sign Constrained Synapses, *J. Phys.* A **22**, 4687 (1989)

[Am89d] D.J. Amit, K.Y.M. Wong, and C. Campbell: Perceptron Learning with Sign-Constrained Weights, *J. Phys.* A **22**, 2039 (1989)

[Am90] S. Amari: Mathematical Foundations of Neurocomputing, *Proc. IEEE* **78**, 1443 (1990)

[An72] J.A. Anderson: A Simple Neural Network Generating Interactive Memory, *Math. Biosci.* **14**, 197 (1972)

[An88a] D.Z. Anderson (ed.): *Neural Information Processing Systems*, American Institute of Physics, New York (1988)

[An88b] A. Anderson: Learning from a Computer Cat, *Nature* **331**, 657 (1988)

[An89a] J.K. Anlauf and M. Biehl: The AdaTron: An Adaptive Perceptron Algorithm, *Europhys. Lett.* **10**, 687 (1989)

[An89b] V.V. Anshelevich, B.R. Amirikian, A.V. Lukashin, and M.D. Frank-Kamenetskii: On the Ability of Neural Networks to Perform Generalization by Induction, *Biol. Cybern.* **61**, 125 (1989)

[An90] J. Anderson, A. Pellionisz, and E. Rosefeld (eds): *Neurocomputing 2: Directions for Research*, MIT Press: Cambridge, MA (1990)

[An94] P.J. Angeline, G.M. Saunders, and J.B. Pollack: An Evolutionary Algorithm that Constructs Recurrent Neural Networks, *IEEE Transactions on Neural Networks* **5**, 54 (1994)

[As89] T. Ash: Dynamic Node Creation in Back-propagation Networks, *Connection Science* **1**, 365 (1989)

[Ba72] L.E. Baum: An Inequality and Associated Maximization Technique in Statistical Estimation for Probabilistic Functions of Markov Processes, *Inequalities* **3**, 1 (1972)

[Ba80] H.B. Barlow: The Absolute Efficiency of Perceptual Decisions, *Phil. Trans. Roy. Soc. London* B **290**, 71 (1980)

[Ba85a] A.G. Barto: Learning by Statistical Cooperation of Self-Interested Neuron-like Computing Elements, *Human Neurobiology* **4**, 229 (1985)

[Ba85b] A.G. Barto and P. Anandan: Pattern-Recognizing Stochastic Learning Automata, *IEEE Trans. Syst. Man Cybern.* **15**, 360 (1985)

[Ba87a] K.L. Babcock and R.M. Westervelt: Dynamics of Simple Electronic Neural Networks, *Physica* D **28**, 305 (1987)

[Ba87b] P. Baldi and S.S. Venkatesh: Number of Stable Points for Spin-Glasses and Neural Networks of Higher Orders, *Phys. Rev. Lett.* **58**, 913 (1987)

[Ba87c] A.G. Barto and M.I. Jordan: Gradient Following without Back-propagation in Layered Networks, *IEEE First Annual Int. Conf. on Neural Networks* II, 629 (1987)

[Ba89] M. Bauer and W. Martienssen: Quasi-Periodicity Route to Chaos in Neural Networks, *Europhys. Lett.* **10**, 427 (1989)

[Ba92] B. Bai and N.H. Farhat: Learning Networks for Extrapolation and Radar Target Identification, *Neural Networks* **5**, 507 (1992)

[Ba93] H. Baird: Recognition Technology Frontiers, *Pattern Recognition Letters* **14**, 327 (1993)

[Ba94a] I.A. Bachelder and A.M. Waxman: Mobile Robot Visual Mapping and Localization: A View-based Neurocomputational Architecture, *Neural Networks* **7**, 1083 (1994)

[Ba94b] R.J. Bauer Jr. : *Genetic Algorithms and Investment Strategies*, John Wiley & Sons, (1994)

[Be89] A. Bergman: Self-Organization by Simulated Evolution, *Lectures in Complex Systems, Proceedings of the 1989 Complex Systems Summer School*, Sante Fe, E. Jen ed., (1989)

[Be91] L. Bellantoni, J.S. Conway, J.E. Jacobsen, Y.B. Pan, and Sau Lan Wu: Using Neural Networks with Jet Shapes to Identify b Jets in e^+e^- Interactions, *Nucl. Inst. and Meth.* **A310**, 618 (1991)

[Be92] A. Bertoni and M. Dorigo: In Search of a Good Crossover Between Evolution and Optimization, in: *Parallel Problem Solving from Nature* 2b, B. Manderick and R. Männer, eds., Amsterdam: Elsevier, p. 479 (1992)

[Be93] K.H. Becks, F. Block, J. Drees, P. Langefeld, and F. Seidel: b-Quark Tagging Using Neural Networks and Multivariate Statistical Methods: A Comparison of Both Techniques, *Nucl. Inst. and Meth.* **A329**, 501 (1993)

[Bi86a] E. Bienenstock, F. Fogelman Soulié, and G. Weisbuch (eds.): *Disordered Systems and Biological Organization*, NATO ASI Ser. F 20, Springer, Berlin, Heidelberg (1986)

[Bi86b] K. Binder and A.P. Young: Spin Glasses: Experimental Facts, Theoretical Concepts, and Open Questions, *Rev. Mod. Phys.* **58**, 801 (1986)

[Bi87a] W. Bialek and A. Zee: Statistical Mechanics and Invariant Perception, *Phys. Rev. Lett.* **58**, 741 (1987)

[Bi87b] E. Bienenstock and C. von der Malsburg: A Neural Network for Invariant Pattern Recognition, *Europhys. Lett.* **4**, 121 (1987)

[Bi92] A. Bischoff, B. Schürmann, J. Maruhn, and J. Reinhardt: Characteristic Properties of Stable Recurrent Higher-Order Neural Networks, in: *Artificial Neural Networks* **2**, I. Aleksander and J. Taylor (eds.), p. 325 (1992)

[Bi94] A. Bischoff: Unfolding Supervised Learning, Technical Report, University of Frankfurt, April 1994

[Bl62a] H.D. Block: The Perceptron: A Model for Brain Functioning. I, *Rev. Mod. Phys.* **34**, 123 (1962)

[Bl62b] H.D. Block, B.W. Knight Jr., and F. Rosenblatt: Analysis of a Four-Layer Series-Coupled Perceptron. II, *Rev. Mod. Phys.* **34**, 135 (1962)

[Bo88] H. Bohr, J. Bohr, S. Brunak, R.M.J. Cotterill, B. Lautrup, L. Nørskov, O.H. Olsen, and S.B. Petersen: Protein Secondary Structure and Homology by Neural Networks, *FEBS Lett.* **241**, 223 (1988)

[Bo89] H. Bohr: Scientific Applications of Neural Networks, Proc. IECON-88 (1989)

[Bo91] C. Bortolotto, A. De Angelis, and L. Lanceri: Tagging the Decays of the Z^0 Boson into b Quark Pairs with a Neural Network Classifier, *Nucl. Inst. and Meth.* **A306**, 459 (1991)

[Br82] C. Braccini, G. Gambardella, G. Sandini, and V. Tagliasco: A Model of the Early Stage of the Human Visual System: Functional and Topological Transformations Performed in the Peripheral Visual Field, *Biol. Cybern.* **44**, 47 (1982)

[Br88a] S. Brunak and H. Bohr: NBI preprint HI-88 (1988)

[Br88b] J. Bruck and J.W. Goodman: On the Power of Neural Networks for Solving Hard Problems, in: [An88a], p. 137

[Br88c] J. Bruck and J.W. Goodman: A Generalized Convergence Theorem for Neural Networks and Its Applications in Combinatorial Optimization, *IEEE Trans. Inform. Theory* **34**, 1089 (1988)

[Bu77] T.H. Bullock, R. Orkand, and A. Grinnell: *Introduction to Nervous Systems*, Freeman, San Francisco (1977)

[Bu87a] J. Buhmann and K. Schulten: Influence of Noise on the Function of a "Physiological" Neural Network, *Biol. Cybern.* **56**, 313 (1987)

[Bu87b] J. Buhmann and K. Schulten: Noise-Driven Temporal Association in Neural Networks, *Europhys. Lett.* **4**, 1205 (1987)

[Bu89] J. Buhmann, R. Divko, and K. Schulten: Associative Memory with High Information Content, *Phys. Rev. A* **39**, 2689 (1989)

[Ca61] E.R. Caianiello: Outline of a Theory of Thought Processes and Thinking Machines, *J. Theor. Biol.* **1**, 204 (1961) {Reprinted in [Sh88]}

[Ca87b] P. Carnevali and S. Patarnello: Exhaustive Thermodynamical Analysis of Boolean Learning Networks, *Europhys. Lett.* **4**, 1199 (1987)

[Ca87c] G.A. Carpenter and S. Grossberg: A Massively Parallel Architecture for a Self-Organizing Neural Pattern Recognition Machine, *Computer Vision, Graphics and Image Processing*, **37**, 54 (1987)

[Ca89a] H.C. Card and W.R. Moore: VLSI Devices and Circuits for Neural Networks, *Int. Journ. of Neural Syst.* **1**, 149 (1989)

[Ca89b] M. Casdagli: Nonlinear Prediction of Chaotic Time Series, *Physica* D **35**, 335 (1989)

[Ch78] P.Y. Chou and G.D. Fasman: *Ann. Rev. Biochem.* **7**, 251 (1978)

[Ch83] J. P. Changeux: *L'homme neuronal*, Librairie Arthème Fayard, Paris (1983) [*Der neuronale Mensch*, Rowohlt, Hamburg (1984)]

[Ch90] D. Chalmers: The Evolution of Learning: An Experiment in Genetic Connectionism, *Proceedings of the 1990 Connectionist Models Summer School*, D.S Touretsky, J.L. Elman, T.J. Sejnowski, and G.E. Hinton eds., Morgan Kaufmann, San Mateo (1990)

[Ch92] K. Chakraborty, K. Mehrotra, C. Mohan, and S. Ranka: Forecasting the Behavior of Multivariate Time Series using Neural Networks, *Neural Networks* **5**, 961 (1992)

[Cl85] J.W. Clark, J. Rafelski, and J.V. Winston: Brain without Mind: Computer Simulation of Neural Networks with Modifiable Neuronal Interactions, *Phys. Rep.* **123**, 215 (1985)

[Cl88] J.W. Clark: Statistical Mechanics of Neural Networks, *Phys. Rep.* **158**, 91 (1988)

[Co65] T. Cover: Geometrical and Statistical Properties of Systems of Linear Inequalities with Applications in Pattern Recognition, *IEEE Trans. Electr. Comp.* **14**, 326 (1965)

[Co72] I.M. Cooke and M. Lipkin: *Cellular Neurophysiology: A Source Book*, Holt, Rinehart & Winston, New York (1972) [A collection of reproduced classical original papers]

[Co83] M. Cohen and S. Grossberg: Absolute Stability of Global Pattern Formation and Parallel Memory Storage by Competitive Neural Networks, *IEEE Trans. Syst. Man Cybern.* **13**, 815 (1983)

[Co88a] E. Collins, S. Ghosh, and S. Scofield: Risk Analysis, in: [DA88], p. 429

[Co88b] A.C.C. Coolen and T.W. Ruijgrok: Image Evolution in Hopfield Networks, *Phys. Rev.* A **38**, 4253 (1988)

[Co92] J.E. Collard: Commodity Trading with a Three Year Old, in: *Neural Networks in Finance and Investment*, R. Trippi and E. Turban eds, Probus Publishing, 411 (1992)

[Cr83] F.C. Crick and G. Mitchison: The Function of Dream Sleep, *Nature* **304**, 111 (1983)

[Cr86] A. Crisanti, D.J. Amit, and H. Gutfreund: Saturation Level of the Hopfield Model for Neural Network, *Europhys. Lett.* **2**, 337 (1986)

[Cr89] F. Crick: The Recent Excitement about Neural Networks, *Nature* **337**, 129 (1989)

[Cu86] Y. Le Cun: Learning Process in an Asymmetric Threshold Network, in: [Bi86a], p.234

[Cy89] G. Cybenko: Approximation by superpositions of sigmoidal functions, *Math. of Control, Signals, and Systems* **2**, 303 (1989)

[DA82] D. D'Amato, L. Pintsov, H. Koay, D. Stone, J. Tan, K. Tuttle, and D. Buck, High Speed Pattern Recognition System for Alphanumeric Handprinted Characters, *Proceeding of the IEEE-CS Conference on Pattern Recognition and Image Processing*, Las Vegas, 165 (1982)

[DA88] *DARPA Neural Network Study*, AFCEA International Press, Fairfax, Virginia (1988)

[Da91] T.E. Davis: Toward an Extrapolation of the Simulated Annealing Convergence Theory onto the Simple Genetic Algorithm, Ph.D. thesis, University of Florida, (1991)

[Da93] S.P. Day: Continuous-Time Back-Propagation with Adaptable Time Delays, Submitted to *IEEE Transactions on Neural Networks*, (1993)

[De86] J.S. Denker: Neural Network Models of Learning and Adaptation, *Physica* D **22**, 216 (1986)

[De87a] J.S. Denker (ed.): *Neural Networks for Computing*, American Inst. of Physics Conf. Proc. Vol. 151 (1987)

[De87b] J.S. Denker, D. Schwartz, B. Wittner, S. Solla, R. Howard, L. Jackel, and J. Hopfield: *Complex Systems* **1**, 877 (1987)

[De87c] B. Derrida, E. Gardner, and A. Zippelius: An Exactly Solvable Asymmetric Neural Network Model, *Europhys. Lett.* **4**, 167 (1987)

[De88a] B. Denby: Neural Networks and Cellular Automata in Experimental High Energy Physics, *Comp. Phys. Comm.* **49**, 429 (1988)

[De88b] B. Derrida and R. Meir: Chaotic Behavior of a Layered Neural Network, *Phys. Rev.* A **28**, 3116 (1988)

[De89] F. Del Giudice, S. Franz, and M.A. Virasoro: Perceptron Beyond the Limit of Capacity, *J. Phys. France* **50**, 121 (1989)

[DE89] The DELPHI Collaboration: DELPHI 89-68 PROG-143

[De90a] B. Denby, M. Campbell, F. Bedeschi, N. Chriss, C. Bowers, and F. Nesti: Neural Networks for Triggering, *1989 IEEE Nuclear Science Symposium*, San Francisco, Fermilab Report Conf-90/20, (1990)

[De90b] B. Denby, E. Lessner, and C. Lindsey: Tests of Track Segment and Vertex Finding with Neural Networks, *1990 Conference on Computing in High Energy Physics*, Sante Fe, NM, Fermilab Report Conf-90/68 (1990)

[De90c] L. Deng P. Kenny, M. Lennig, and P. Mermelstein: Modeling Micro-Segments of Stop Consonants in a Hidden Markov Model Based Word Recognizer, *Journal of the Acoustical Society of America* **87**, 2738 (1990)

[DE91] The DELPHI Collaboration: *Nucl. Inst. and Meth.* **A303**, 233 (1991)

[De91] L. Deng P. Kenny, M. Lennig, and P. Mermelstein: Phonemic Hidden Markov Models with Continuous Mixture Output Densities for Large Vocabulary Recognition, *IEEE Transactions on Signal Processing* **40**, 265 (1991)

[DE92] The DELPHI Collaboration: Physics Letters B **276**, 383 (1992)

[De94a] L. Deng, K. Hassanein, and M. Elmasry: Analysis of the Correlation Structure for a Neural Predictive Model with Application to Speech Recognition, *Neural Networks* **7**, 331 (1994)

[De94b] G.J. Deboeck ed: *Trading on the Edge: Neural, Genetic, and Fuzzy Systems for Chaotic Financial Markets*, John Wiley & Sons, (1994)

[Di87] S. Diederich and M. Opper: Learning of Correlated Patterns in Spin-Glass Networks by Local Learning Rules, *Phys. Rev. Lett.* **58**, 949 (1987)

[Do86] V.S. Dotsenko: Hierarchical Model of Memory, *Physica* A **140**, 410 (1986)

[Do88] V.S. Dotsenko: Neural Networks: Translation, Rotation and Scale-Invariant Pattern Recognition, *J. Phys.* A **21**, L783 (1988)

[Do92] K. Doya: Bifurcation in the Learning of Recurrent Neural Networks, *Proc. IEEE Intl. Symposium on Circuits and Systems*, **6**, 2777 (1992)

[Dr95] I.E. Dror, M. Zagaeski, and C.F. Moss: Three-Dimensional Target Recognition via Sonar: A Neural Network Model, *Neural Networks* **8**, 149 (1995)

[Du89] R. Durbin, C. Miall, and G. Mitchison (eds.): *The Computing Neuron*, Addison-Wesley, Wokingham (1989)

[Du80] B. Duerr, W. Haettich, H. Tropf, and G. Winkler: A combination of Statistical and Syntactical Pattern Recognition Applied to Classification of Unconstrained Handwritten Numerals, *Pattern Recognition* **12**, 189 (1980)

[Ec77] J.C. Eccles: *The Understanding of the Brain*, McGraw-Hill, New York (1977)

[Ec89] R. Eckmiller: Generation of Movement Trajectories in Primates and Robots, in: [Al89a], p. 305

[Ed75] S.F. Edwards and P.W. Anderson: Theory of Spin Glasses, *J. Phys.* F **5**, 965 (1975)

[Ei90] A.E. Eiben, E.H.L. Aarts, and K.M. Van Hee: Global Convergence of Genetic Algorithms: a Markov Chain Analysis, *Proc. First Workshop on Problem Solving from Nature*, H.P. Schwefel and R. Männer eds., Berlin, Heidelberg: Springer-Verlag, p. 4, 1990.

[En84] A.C.D. van Enter and J.L. van Hemmen: Statistical-Mechanical Formalism for Spin Glasses, *Phys. Rev.* A **29**, 355 (1984)

[Fa88] D. Farmer and J. Sidorowich: Exploiting Chaos to Predict the Future and Reduce Noise, in: [Le88], p. 277

[Fa90] S.E. Fahlman: The Cascade Correlation Algorithm, in: *Neural Information Processing Systems 2*, D.E. Touretzky (ed.), Morgan Kaufmann, Palo Alto (1990)

[Fe72] R.P. Feynman: *Statistical Mechanics: A Set of Lectures*, Benjamin, Reading (1972)

[Fe87] M.V. Feigelman and L.B. Ioffe: The Augmented Models of Associated Memory: Asymmetric Interaction and Hierarchy of Patterns, *J. Mod. Phys.* B **1**, 51 (1987)

[Fi73] B. Fischer: Overlap of Receptive Field Centres and Representation of the Visual Field in the Cat's Optic Tract, *Vision Res.* **13**, 2113 (1973)

[Fi80] R.J.P. de Figueiredo: Implications and Applications of Kolmogorov's Superposition Theorem, *J. Math. Anal. Appl.* **38**, 1227 (1980)

[Fo88a] J.F. Fontanari and R. Köberle: Information Processing in Synchronous Neural Networks, *J. Phys. France* **49**, 13 (1988)

[Fo88b] B.M. Forrest: Content-addressability and Learning in Neural Networks, *J. Phys.* A **21**, 245 (1988)

[Fo88c] J.C. Fort: Solving a Combinatorial Problem via Self-organizing Process: An Application of the Kohonen Algorithm to the Traveling Salesman Problem, *Biol. Cybern.* **59**, 33 (1988)

[Fo92] D.B. Fogel: Evolving Artificial Intelligence, Ph.D. Thesis, University of California, San Diego (1992)

[Fr94] J. Freeman: *Simulating Neural Networks with Mathematica*, Addison-Wesley, (1994)

[Fu75] K. Fukushima: Cognitron: A Self-Organizing Multilayered Neural Network, *Biol. Cybern.* **20**, 121 (1975)

[Fu80] K. Fukushima: Neocognitron: A Self-Organizing Neural Network Model for a Mechanism of Pattern Recognition Unaffected by Shift in Position, *Biol. Cybern.* **36**, 193 (1980)

[Fu88a] K. Fukushima: Neocognitron: A Hierarchical Neural Network Capable of Visual Pattern Recognition, *Neural Networks* **1**, 119 (1988)

[Fu88b] A. Fuchs and H. Haken: Pattern Recognition and Associative Memory as Dynamical Processes in a Synergetic System, *Biol. Cybern.* **60**, 17 and 107 (1988) [Erratum: p. 476]

[Ga79] M.R. Garey and D.S. Johnson: *Computers and Intractability: A Guide to the Theory of NP-Completeness*, Freeman, San Francisco (1979)

[Ga87] S.I. Gallant: Optimal Linear Discriminants, IEEE Proc. 8th Conf. on Pattern Recognition, Paris (1987)

[Ga88a] E. Gardner: The Space of Interactions in Neural Network Models, *J. Phys.* A **21**, 257 (1988)

[Ga88b] E. Gardner and B. Derrida: Optimal Storage Properties of Neural Network Models, *J. Phys.* A **21**, 271 (1988)

[Ga89a] E. Gardner and B. Derrida: Three Unfinished Works on the Optimal Storage Capacity of Networks, *J. Phys.* A **22**, 1983 (1989)

[Ga89b] E. Gardner, H. Gutfreund, and I. Yekutieli: The Phase Space of Interactions in Neural Networks with definite Symmetry, *J. Phys.* A **22**, 1995 (1989)

[Ge87] E. Geszti and F. Pázmándi: Learning within Bounds and Dream Sleep, *J. Phys.* A **20**, L1299 (1987)

[Ge89] E. Geszti and F. Pázmándi: Modeling Dream and Sleep, *Physica Scripta.* T **25**, 152 (1989)

[Ge93] K.A. Gernoth, J.W. Clark, J.S. Prater, and H. Bohr: Neural Network Models of Nuclear Systematics, *Physics Letters B* **300**, 1 (1993)

[Gi88] C.L. Giles, R.D. Griffin, and T. Maxwell: Encoding Geometric Invariances in Higher-Order Neural Networks, in: [An88a], p. 301

[Gi93] M. Gilloux: Research into the New Generation of Character and Mailing Address Recognition Systems at the French Post Office Research Center, *Pattern Recognition Letters* **14**, 267 (1994)

[Gl63] R. Glauber: Time-Dependent Statistics of the Ising Model, *J. Math. Phys.* **4**, 294 (1963)

[Gl94] M. Glesner and W. Pöchmüller: *Neurocomputers: An Overview of Neural Networks in VLSI*, Chapman & Hall, London (1994)

[Go88] D. Goldberg: *Genetic Algorithms in Machine Learning, Optimization, and Search*, Addison-Wesley, Reading (1988)

[Gr76] S. Grossberg: Adaptive Pattern Classification and Universal Recoding: I. Parallel Development and Coding of Neural Feature Detectors, *Biol. Cybern.* **23**, 121 (1976)

[Gr87] H.P. Graf, L.D. Jackel, R.E. Howard, B. Straughn, J.S. Denker, W. Hubbard, D.M. Tennant, and D. Schwarz: VLSI Implementation of a Neural Network Memory with Several Hundreds of Neurons, in: [De87a], p. 182

[Gu88] H. Gutfreund and M. Mézard: Processing of Temporal Sequences in Neural Networks, *Phys. Rev. Lett.* **61**, 235 (1988)

[Ha87] H. Haken: in: *Computational Systems – Natural and Artificial*, H. Haken (ed.), Springer Series in Synergetics, Vol. 38, Springer, Berlin, Heidelberg (1987)

[Ha94a] H. Hayakawa, S. Nishida, Y. Wada: A Computational Model for Shape Estimation by Integration of Shading and Edge Information, *Neural Networks* **7**, 1193 (1994)

[Ha94b] S. Haykin: *Neural Networks, a Comprehensive Foundation*, Macmillan, New York (1994)

[He49] D.O. Hebb: *The Organization of Behavior: A Neurophysiological Theory*, Wiley, New York (1949)

[He82] J.L. van Hemmen: Classical Spin-Glass Model, *Phys. Rev. Lett.* **49**, 409 (1982)

[He86] J.L. van Hemmen and R. Kühn: Nonlinear Neural Networks, *Phys. Rev. Lett.* **57**, 913 (1986)

[He87a] R. Hecht-Nielsen: Counterpropagation Networks, *Appl. Optics* **26**, 4979 (1987)

[He87b] R. Hecht-Nielsen: Kolmogorov's Mapping Neural Network Existence Theorem, *IEEE First Annual Int. Conf. on Neural Networks* , paper III-11

[He87c] J.L. van Hemmen: Nonlinear Neural Networks Near Saturation, *Phys. Rev.* A **36**, 1959 (1987)

[He87d] R.D. Henkel and W. Kinzel: Metastable States of the SK Model of Spin Glasses, J. Phys. A **20**, L727 (1987)

[He88a] J.L. van Hemmen, D. Grensing, A. Huber, and R. Kühn: Nonlinear Neural Networks, *Journ. Stat. Phys.* **50**, 231 and 259 (1988)

[He88b] R. Hecht-Nielsen, Theory of the Backpropagation Neural Network, *Neural Networks* **1**, 131 (1988)

[He88c] J.L. van Hemmen, G. Keller, and R. Kühn: Forgetful Memories, *Europhys. Lett.* **5**, 663 (1988)

[He89] H.G.E. Hentschel and A. Fine: Statistical Mechanics of Stereoscopic Vision, *Phys. Rev.* A **40**, 3983 (1989)

[He90] R. Hecht-Nielsen: *Neurocomputing*, Addison Wesley, (1990)

[He91] J.A. Hertz, A.S. Krogh, and R.G. Palmer: *Introduction to the Theory of Neural Computation*, Addison-Wesley, New York (1991).

[Hi83] G.E. Hinton and T.J. Sejnowski: Proc. IEEE Comp. Soc. Conf. Computer Vision and Pattern Recognition, Washington, D.C. (1983), p. 448

[Ho75] J. Holland: *Adaptation in Natural and Artificial Systems*, University of Michigan Press (1988)

[Ho82] J.J. Hopfield: Neural Networks and Physical Systems with Emergent Collective Computational Abilities, *Proc. Natl. Acad. Sci. USA* **79**, 2554 (1982)

[Ho83] J.J. Hopfield, D.I. Feinstein, and R.G. Palmer: "Unlearning" Has a Stabilizing Effect in Collective Memories, *Nature* **304**, 158 (1983)

[Ho84] J.J. Hopfield: Neurons with Graded Response Have Collective Computational Properties Like Those of Two-State Neurons, *Proc. Natl. Acad. Sci. USA* **81**, 3088 (1984)

[Ho85] J.J. Hopfield and D.W. Tank: 'Neural' Computation of Decisions in Optimization Problems, *Biol. Cybern.* **52**, 141 (1985)

[Ho86] J.J. Hopfield and D.W. Tank: Computing with Neural Circuits: A Model, *Science* **233**, 625 (1986)

[Ho88a] J. Holland: The Dynamics of Searches Directed by Genetic Algorithms, in: [Le88], p. 111

[Ho88b] D. Horn and M. Usher: Capacities of Multiconnected Memory Models, *Journ. de Physique* **49**, 389 (1988)

[Ho89a] L.H. Holley and M. Karplus: Protein Secondary Structure Prediction with a Neural Network, *Proc. Natl. Acad. Sci. USA* **86**, 152 (1989)

[Ho89b] D. Horn and M. Usher: Neural Networks with Dynamical Thresholds, *Phys. Rev.* A **40**, 1036 (1989)

[Ho89c] H. Horner: Neural Networks with Low Levels of Activity: Ising vs. McCulloch-Pitts Neurons, *Z. Physik* B **75**, 133 (1989)

[Ho89d] K. Hornik, M Stinchcomb, and H. White: Multilayer FeedForward Networks are Universal Approximators, *Neural Networks* **2**, 359 (1989)

[Ho91] M.A. Holler: VLSI Implementations of Learning and Memory systems, A Review, in: [Li91], p. 993

[Ho93] J. Holt and J. Hwang: Finite Precision Error Analysis of Neural Networks Hardware Implementations, *IEEE Trans. on Computers* **42**, 281 (1993)

[Hr90] G. Hripcsak: Using Connectionist Modules for Decision Support, *Methods of Information in Medicine* **29**, 167 (1990)

[Hu62] D.H. Hubel and T.N. Wiesel: Receptive Fields, Binocular Interaction and Functional Architecture in the Cat's Visual Cortex, *J. Physiol.* **160**, 106 (1962)

[Hu65] D.H. Hubel and T.N. Wiesel: Receptive Fields and Fundamental Architecture in Two Nonstriate Visual Areas (18 and 19) of the Cat, *J. Neurophysiol.* **28**, 229 (1965)

[Hu88] D.H. Hubel: *Eye, Brain and Vision*, Freeman, New York (1988)

[Hu87] J.M. Hutchinson and C. Koch: Simple Analog and Hybrid Networks for Surface Interpolation, in: [De87a], p. 235.

[Ik87] T. Ikegami and M. Suzuki: Spatio-temporal Patterns of a Plastic Neural Network, *Prog. Theor. Phys.* **78**, 38 (1987)

[Is25] E. Ising: Beitrag zur Theorie des Ferromagnetismus, *Z. Physik* **31**, 253 (1925)

[Is90] K. Iso and T. Watanbe: Speaker-Dependent Word Recognition Using a Neural Prediction Model, *Proceedings of the IEEE International Conference on Signal Processing* **1**, 441 (1990)

[Ka85] E. Kandel and J.H. Schwartz: *Principles of Neural Science*, Elsevier-North Holland, San Francisco (1985)

[Ka87] I. Kanter and H. Sompolinski: Associative Recall of Memories without Errors, *Phys. Rev.* A **35**, 380 (1987)

[Ka94] C.E. Kahn Jr.: Artificial Intelligence in Radiology: Decision Support Systems, *Radiographics* **14**, 849 (1994)

[Ke87] J.D. Keeler: Basins of Attraction of Neural Networks Models, in: [De87a], p. 259

[Ke88] T.B. Kepler and L.F. Abbott: Domains of Attraction in Neural Networks, *J. Phys. France* **49**, 1656 (1988)

[Ke91] A. Kehagias: Stochastic Recurrent Networks: Prediction and Classification of Time Series, (unpublished), st401843@brownvm.brown.edu (1991)

[Ki83] S. Kirkpatrick, C.D. Gelatt, Jr., and M.P. Vecchi: Optimization by Simulated Annealing, *Science* **220**, 671 (1983)

[Ki85] W. Kinzel: Learning and Pattern Recognition in Spin Glass Models, *Z. Physik* B **60**, 205 (1985); [Erratum: B **62**, 267 (1986)]

[Ki87] W. Kinzel: Statistical Mechanics of Neural Networks, *Physica Scripta* **35**, 398 (1987)

[Ki94] H.J. Kim and H.S. Yang: A Neural Network Capable of Learning and Inference for Visual Pattern Recognition, *Pattern Recognition Letters* **27**, 1291 (1994)

[Kl82] A.H. Klopf: *The Hedonistic Neuron: A Theory of Memory, Learning, and Intelligence*, Hemisphere, Washington, DC (1982)

[Kl86] D. Kleinfeld: Sequential State Generation by Model Neural Networks, *Proc. Natl. Acad. Sci. USA* **83**, 9469 (1986)

[Kl87] D. Klatt: Review of Text to Speech Conversion for English, *J. Acoustic Soc. Amer.* **82**, 737 (1987)

[Ko57] A.K. Kolmogorov: On the Representation of Continuous Functions of Several Variables by Superposition of Continuous Functions of One Variable and Addition, *Dokl. Akad. Nauk SSSR* **114**, 953 (1957)

[Ko72] T. Kohonen: Correlation Matrix Memories, *IEEE Trans. Electr. Comp.* **21**, 353 (1972)

[Ko82] T. Kohonen: Self-organized Formation of Topologically Correct Feature Maps, *Biol. Cybern.* **43**, 59 (1982)

[Ko84] T. Kohonen: *Self-Organization and Associative Memory*, Springer, Berlin, Heidelberg (1984)

[Ko86] C. Koch, J. Marroquin, and A. Yuille: Analog "Neuronal" Networks in Early Vision, *Proc. Natl. Acad. Sci. USA* **83**, 4263 (1986)

[Ko89a] C. Koch: Seeing Chips: Analog VLSI Circuits for Computer Vision, *Neural Computation* **1**, 184 (1989)

[Ko89b] I.Ya. Korenblit and E.F. Shender: Spin Glasses and Nonergodicity, *Sov. Phys. Usp.* **32**, 139 (1989)

[Ko90] G.A. Kohring: Neural Networks with Many-Neuron Interactions, *J. Phys. France* **51**, 145 (1990)

[Kr87a] W. Krauth and M. Mézard: Learning Algorithms with Optimal Stability in Neural Networks, *J. Phys. A* **20**, L745 (1987)

[Kr87b] R. Kree and A. Zippelius: Continuous-Time Dynamics of Asymmetrically Diluted Neural Networks, *Phys. Rev.* A **36**, 4421 (1987)

[Kr88] R. Kree and A. Zippelius: Recognition of Topological Features of Graphs and Images in Neural Networks, *J. Phys.* A **21**, L813 (1988)

[Kr89] W. Krauth and M. Mézard: Storage Capacity of Networks with Binary Couplings, *J. Phys.* A **22**, 3057 (1989)

[Kr94] D. DeKruger and B. Hunt: Image Processing and Neural Networks for Recognition of Cartographic Features, *Pattern Recognition Letters* **27**, 461 (1994)

[Ku91] C.-C. Kuan, J.J. Hull, and S.N. Srihari: Method and Apparatus for Hand-written Character Recognition, United States Patent No. 5 058, 182 (1991)

[Kü88] K.E. Kürten: Critical Phenomena in Model Neural Networks, *Phys. Lett.* A **129**, 157 (1988)

[La81] S. Lakshmivarahan: *Learning Algorithms and Applications*, Springer, New York (1981)

[La86] A. Lapedes and R.M. Farber: A Self-optimizing, Nonsymmetrical Neural Net for Content Addressable Memory and Pattern Recognition, *Physica* D **22**, 247 (1986)

[La87a] P.J.M. van Laarhoven and E.H.L. Aarts: *Simulated Annealing: Theory and Applications*, Reidel, Dordrecht (1987)

[La87b] A.S. Lapedes and R.M. Farber: Nonlinear Signal Processing using Neural Networks: Prediction and System Modelling, Los Alamos technical report LA-UR-87-2662 (1987)

[La87c] A.S. Lapedes and R.M. Farber: Programming a Massively Parallel Computation Universal System: Static Behaviour, in: [De87a], p. 283

[La88] A.S. Lapedes and R.M. Farber: How Neural Nets Work, in: [Le88], p. 331

[La89] S. Laughlin: The Reliability of Single Neurons and Circuit Design: A Case Study, in: [Du89], p.322

[Le88] Y.S. Lee (ed.): *Evolution, Learning and Cognition*, World Scientific, Singapore (1988)

[Le90] E. Levin: Word Recognition Using Hidden Control Neural Architecture, *Proc. of the IEEE Intern. Conf. on Signal Processing* **1**, 433 (1990)

[Le94] B. LeBaron and A.S. Weigend: Evaluating Neural Network Predictors by Bootstrapping, Proceedings of International Conference on Neural Information Processing (ICONIP'94), andreas@cs.colorado.edu (1994)

[Li73] S. Lin and B.W. Kernighan: An Effective Heuristic Algorithm for the Traveling-Salesman Problem, *Operations Research* **21**, 498 (1973)

[Li74] W.A. Little: The Existence of Persistent States in the Brain, *Math. Biosci.* **19**, 101 (1974) {Reprinted in [Sh88]}

[Li78] W.A. Little and G.L. Shaw: Analytic Study of the Memory Capacity of a Neural Network, *Math. Biosci.* **39**, 281 (1978) {Reprinted in [Sh88]}

[Li87] R.P. Lippmann: An Introduction to Computing with Neural Nets, *IEEE ASSP Magazine*, April 1987, p. 4

[Li91] R.P. Lippmann, J.E. Moody, and D.S. Touretzky (eds.): *Advances in Neurel Information Processing Systems 3*, Morgan Kaufmann, San Mateo (1991)

[Li93] F. Lin and R.D. Brandt: Towards Absolute Invariants of Images under Translation, Rotation, and Dilation, *Pattern Recognition Letters* **14**, 369 (1993)

[Li95] C.S. Lindsey and T. Lindblad: *Review of Hardware Neural Networks: a User's Perspective*, hypertext manuscript accessible at http://msia02.msi.se/~lindsey/elba2html/elba2html.html

[Lo89] S.R. Lockery, G. Wittenberg, W.B. Kristan, G.W. Cottrell: Function of Identified Interneurons in the Leech Elucidated Using Neural Networks Trained by Back-Propagation, *Nature* **340**, 468 (1989)

[Lo90] L. Lönnblad, C. Peterson, T. Rögnvaldsson: Finding Gluon Jets with a Neural Trigger, *Phys. Rev. Lett.* **65**, 1321 (1990)

[Lo91] L. Lönnblad, C. Peterson, T. Rögnvaldsson: Using Neural Networks to Identify Jets, *Nuclear Physics* **B349**, 675 (1991)

[Lo92] L. Lönnblad, C. Peterson, T. Rögnvaldsson: Mass Reconstruction with a Neural Network, *Physics Letters B* **278**, 181 (1992)

[Lo94] L. Lönnblad, C. Peterson, T. Rögnvaldsson: JETNET 3.0 - A Versatile Artificial Neural Network Package, *Comp. Phys. Comm.* **81**, 185 (1994)

[Ma69] D. Marr: A Theory of Cerebellar Cortex, *J. Physiol. London* **202**, 437 (1969)

[Ma73] C. von der Malsburg: Self-organization and Orientation Sensitive Cells in the Striate Cortex, *Kybernetik* **14**, 85 (1973)

[Ma77] M.C. Mackey and L. Glass: Oscillation and Chaos in Physiological Control Systems, *Science* **197**, 287 (1977)

[Ma87] C. von der Malsburg and E. Bienenstock: A Neural Network for the Retrieval of Superimposed Connection Patterns, *Europhys. Lett.* **3**, 1243 (1987)

[Ma89a] C.M. Marcus and R.M. Westervelt: Stability of Analog Neural Networks with Delay, *Phys. Rev.* A **39**, 347 (1989)

[Ma89b] C.M. Marcus and R.M. Westervelt: Dynamics of Iterated-Map Neural Networks, *Phys. Rev.* A **40**, 501 (1989)

[Ma91] L. Marquez et al.: Neural Network Models as an Alternative to Regression, *Proceedings of the IEEE 24th Annual Hawaii International Conference on Systems Sciences* **VII**, IEEE Computer Society Press, 129 (1991)

[Ma94a] V. Maniezzo: Genetic Evolution of the Topology and Weight Distribution of Neural Networks, *IEEE Transactions on Neural Networks* **5**, p. 39 (1994)

[Ma94b] B. Manly: *Multivariate Statistical Methods: a Primer*, Chapman & Hall, London (1994)

[Ma94c] T. Masters: *Practical Neural Network Recipes in C++*, Academic Press, (1994)

[Mc43] W.S. McCulloch and W. Pitts: A Logical Calculus of the Ideas Immanent in Nervous Activity, *Bull. Math. Biophys.* **5**, 115 (1943) {Reprinted in [Sh88]}

[Mc87] R.J. McEliece, E.C. Posner, E.R. Rodemich, S.S. Venkatesh: The Capacity of the Hopfield Associative Memory, *IEEE Trans. Inform. Theory* **33**, 461 (1987)

[Mc91] H. McCartor: Back Propagation Implementation on the Adaptive Solutions CNAPS Neurocomputer Chip, in: [Li91], p. 1028

[Mc94] J.R. McDonnell and D. Waagen: Evolving Recurrent Perceptrons for Time-Series Modeling, *IEEE Transactions on Neural Networks* **5**, 24 (1994)

[Mc95a] P.J. McCann and B.L. Kalman: A Neural Network Model for the Gold Market, unpublished, pjm3@cs.wustl.edu (1994)

[Mc95b] J.R. McDonnell and D.E. Waagen: Evolving Cascade-Correlations Networks for Time-Series Forecasting, *Int. Journal of AI Tools*, March (1995)

[Me53] N. Metropolis, A.W. Rosenbluth, M.N. Rosenbluth, A.H. Teller, and E. Teller: Equation of State Calculations for Fast Computing Machines, *J. Chem. Phys.* **21**, 1087 (1953)

[Me80] C. Mead and L. Conway, eds.: *Algorithms for VLSI Processor Arrays*, Addison-Wesley (1980)

[Me87] M. Mézard, G. Parisi, and M.A. Virasoro: *Spin-Glass Theory and Beyond*, World Scientific, Singapore (1987)

[Me88] R. Meir and E. Domany: Layered Feed-Forward Neural Networks with Exactly Soluble Dynamics, *Phys. Rev.* A **37**, 608 (1988)

[Me89a] C. Mead: *Analog VLSI and Neural Systems*, Addison-Wesley, Menlo Park (1989)

[Me89b] M. Mézard: The Space of Interactions in Neural Networks: Gardner's Computation with the Cavity Method, *J. Phys.* A **22**, 2181 (1989)

[Me89c] M. Mézard and J.-P. Nadal: Learning in Feedforward Layered Neural Networks: The Tiling Algorithm, *J. Phys.* A **22**, 2191 (1989)

[Mi69] M. Minsky and S. Papert: *Perceptrons: An Introduction to Computational Geometry*, MIT Press, Cambridge (1969); 2nd ed. (1988)

[Mi89] J.E. Midwinter and D.R. Selviah: Digital Neural Networks, Matched Filters and Optical Implementations, in: [Al89a], p. 258

[Mi90] R.M. Miller: *Computer-Aided Financial Analysis*, Addison-Wesley, (1990)

[Mi91] S. Miesbach: Efficient Gradient Computation for Continuous and Discrete Time-Dependent Neural Networks, *Proc. IJCNN*, Singapore, p. 2337 (1991)

[Mi92] A.S. Miller, B.H. Blott, and T.K. Hames: Review of Neural Network Applications in Medical Image and Signal Processing, *Medical & Biological Engineering & Computing* **30**, 449 (1992)

[Mi94] Z. Michalewicz: *Genetic algorithms + data structures = evolution programs*, second edition, Springer-Verlag (1994)

[Mo87] I. Morgenstern in: *Proceedings of the Heidelberg Colloquium on Glassy Dynamics and Optimization*, J.L. van Hemmen and I. Morgenstern (eds.), Lecture Notes in Physics, Vol. 275, Springer, Berlin, Heidelberg, New York (1987)

[Mo88] R.J.T. Morris and W.S. Wong: A Short-Term Neural Network Memory, *SIAM J. Comput.* **17**, 1103 (1988)

[Mo89] D.J. Montana and L. Davis: Training Feedforward Neural Networks Using Genetic Algorithms, in: *Eleventh International Joint Conference on Artificial Intelligence*, ed. N.S. Sridharan, Morgan Kaufmann, San Mateo (1989), p. 762

[Mo95] N. Morgan and H.A. Bourlard: Neural Networks for Statistical Recognition of Continuous Speech, *Proc. IEEE* **83**, 742 (1995)

[Na72] K. Nakano: Associatron – A Model of Associative Memory, *IEEE Trans. Syst. Man Cybern.* **2**, 381 (1972)

[Na86] J.P. Nadal, G. Toulouse, J.P. Changeux, and S. Dehane: Networks of Formal Neurons and Memory Palimpsests, *Europhys. Lett.* **1**, 535 (1986)

[Na89] L. Nadel, L.A. Cooper, P. Culicover, and R.M. Harnish (eds.): *Neural Connections, Mental Computation*, MIT Press, Cambridge (1989)

[Na94] N. Nandhakumar, S. Karthik, and J.K. Aggarwal: Unified Modeling of Non-Homogenous 3D Objects for Thermal and Visual Image Systems, *Pattern Recognition* **27**, 1303 (1994)

[Ol92] P. Olmos, J.C. Diaz, G. Garcia-Belmonte, P. Gomez, and V. Rodellar: Application of Neural Network Techniques in Gamma Spectroscopy, *Nucl. Inst. and Meth.*, **A312**, 167 (1992)

[Op88] M. Opper: Learning Times of Neural Networks: Exact Solution for a PERCEPTRON Algorithm, *Phys. Rev.* A **38**, 3824 (1988)

[Pa80] G. Parisi: The Order Parameter for Spin Glasses: A Function on the Interval 0–1, *J. Phys.* A **13**, 1101 (1980)

[Pa81] G. Palm: On the Storage Capacity of an Associative Memory with Randomly Distributed Storage Elements, *Biol. Cybern.* **39**, 125 (1981)

[Pa85] D.B. Parker: Learning-Logic: Casting the Cortex of the Human Brain in Silicon, MIT Techn. Rep. TR-47 (1985)

[Pa86a] N. Parga and M.A. Virasoro: The Ultrametric Organization of Memories in a Neural Network, *J. Phys. France* **47**, 1857 (1986)

[Pa86b] G. Parisi: A Memory which Forgets, *J. Phys.* A **19**, L617 (1986)

[Pa87a] M. Padberg and G. Rinaldi: *Oper. Res. Lett.* **6**, 1 (1987)

[Pa87b] S. Patarnello and P. Carnevali: Learning Networks of Neurons with Boolean Logic, *Europhys. Lett.* **4**, 503 (1987)

[Pa88] S. Patarnello and P. Carnevali: Learning Capabilities of Boolean Networks, *Neural Networks, from Models to Applications*, L. Personnaz and G. Dreyfus eds., (1988)

[PDP86] J.L. McClelland, D.E. Rumelhart (eds.): *Parallel Distributed Processing*, 2 vols., MIT Press, Cambridge (1986)

[Pe86a] P. Peretto and J.J. Niez: Long Term Memory Storage Capacity of Multi-connected Neural Networks, *Biol. Cybern.* **54**, 53 (1986)

[Pe86b] L. Personnaz, I. Guyon, and G. Dreyfus: Collective Computational Properties of Neural Networks: New Learning Mechanisms, *Phys. Rev.* A **34**, 4217 (1986)

[Pe87] C. Peterson and J.R. Anderson: A Mean Field Theory Learning Algorithm for Neural Networks, *Complex Systems* **1**, 995 (1987)

[Pe89a] B.A. Pearlmutter: Learning State Space Trajectories in Recurrent Neural Networks, *Neural Computation* **1**, 263 (1989)

[Pe89b] P. Perez: Memory Melting in Neural Networks, *Phys. Rev.* A **39**, 4303 (1989)

[Pe89c] C.J. Perez-Vicente: Sparse Coding and Information in Hebbian Neural Networks, *Europhys. Lett.* **10**, 621 (1989)

[Pe89d] C. Peterson and B. Söderberg: A New Method for Mapping Optimization Problems onto Neural Networks, *Int. Journ. of Neural Syst.* **1**, 3 (1989)

[Pe90] B.A. Pearlmutter: Dynamic Recurrent Neural Networks, Carnegie-Mellon University Technical Report CMU-CS-90-196, Dec. 1990

[Pi47] W. Pitts and W.S. McCulloch: How We Know Universals. The Perception of Auditory and Visual Forms, *Bull. Math. Biophys.* **9**, 127 (1947)

[Pi87] F.J. Pineda: Generalization of Back-Propagation to Recurrent Neural Networks, *Phys. Rev. Lett.* **59**, 2229 (1987)

[Pl88] J.C. Platt and A.H. Barr: Constrained Differential Optimization, in: [An88a] p. 612

[Po77] K.R. Popper and J.C. Eccles: *The Self and Its Brain*, Springer, Berlin, Heidelberg (1977)

[Po85] T. Poggio and C. Koch: Ill-Posed Problems in Early Vision: From Computational Theory to Analogue Networks, *Proc. Roy. Soc. Lond.* B **226**, 303 (1985)

[Pö87] G. Pöppel and U. Krey: Dynamical Learning Process for Recognition of Correlated Patterns in Symmetric Spin Glass Models, *Europhys. Lett.* **4**, 979 (1987)

[Po88] A.B. Poritz: Hidden Markov Models: A Guided Tour, *Proceedings of the IEEE International Conference on Signal Processing* **1**, 7 (1988)

[Po91] D.A. Pomerleau: Rapidly Adapting Artificial Neural Networks for Autonomous Navigation, in: [Li91], p. 429

[Po93] D.A. Pomerleau: Reliability Estimation for Neural Network Based Autonomous Driving, *Robotics and Autonomous Systems* **12**, 113 (1994)

[Pr83] J.P. Provost and G. Vallee: Ergodicity of the Coupling Constants and the Symmetric n-Replicas Trick for a Class of Mean-Field Spin-Glass Models, *Phys. Rev. Lett.* **50**, 598 (1983)

[Pr86] W.H. Press, B.P. Flannery, S.A. Teukolsky, and W.T. Vetterling: *Numerical Recipes: The Art of Scientific Computing*, Cambridge University Press (1986)

[Pr88] M.B. Priestly: *Non-Linear and Non-Stationary Time Series Analysis*, Academic Press, (1988)

[Pr90] J. Proriol et al., *New Computing Techniques in Physics Research*, Editions du Centre National de la Recherche Scientifique (1990)

[Pr94] A. Prügel-Bennett and J.L. Shapiro: Analysis of Genetic Algorithms Using Statistical Mechanics, *Phys. Rev. Lett.* **72**, 1305 (1994)

[Qi88] N. Qian und T.J. Sejnowski: Learning to Predict the Secondary Structure of Globular Proteins, in: [Le88], p. 257

[Qi94] X. Qi: Theoretical Analysis of Evolutionary Algorithms With an Infinite Population Size in Continous Space, *IEEE Transactions on Neural Networks* **5**, p. 102 (1994)

[Ra86] R. Rammal, G. Toulouse, and M.A. Virasoro: Ultrametricity for Physicists, *Rev. Mod. Phys.* **58**, 765 (1986)

[Ra91] U. Ramacher and U. Rückert: *VLSI Design of Neural Networks*, Kluver Academic Publishers (1991)

[Ra92] U. Ramacher: SYNAPSE – A Neurocomputer that Synthesizes Neural Algorithms on a Parallel Systolic Engine, *Journ. of Parallel and Distributed Computing* **14**, 306 (1992)

[Re80] L.E. Reichl: *A Modern Course in Statistical Mechanics*, University of Texas Press, Austin (1980)

[Re84] H.J. Reitboek and J. Altmann: A Model for Size- and Rotation-Invariant Pattern Processing in the Visual System, *Biol. Cybern.* **51**, 113 (1984)

[Re90] H. Reininger and D. Wolf: Nonlinear Prediction of Stochastic Processes Using Neural Networks in: *Proc. V European Signal Processing Conf.* (1990)

[Re94] A.N. Refenes, A. Zaparanis, and G. Francis, Stock Performance Modeling using Neural Networks: A Comparative Study with Regression Models, *Neural Networks* **7**, 375 (1994)

[Ri86] H. Ritter and K. Schulten: On the Stationary State of Kohonen's Self-organizing Sensory Mapping, *Biol. Cybern.* **54**, 99 (1986)

[Ri88a] U. Riedel, R. Kühn, J.L. van Hemmen: Temporal Sequences and Chaos in Neural Nets, *Phys. Rev. A* **38**, 1105 (1988)

[Ri88b] H. Ritter and K. Schulten: Convergence Properties of Kohonen's Topology Conserving Maps: Fluctuations, Stability, and Dimension Selection, *Biol. Cybern.* **60**, 59 (1988)

[Ri92] H. Ritter, T. Martinetz, and K. Schulten: *Neural Computation and Self-Organizing Maps*, Addison-Wesley (1992)

[Ro62] F. Rosenblatt: *Principles of Neurodynamics*, Spartan, New York (1962)

[Ru86a] D.E. Rumelhart, G.E. Hinton, and R.J. Williams: Learning Representations by Back-propagating Errors, *Nature* **323**, 533 (1986)

[Ru86b] D.E. Rumelhart, G.E. Hinton, and R.J. Williams: Learning Internal Representations by Error Propagation, in: [PDP86], Vol. 1, p. 318

[Sa87] O. Sacks: *The Man Who Mistook His Wife for a Hat*, Harper & Row, New York (1987)

[Sa89] D. Sanger: Contribution Analysis: A Technique for Assigning Responsibilities to Hidden Units in Connectionist Networks, *Connection Science* **1**, 115 (1989)

[Sc77] E.L. Schwarz: Spatial Mapping in the Primate Sensory Projection: Analytic Structure and Relevance to Perception, *Biol. Cybern.* **25**, 181 (1977)

[Sc78] J. Schurmann: A Multifont Word Recognition System for Postal Address Reading, *IEEE Transactions in Computing* **27**, 721 (1978)

[Sc84] H.G. Schuster: *Deterministic Chaos: An Introduction*, Physik-Verlag, Weinheim (1984)

[Sc87] R. Scalettar and A. Zee: Three Issues on Generalization in Feed-Forward Nets, preprint NSF-ITP-87-150

[Sc88] R. Scalettar and A. Zee: Perception of Left and Right by a Feed Forward Net, *Biol. Cybern.* **58**, 193 (1988)

[Sc89] B. Schürmann: Stability and Adaptation in Artificial Neural Systems, *Phys. Rev. A* **40**, 2681 (1989)

[Sc90] E. Schöneburg: Stock Price Prediction using Neural Networks: A Project Report, *Neurocomputing 2*, 17 (1990)

[Sc94] H. Schmid: Part-of-Speech Tagging with Neural Networks, Unpublished, schmid@ims.uni-stuttgart.de (1994)

[Se86] T.J. Sejnowski, P.K. Kienker, and G.E. Hinton: Learning Symmetry Groups with Hidden Units: Beyond the Perceptron, *Physica* D **22**, 2060 (1986)

[Se87] T.J. Sejnowski und C.R. Rosenberg: Parallel Networks that Learn to Pronounce English Text, *Complex Systems* **1**, 145 (1987)

[Se89] A. Selverston and P. Mazzoni: Flexibility of Computational Units in Invertebrate CPGs, in: [Du89], p. 205

[Sh75] D. Sherrington and S. Kirkpatrick: Solvable Model of a Spin-Glass, *Phys. Rev. Lett.* **35**, 1792 (1975)

[Sh78] D. Sherrington and S. Kirkpatrick: Infinite-Ranged Models of Spin-Glasses, *Phys. Rev.* B **17**, 4384 (1978)

[Sh79] G.M. Shepherd: *The Synaptic Organization of the Brain*, Oxford University Press, New York (1979)

[Sh87] S. Shinomoto: A Cognitive and Associative Memory, *Biol. Cybern.* **57**, 197 (1987)

[Sh88] G.L. Shaw and G. Palm: *Brain Theory*, Reprint volume, World Scientific, Singapore (1988)

[Sh90] R. Sharda, and R.B. Patil: Neural Networks as Forecasting Experts: an Empirical Test, *Proceedings of the IJCNN* Washington, 491 (1990)

[Si87] M.A. Sivilotti, M.R. Emerling and C.A. Mead: VLSI Architectures for Implementation of Neural Networks, in: [De87a] p. 408

[Sj82] T. Sjöstrand: The LUCLUS Algorithm, *Comp. Phys. Comm.* **27**, 229 and 242 (1982)

[Sj87] T. Sjöstrand and H.-U. Bengtsson: The LUCLUS Algorithm, *Comp. Phys. Comm.* **43** (1987)

[Sj89] T. Sjöstrand: JETSET 7.2 Program and Manual; see B. Bambad et al.: CERN Report No. CERN-TH.5466/89 (unpublished), (1989)

[So86a] H. Sompolinsky and I. Kanter: Temporal Association in Asymmetric Neural Networks, *Phys. Rev. Lett.* **57**, 2861 (1986)

[So86b] H. Sompolinski: Neural Networks with Nonlinear Synapses and a Static Noise, *Phys. Rev.* A **34**, 2571 (1986)

[So88] S.A. Solla, E. Levin, and M. Fleisher: Accelerated Learning in Layered Neural Networks, *Complex Systems* **2**, 625 (1988)

[So89] S.A. Solla: Learning and Generalization in Layered Neural Networks: The Contiguity Problem, in: *Neural Networks: From Models to Applications*, L. Personnaz and G. Dreyfus (eds.), I.D.S.E.T., Paris (1989), p. 168

[Sp89] P. Spitzner and W. Kinzel: Hopfield Network with Directed Bonds, *Z. Physik* B **74**, 539 (1989)

[Sr83] M.S. Srivastava and E.M. Carter: *An Introduction to Applied Multivariate Statistics*, North Holland, (1983)

[Sr93] S. N. Srihari: Recognition of Handwritten and Machine-Printed Text for Postal Address Implementation, *Pattern Recognition Letters* **14**, 291 (1993)

[St80] J. Stoer and R. Bulirsch, *Introduction to Numerical Analysis*, Springer, New York (1978)

[St90] D.F. Stubbs: Multiple Neural Network Approaches to Clinical Expert Systems, *SPIE Applications of Artificial Neural Networks* **1294**, 433 (1990)

[St94] C. Stewart, Y.C. Lu, and C. Larson: A Neural Clustering Approach for High Resolution Radar Target Classification, *Pattern Recognition* **27**, 503 (1994)

[Su88] J.P. Sutton, J.S. Beis, and L.E.H. Trainor: Hierarchical Model of Memory and Memory Loss, *J. Phys.* A **21**, 4443 (1988)

[Su93] C.Y. Suen, R. Legault, C. Nadal, M. Cheriet, and L. Lam: Building a New Generation of Handwriting Recognition Systems, *Pattern Recognition Letters* **14**, 303 (1993)

[Sv92] C. Svarer, L.K. Hansen, and J. Larsen: On Design and Evaluation of Tapped-Delay Neural Network Architectures, *IEEE Int. Conf. on Neural Networks*, San Francisco, (1992)

[Ta80] F. Tanaka and S.F. Edwards: Analytic Theory of the Ground State Properties of a Spin Glass, *J. Phys.* F10, 2769 (1980)

[Ta91] Z. Tang et al.: Time Series Forecasting using Neural Networks vs. Box-Jenkins Methodology, *Simulation*, 303 (1991)

[Ta95] J.G. Taylor and D. Husmeier: Improving Time Series Prediction by Learning Time Lags, unpublished, UDAH222@bay.cc.kcl.ac.uk (1995)

[Te88] G. Tesauro and T.J. Sejnowski: A "Neural" Network that Learns to Play Backgammon, in: [An88a], p. 794

[Te90] J. Tebelskis and A. Waibel: Large Vocabulary Recognition using Linked Predictive Neural Networks, *Proceedings of the IEEE International Conference on Signal Processing* **1**, 437 (1990)

[Te93] B.A. Telfer, and D.P. Casasent: Minimum-Cost Associate Processor for Piecewise-Hyperspherical Classification, *Neural Networks* **6**, 1117 (1993)

[Th94] D.S. Thomas and A. Mitiche: Asymptotic Optimality of Pattern Recognition by Regression Analysis, *Neural Networks* **7**, 313 (1994)

[Ti89] N. Tishby, E. Levin, and S.A. Solla: Consistent Inference of Probabilities in Layered Neural Networks: Predictions and Generalization, in: Proc. Intern. Joint Conf. on Neural Networks, IEEE, Washington, DC, **2**, 403 (1989)

[To79] G. Toulouse: Symmetry and Topology Concepts for Spin Glasses and Other Glasses, *Phys. Rep.* **49**, 267 (1979)

[To80] H. Tong and K.S. Lim: Threshold Autoregression, Limit Cycles and Cyclical Data, *Journal of the Royal Statistical Society* **42B**, (1980)

[Ts88a] M.V. Tsodyks and M.V. Feigel'man: The Enhanced Storage Capacity in Neural Networks with Low Activity Level, *Europhys. Lett.* **6**, 101 (1988)

[Ts88b] M.V. Tsodyks: Associative Memory in Asymmetric Diluted Network with Low Level of Activity, *Europhys. Lett.* **7**, 203 (1988)

[Tu92] E. Turban and R. Trippi eds.: *Neural Network Applications in Investment and Finance Services*, Probus Publishing, (1992)

[Va84] L.G. Valiant: A Theory of the Learnable, *Comm. ACM* **27**, 1134 (1984)

[Va89] C. de la Vaissaiére and S. Palma-Lopez, DELPHI 89-32 PHYS 38 (1989)

[Va94] D. Valentin, H. Abdi, A. O'Toole, G. Cottrell: Connectionist Models of Face Processing - A Survey, *Pattern Recognition Letters* **27**, 1209 (1994)

[Ve86] S.S. Venkatesh: Epsilon Capacity of Neural Networks, in: [De87a] p. 440

[Vi88] M.A. Virasoro: The Effect of Synapses Destruction on Categorization by Neural Networks, *Europhys. Lett.* **7**, 293 (1988)

[Vs93] G.V. Vstovsky and A.V. Vstovskaya: A Class of Hidden Markov Models for Image Processing, *Pattern Recognition Letters* **14**, 391 (1994)

[Wa89a] E. Wacholder, J. Han, and R.C. Mann: A Neural Network Algorithm for the Multiple Traveling Salesmen Problem, *Biol. Cybern.* **61**, 11 (1989)

[Wa89b] A. Waibel, T. Hanazawa, G. Hinton, K. Shikano, and K.J. Lang: Phoneme Recognition using Time-delay Neural Networks, *IEEE Trans. ASSP* 37, 328 (1989)

[Wa91] A. Waibel, A. Jain, A. McNair, K. Saito, A. Hauptmann, and J. Tebelskis: Janus: A Speech-to-Speech Translation System Using Connectionist and Symbolic Processing Strategies, *Proceedings of the 1991 ICASSP*, (1991)

[Wa93a] E.A. Wan: Finite Impulse Response Neural Networks with Applications in Time Series Prediction, Ph.D. dissertation, Stanford University, (1993)

[Wa93b] T.L.H. Watkin, A. Rau, and M. Biehl: The Statistical Mechanics of Learning a Rule, *Rev. Mod. Phys.* **65**, 499 (1993)

[Wa94] S.R. Waterhouse and A.J. Robinson: Non-Linear Prediction of Acoustic Vectors using Hierarchical Mixtures of Experts, to appear in *Neural Information Processing Systems* **7**, Morgan Kaufmann, (1994)

[We74] P. Werbos: Beyond Regression: New Tools for Prediction and Analysis in the Behavioral Sciences, Ph.D. thesis, Harvard University (1974)

[We85] G. Weisbuch and F. Fogelman Soulié: Scaling Laws for the Attractors of Hopfield Networks, *J. Physique Lett.* **46**, 623 (1985)

[We90] A. Weigend, D.E. Rumelhart, and B.A. Huberman: Predicting the Future: A Connectionist Approach, *Int. Journ. of Neural Systems* **1**, 193 (1990)

[We94a] A. Weigend and N. Gershenfeld: *Time Series Prediction: Forecasting the Future and Understanding the Past*, Addison-Wesley, Sante Fe (1994)

[We94b] W.R. Webber, B. Litt, K. Wilson, and R.P. Lesser: Practical Detection of Epileptiform Discharges in the EEG using an Artificial Neural Network: A Comparison of Raw and Parameterized EEG Data, *Electroencephalography & Clinical Neurophysiology* **91**, 194 (1994)

[Wi69] D.J. Willshaw, O.P. Buneman, H.C. Longuet-Higgins: Nonholographic associative memory, *Nature* **222**, 960 (1969)

[Wi73] B. Widrow, N.K. Gupta, and S. Maitra: *IEEE Trans. Syst. Man Cybern.* **5**, 455 (1973)

[Wi86a] F.W. Wiegel: *Introduction to Path-Integral Methods in Physics and Polymer Science*, World Scientific, Singapore (1986)

[Wi86b] R.J. Williams: Reinforcement Learning in Connectionist Networks: A Mathematical Analysis, UC San Diego Technical Report ICS-8605 (1986)

[Wi87] R.J. Williams: Reinforcement-Learning Connectionist Systems, Northeastern University Report NU-CCS-87-3 (1987)

[Wi88] G.V. Wilson and G.S. Pawley: On the Stability of the Travelling Salesman Problem Algorithm of Hopfield and Tank, *Biol. Cybern.* **58**, 63 (1988)

[Wi89] R.J. Williams and D. Zipser: A Learning Algorithm for Continually Running Fully Recurrent Neural Networks, *Neural Computation* **1**, 270 (1989)

[Wo63] D.E. Wooldridge: *The Machinery of the Brain*, McGraw-Hill, New York (1963)

[Wo90] F.S. Wong and P.Z. Wang: A Stock Selection Strategy using Fuzzy Neural Networks, *Neurocomputing* **2**, 233 (1990)

[Wo92] F. Wong and P.Y. Tan: Neural Networks and Genetic Algorithms for Economic Forecasting, *AI in Economics and Business Administration*, (1992)

[Wo94] M. Woszcyna, N. Aoki-Waibel, F.D. Buo, N. Coccaro, K. Horiguchi, T. Kemp, A. Lavie, A. McNair, T. Polzin, I. Rogina, C.P. Rose, T. Schultz, B. Suhm, M. Tomita, and A. Waibel: Janus 93: Towards Spontaneous Speech Translation, *Proceedings of the 1994 ICASSP*, (1994)

[Wu83] F.Y. Wu: The Potts Model, *Rev. Mod. Phys.* **54**, 235 (1983)

[Za89] V.A. Zagrebnov and A.S. Chyvrov: The Little–Hopfield Model: Recurrence Relations for Retrieval-Pattern Errors, *Sov. Phys. JETP* **68**, 153 (1989)

[Ze88] A. Zee: Some Quantitative Issues in the Theory of Perception, in: [Le88], p. 183

[Zi88] D. Zipser and R.A. Andersen: A Back-propagation Programmed Network that Simulates Response Properties of a Subset of Posterior Parietal Neurons, *Nature* **331**, 679 (1988)

Index

Physics of Neural Networks

Springer-Verlag
and the Environment

We at Springer-Verlag firmly believe that an international science publisher has a special obligation to the environment, and our corporate policies consistently reflect this conviction.

We also expect our business partners – paper mills, printers, packaging manufacturers, etc. – to commit themselves to using environmentally friendly materials and production processes.

The paper in this book is made from low- or no-chlorine pulp and is acid free, in conformance with international standards for paper permanency.